Government
and
Technical Progress

The Technology Policy and Economic Growth Series

Herbert I. Fusfeld and Richard R. Nelson, Editors

Fusfeld/Haklisch INDUSTRIAL PRODUCTIVITY AND
INTERNATIONAL TECHNICAL COOPERATION
Fusfeld/Langois UNDERSTANDING R&D PRODUCTIVITY
Hazewindus THE U.S. MICROELECTRONICS INDUSTRY:
Technical Change, Industry Growth, and Social Impact
Nelson GOVERNMENT AND TECHNICAL PROGRESS:
A Cross–Industry Analysis

Pergamon Titles of Related Interest

Dewar INDUSTRY VITALIZATION: Toward a
National Industrial Policy
Hill/Utterback TECHNOLOGICAL INNOVATION FOR A
DYNAMIC ECONOMY
Lundstedt/Colglazier MANAGING INNOVATION: The Social
Dimensions of Creativity
Perlmutter/Sagafi-nejad INTERNATIONAL TECHNOLOGY
TRANSFER
Sagafi-nejad/Moxon/Perlmutter CONTROLLING INTERNATIONAL
TECHNOLOGY TRANSFER
Sagafi-nejad/Belfield TRANSNATIONAL CORPORATIONS,
TECHNOLOGY TRANSFER AND DEVELOPMENT

Related Journals*

BULLETIN OF SCIENCE, TECHNOLOGY AND SOCIETY
COMPUTERS AND INDUSTRIAL ENGINEERING
COMPUTERS AND OPERATIONS RESEARCH
SOCIO-ECONOMIC PLANNING SCIENCES
TECHNOLOGY IN SOCIETY
WORK IN AMERICA INSTITUTE STUDIES IN PRODUCTIVITY

*Free specimen copies available upon request.

Government
and
Technical Progress
A Cross-Industry Analysis

edited by
Richard R. Nelson

The Technology Policy and Economic Growth Series,
Herbert I. Fusfeld and Richard R. Nelson, Editors

Published in cooperation with the Center for Science and Technology Policy,
Graduate School of Business Administration, New York University

Pergamon Press

New York   Oxford   Toronto   Sydney   Paris   Frankfurt

Pergamon Press Offices:

U.S.A.	Pergamon Press Inc.. Maxwell House. Fairview Park. Elmsford. New York 10523. U.S.A.
U.K.	Pergamon Press Ltd.. Headington Hill Hall. Oxford OX3 OBW. England
CANADA	Pergamon Press Canada Ltd.. Suite 104. 150 Consumers Road. Willowdale. Ontario M2J 1P9. Canada
AUSTRALIA	Pergamon Press (Aust.) Pty. Ltd.. P.O. Box 544. Potts Point. NSW 2011. Australia
FRANCE	Pergamon Press SARL. 24 rue des Ecoles. 75240 Paris. Cedex 05. France
FEDERAL REPUBLIC OF GERMANY	Pergamon Press GmbH. Hammerweg 6 6242 Kronberg/Taunus. Federal Republic of Germany

Library of Congress Cataloging in Publication Data

Main entry under title:

Government and technical progress.

   (The Technology policy and economic growth
series)
      Includes index.
      1. Technological innovations--Government policy--
United States. 2. Technology and state--United States.
3. Industry and state--United States. I. Nelson,
Richard R. II. Series.
HC110.T4G68   1982   338'.06   82-3869
ISBN 0-08-028837-5   AACR2

Printed in the United States of America

Contents

v

Preface

This study was conceived and begun in late 1979. The Carter Administration was then conducting its Domestic Policy Review on industrial innovation, and it appeared that a number of proposals for a more active federal role in the R&D process were about to emanate from that initiative.

Both Herbert Fusfeld and I were keenly struck by a sense of de novo about the review. It seemed to us that the discussion did not adequately recognize that the United States has had a long history of policies aimed at stimulating innovation in various economic sectors. Some of these policies, we knew, had been very successful; others had been unsuccessful or worse. Furthermore, it seemed to us that this legacy was a diverse one, with the nature of the innovation policies varying greatly from sector to sector.

What was needed, we concluded, was an analytic review of the historical experience of government involvement in technical change in the United States. This book is the result.

Our approach, as is appropriate to historical analysis, is one of case studies. After designing and outlining the study in the rough, I set about assembling what I hoped would be a first-rate team of economic historians to tackle a set of seven industrial sectors. And I wasn't disappointed: my first choices for each of the sector cases accepted our invitation. The study team met twice - once early on to discuss broad strategy and to exchange basic information and viewpoints, and once again after first drafts of the chapters had been written. It was after this second meeting that I set about writing chapter 9, an integrative summary that attempts to draw and mesh conclusions from the seven case studies.

While the individual industry chapters are the responsibility of the authors who signed them, all the authors acknowledge the useful feedback they received from one another while the chapters were in draft. I would also like to express my thanks to these colleagues for the comments they made on drafts of my own chapter - although I retain responsibility for the conclusions expressed.

The authors as a group would like to thank Mr. Leonard A. Ault, Acting Chief, Dissemination and Analysis Branch, NASA; Dr. John D. Holmfeld, Science Consultant, House Committee on Science, Research, and Technology; and Mr. Steven A. Merrill, Staff Member, Subcommittee on Commerce, Science, and Space, Senate Committee on Commerce, Science, and Transportation, for their support of this project.

We would also like to express our appreciation to Dr. Herbert I. Fusfeld, Director of the Center for Science and Technology Policy, under whose auspices the project was undertaken, for his support and advice throughout. In the early days of the project, Dr. Hedvah L. Schuchman helped to guide the project, both administratively and intellectually. Dr. Richard Langlois of the Center has helped us in innumerable ways throughout the project.

This work was supported by the National Aeronautics and Space Administration under grant NSG7636. The opinions, conclusions, and recommendations in this book are, of course, those of the authors, and do not necessarily reflect the views of NASA or its employees.

Government
and
Technical Progress

1

Public Policy and Technical Progress: A Cross-Industry Analysis

Richard R. Nelson

This study describes the nature of the public policies that have influenced the pace and pattern of technical progress in seven key American industries, and tries to assess the broad effects of these policies. The policies considered include funding or subsidy of certain kinds of research and development, but attention is also directed to government procurement, policies regarding education and training, information dissemination, patent protection and licensing, and, where germane, regulatory and antitrust policies. The industries studied are agriculture, pharmaceuticals, semiconductors, computers, civil aircraft, automobiles, and residential construction. These industries vary significantly in the pace and character of technical progress that has been achieved, institutional structure, and the government policies that have had the most important effects.

The present seems a particularly appropriate time for such a study. When there is an active search for new policies and a sense of urgency about the matter, there is little time or patience with broad historical and analytic reflection. Over the past two decades there have been three occasions of active policy interest. Only a short time ago, the Carter administration had a domestic policy review on industrial innovation, in search for policies that could restore America's lagging productivity growth and international competitiveness. Nearly a decade earlier, the Nixon administration engaged in a similar review of how federal policy could better spur industrial innovation, motivated by similar concerns that America was losing its position of technological leadership. In the early 1960s, the Kennedy administration attempted to mount a civilian technology program as part of its package of policies to lift the economy from the doldrums of the late 1950s. It is perhaps revealing that on none of these occasions did the government

1

agencies involved engage in thoughtful review of past government policies that have affected industrial innovation. Indeed, many of the documents read as if there were no such experience. Perhaps relatedly, the arguments, both pro and con, about policies tended to be global. They proceeded as if structural differences among sectors in industries of the American economy were slight, or as if feasible or appropriate policies were independent of these differences. In fact, past policies have differed significantly from sector to sector, and in ways that seem appropriately tailored to differences in economic structures or purposes or both. A central premise behind this study is that, if they are to be successful, public policies to stimulate technical progress need to be nicely tuned to the particulars of the different economic sectors.

Perhaps because there was no such historical reflection and analysis, few of the proposals that emanated from the forementioned attempts to formulate a policy were presented forcefully enough to persuade both the president and the Congress. Of those that were initiated, many were abandoned after a few years. The present, when there is little political pressure to find effective active policies to spur industrial innovation, seems an appropriate time for historical scrutiny and reflection.

In treating the question of appropriate government policy to support industrial innovation as an empirical one, we are in effect dismissing as uniformed the sometimes articulated position that government involvement in the innovation process is virtually always expensive folly. There are indeed many instances where government programs were just that. But, as the case studies we present will testify, there are other instances where the success of such programs has been outstanding. It is just this variation that calls for analysis.

In treating the question as one that warrants detailed empirical exploration, we acknowledge, reluctantly, that the general theoretical analyses and empirical observations of economists provide only limited and incomplete guidance regarding the kinds of policies that will pay off under different circumstances. Indeed the economic literature on this subject has grown progressively less conclusive.

A decade ago economists writing on the subject were stressing the limits of the ability of a business firm that finances an R&D project to appropriate and profit from the benefits that flow from that project, and the uncertainties that often are entailed in R&D seeking major technological advance. The former appeared to point toward the desirability of government policy to subsidize or supplement private R&D which would otherwise be conducted at less than the socially optimal level. The latter seemed to call for mechanisms for government sharing of risks on large and adventuresome projects. Over the past several years economists have come to recognize that the situation is much more complicated.

In the first place, it is now better understood that the protection of an invention by a patent or industrial secrecy leads not only to some restriction of its use (economists had long understood that) but also in some cases to duplicate or near duplicate R&D efforts by firms, which yield little net social value. This phenomenon casts doubt on the earlier logic that unaided private enterprise will spend "too little" on R&D, and calls attention to inefficiencies of the allocation of R&D among different kinds of projects that the industrial R&D system will generate. In the second place, economists now better recognize that the uncertainty surrounding efforts to advance a technology significantly call for the exploration of a variety of different approaches without premature heavy financial commitment to any, and warns that large-scale concerted efforts are, in general, inadvisable until the uncertainties have been significantly reduced. Again, the policy problem is better described in terms of a possible failure of the market to spawn the appropriate portfolio of projects than in terms of private expenditures being too small in the absence of government assistance.

While economists tended to diverge significantly a decade ago, about the appropriate roles of government in industrial R&D, there was consensus about the appropriateness, indeed the necessity, of governmental support of basic scientific research. That consensus has not become unglued, but it is now better recognized than was the case earlier that the simple statement masks an important policy issue. What is treated as basic research, with the proposed funding requests subject to peer review, and the research findings openly disseminated, is itself a matter of policy choice. While most of the R&D done by private for-profit business firms aimed at enhancing the design of their products is going to be treated by them as proprietary, and the research done on a basic theoretical problem by a physicist in the university is going to be treated by the researcher as contributing to public knowledge, research to improve seed varieties, or to discover a cure for a particular disease, or to identify and measure the properties of certain materials may eventuate in public or proprietary knowledge depending on who does it, the sources of the financing, and the precise form that the findings take on. As we shall see in the following studies, in several industries (agriculture, pharmaceuticals, computers, aviation), what the government in effect did was to define certain areas as basic and nonproprietary, and proceeded to fund research in these areas.

It has also been proposed that the government should fund R&D aimed at meeting public-sector needs, but should stay out of funding R&D on private-sector technologies. It turns out that this adage too provides little guidance. Regarding needs of the public sector, the government certainly can, and in many cases has, funded or even undertaken R&D

aimed to meet them better. But the fact that a demand is governmental does not automatically signal that government R&D is needed if innovation is to occur. For many public-sector needs, the government has not funded significant R&D. In many of these cases, private firms have funded R&D in order to create products that governments would find attractive and would buy. It is interesting that prior to World War II, much of R&D on military aircraft was privately funded.

To complicate the picture further, often no clean lines can be drawn between a technology or industry devoted to private needs, and one devoted to public needs. The most general case is overlap. Aircraft, computers, semiconductor devices that are used in computers, and more broadly, medicines, and buildings are inputs into both private and public-sector activity. As we shall see in the following chapters, federal support of the development of a technology for public-sector purposes often has led to capabilities that meet private demands as well.

Similarly, economists studying the relationships between economic structure and technological innovation, and speculating upon how the structure-innovation links might bear on government policy, now recognize better the complexities involved. Two decades ago the focus was on the proposition put forth by Schumpeter, and later echoed by Galbraith, that industries composed of large firms with significant market power tended to be significantly more progressive technologically than industries more atomistically organized. The implications of the hypothesis seemed to be twofold. First, government R&D might be needed in industries where the bulk of the firms were small. Second, a tough antitrust policy might be antithetical to technological progress.

Empirical research has revealed a more complicated picture than suggested by the simple Schumpeterian hypothesis. Some industries, dominated by large firms, are not technologically progressive. Some industries, populated by small and medium-sized firms, are very technologically progressive. The early days of the semiconductor industry provides a good case in point. The fact that firms are small does not automatically indicate that the industry can benefit from or even tolerate government R&D support. While government R&D support for agriculture, where the farms are small, is a success story, government attempts to advance house construction technologies have not been particularly fruitful. Nor does the fact that firms are large indicate that government R&D support will not be fruitful. Aviation is a case in point. Similarly, there are no simple implications for antitrust policy.

Further, industry structure and character of fruitful R&D tend to change over time. It is common, if not universal, for new industries to begin as a collection of many small firms with important technological developments coming from individuals or small groups of scientists and engineers. In many cases such an initial configuration tends to evolve over time into one in which viable firms are much larger and R&D projects much more costly. This seems to have happened in the semiconductor industry during the 1970s. Relatedly, government policies that are appropriate and feasible at one stage in an industry's history may not be appropriate or feasible at another stage.

Industry structure limits what government can do. Whether a government policy will be effective or not depends at least as much on the changes in the allocation of R&D it stimulates as on whether or not total R&D spending rises. In designing a program, or in evaluating one, the allocating mechanism is of central concern. Government agencies, however, in some circumstances are quite constrained regarding the range of allocation mechanisms they can employ effectively. In particular, there may be limitations on the information to which public officials have access. For example, if much of the information needed to make effective R&D decisions is proprietary, government officials are unlikely to be in a position to make detailed judgments. And, in a large pluralistic democracy like the United States, there are also likely to be political constraints on what governments can do. For example, the government is likely to be attacked as unfair if it pushes a program that obviously benefits one part of an industry at the expense of another part. On the other hand, where firms do not consider each other rivals (as in farming, or the practice of medicine), there are fewer constraints on governmental access and action. A public-sector mission, as in aviation, and computers, also can relax constraints.

The foregoing comments were designed to help the reader of the following seven chapters know what to watch for. These chapters were researched and written by the scholars who signed them. They differ considerably in scope and detail, reflecting the inclinations of the scholars and the particular salient features of the industries. All are organized, however, according to a common format. Each of the chapters describes the industry in question, and its evolution over time. Each chapter presents various descriptions, quantitative and qualitative, of the technological advances that have occurred, and attempts to trace the sources of those advances. The particular focus, of course, is on the government policies that have had the most significant influence. The industries studied and described differ significantly in all of the above respects.

Agriculture, or rather farming, is an industry where active government policies to stimulate technological advance date back to the middle of the nineteenth century. The federal-state-supported experimentation stations, and the agricultural extension services generally are affiliated with land-grant state colleges or universities, still another govern-

ment invention aimed to spur productivity in agriculture.
These programs have been enormously, sometimes embarrass-
ingly, successful. Not so long ago the United States felt it
faced a food glut. Interestingly, the response to that was to
establish a food-price support system, and to try to get land
out of cultivation, rather than to slow down the governmentally
fashioned engine of progress.

Pharmaceuticals is a different story, or rather a set of
different stories. Part of it is the massive government funding of biomedical research and the training of research scientists, largely a post-World War II development. Part of it is
the complicated regulatory structure that has evolved over the
years, first to check on the safety of new pharmaceuticals the
companies proposed to put on the market, later to assess the
efficacy of new drugs, and increasingly, to monitor and constrain the human experimentation parts of the research process. The story also includes antitrust litigation, issues about
patent life, and about whether physicians should be required
to prescribe generically, rather than by brand name.

Aviation is an industry where, from the beginning, a
strong national security interest has spilled over to facilitate
the development of civil aircraft as well as military. The
history contains the aborted, and in our eyes at least miscon-
ceived, supersonic transport effort, but it also contains a well
conceived and effective program under the National Advisory
Commission on Aeronautics, which later gave rise to NASA.
During the 1920 and 1930s, the NACA undertook research and
testing, which played an extremely important role in permitting
the development of the modern passenger airliner. Also, in
subsidizing the airlines (and the development of aviation)
through the Airmail Act of 1930, the government required that
airlines and airframe producers stand as separate corporate
entities. (Until that time there was a considerable degree of
vertical integration.) This structural policy has had, as we
shall see, a profound effect on technological advance in civil
aviation.

The semiconductor industry shares with aviation both the
fact of government national security interest, and the strong
influences of government policy with respect to structure.
While Bell Laboratories (where the transistor was invented)
and Western Electric earlier had decided not to produce transistors commercially and to make the technology generally
available, an antitrust decree blocked reversal of these
decisions. The semiconductor industry, like the airframe
industry, in its early days sold mostly to a government-made
market. The industry benefited greatly from the support of
research in basic physics and materials research sponsored by
agencies ranging from the NSF to the DOD. Likewise, the
industry was a beneficiary of a strong federal support given
during the 1960s to advance scientific and engineering education.

The first operational computers were developed on government contract. The early market for computers was largely governmental. The computer story and the semiconductor story are, of course, closely intertwined. But while in the semiconductor case, government policy led to an industry consisting initially of many firms no one with a major initial headstart over the others, in the computer case a dominant firm came into being very early in the game. (It might be noted that the consent decree kept Bell Labs and Western Electric out of the commercial computer business). Thus, the computer case typifies the antitrust policy dilemma that occurs when a firm comes to dominate an industry because (initially at least) it made shrewd judgments about where the technology and the market were going.

The automobile industry is one where the government's influence on the evolution of technology has been indirect and, until recently at least, unintended. At the present time the story is mainly about clean air and safety regulation, and the effect of these on R&D incentives and constraints. Policies affecting gasoline prices have also been important. The recent, and now aborted, Cooperative Automotive Research Program represents an attempt to define for the automobile industry a range of nonproprietary research for which federal funding would be appropriate. There are some interesting parallels with other industries, like farming, and pharmaceuticals, where a similar "nonproprietary" area has been defined.

Housing, or residential construction, is a sector where by all measures technological progress has been very slow. It often has been alleged that the government, through its building codes, and more recently through other forms of regulation, has been a large part of the problem. As we shall see, that is arguable. Residential construction is interesting for our purposes largely because it is a sector where the federal government has tried several times to organize an R&D support program, each time without much success. Each time, analogies were drawn to agriculture, but apparently the analogies were wrong, or at least incomplete. There clearly are some interesting issues here.

But already I am slipping over into a comparative discussion. The great advantage of collecting a number of different case studies, each of which covers comparable material, is that this permits comparison. The ensuing chapters of this volume will be explicitly comparative in nature, and will attempt to assess what kinds of policies are appropriate to what objectives and what industry structures. The concluding chapter will be explicitly comparative in nature, and will attempt to assess what kinds of policies are appropriate to what objectives and in what industry structures.

2
The Semiconductor Industry
Richard C. Levin

The astounding achievements made possible by advances in semiconductor technology are without parallel in recent industrial history. Computations that required a roomful of electronic equipment at work for minutes 25 years ago, or a roomful of mathematicians at work for days 50 years ago, now can be performed by a single microprocessor the size of a human fingernail in a fraction of a second. Technological improvements in speed and cost on this scale dwarf by orders of magnitude even the phenomenal improvements in transport technology over the past two centuries, and it is a reasonable conjecture that the full economic and social consequences of microelectronics technology will eventually be as profound as those occasioned by the succession of transport innovations from railroad to automobile to jet aircraft.

In this chapter, I shall describe and evaluate the role of public policy in facilitating technical advance in the developing semiconductor industry. In comparison with sectors such as agriculture and aviation, the contribution of public policy in microelectronics has been modest, but nevertheless of considerable significance. Without question, the most important policy instruments influencing technical advance have been the public procurement of electronic components and systems - principally by the military services - and public support for research, development, and production engineering - principal-

*This research was supported by National Science Foundation grant number PRA-8019779. My understanding of this subject has benefited greatly from valuable conversations with Richard R. Nelson, Robert W. Wilson, and executives of several leading semiconductor firms.

9

ly by the military and NASA, with some contribution from the National Science Foundation and the National Bureau of Standards. Accordingly, the use of these policy instruments will warrant careful attention in this discussion. Other policies are of relevance, however, and they will come in for some scrutiny. In particular, patent and antitrust policies have established certain ground rules that govern the competition of private firms for profitable innovations. Manpower polices, especially those involving public support for engineering education, have to some degree influenced the technological capability of the electronics industries. Finally, tax policies – especially those involving the tax treatment of capital gains, stock options, and R&D expenditures – may have influenced technological advance, especially through their effects on the contribution made by small firms and new entrants.

The plan of the chapter is as follows: section I provides a brief sketch of semiconductor technology and of key characteristics of semiconductor manufacture. Section II discusses the market structure of the semiconductor industry: the structure of demand and cost, conditions of entry, vertical integration, and the size distribution of firms. Against this background, Section III attempts to characterize the course of technical progress in microelectronics and to identify the sources of innovation. Section IV takes up the principal subject of the chapter, the contribution of public policy to semiconductor innovation, discussing each of the areas of policy mentioned above. Section V concludes with a brief treatment of policy issues currently under discussion and programs underway that may influence the future course of technical advance. In particular, current issues in tax, trade, and manpower policy will be discussed, and a major new program of military R&D support, the VHSIC (Very High Speed Integrated Circuits) program will be reviewed.

Before proceeding, one caveat is required. As a consequence of its growth and its strategic significance in the U.S. economy, the semiconductor industry has been the subject of extraordinarily intensive study. During the past decade, no less than six reasonably comprehensive industry studies have appeared: Tilton (1971), Golding (1971), Webbink (1977), Braun and Macdonald (1978), Department of Commerce (1979), and Wilson, Ashton, and Egan (1980). These studies differ somewhat in emphasis, but in large part they cover the same ground. In addition, there have been several monographs on the impact of specific government programs on semiconductor innovation: notably Asher and Strom (1977), Utterback and Murray (1977), and Charles River Associates (1981). Moreover, the technological history is also richly documented; a reader without technical training can absorb a great deal of semiconductor technology and its history from perusal of relatively recent special issues of Science (March 18, 1977),

Scientific American (September 1977), and IEEE Transactions on Electron Devices (June 1976). Taken together, this literature contains a reasonably complete record of the technological and economic history of the semiconductor industry, including description and analysis of government policies relevant to the industry's development. This chapter adds little new evidence to the abundant record; instead, it presents a synthesis and an interpretation of the existing literature. Thus, it may serve as an introduction to the role of public policy in the semiconductor industry, but it is unlikely to provide substantial new information to a reader who is already well acquainted with the industry and with the literature surveyed.

I. A SKETCH OF SEMICONDUCTOR TECHNOLOGY

Semiconductor Devices

A semiconductor, as the name suggests, is a material with electrical properties intermediate between those of conductors and insulators. In a conductor, such as a metal, electrons move freely through the material, and hence are free to carry an electric current when one is supplied. In an insulator, electrons are tightly bound to atoms or molecules, and they are therefore not available to carry electric current. In a semiconductor, free charge carriers are not usually present, but they can be generated with application of a modest amount of energy.

An atom of silicon has four electrons in its outermost shell. In the solid state, pairs of electrons are shared by neighboring atoms: in the crystal lattice each atom is thus surrounded by eight shared electrons. All electrons are committed to the bonds between atoms; hence, a crystal of pure silicon is a poor conductor of electricity.

By a process called "doping," a controlled number of atoms of another substance may be introduced into a silicon crystal, altering its electrical properties. For example, a phosphorus atom with five electrons in its outer shell may replace a silicon atom without disturbing the crystal structure, but its extra electron has no role in bonding the adjacent atoms. In the absence of external stimulus, the free electron remains in the vicinity of the phosphorus atom, and the material remains an insulator. But application of a small voltage across the crystal can mobilize the free electron, making it available to carry electric current. Similarly, a boron atom with three electrons in its outer shell may be introduced into the silicon crystal, creating a "hole" in the lattice, a deficiency of one electron. Holes are also immobile in the absence of stimulus but a small voltage can mobilize a hole to flow through the material, much as a bubble moves through a liquid. Silicon doped with phosphorus (or any other penta-valent element) is called an n-type semiconductor, where n stands for the negative charge of the free electrons. Silicon doped with boron (or another trivalent element) is called a p-type semiconductor, representing the positive charge of the holes.

Semiconductor devices rely on the properties of adjacent regions of n-type and p-type material to produce desired electrical effects. The simplest of semiconductor devices, a diode, consists of adjoining n-type and p-type regions in a single crystal of silicon. The distinctive property of a diode is its asymmetry. A diode conducts current of one polarity, yet offers high resistance to current of the opposite polarity. This asymmetrical response is an essential feature of the junction, or bipolar transistor, which is essentially two diodes joined back to back on a single crystal of silicon. That is, a junction transistor may consist of a region of p-type material sandwiched between two regions of n-type material (an npn transistor), or the reverse (a pnp transistor). As in a diode, current flow through a junction transistor can be switched on and off by reversing polarity. But, more importantly, by applying different voltages to each of the doped regions, current flowing through can be amplified. It was the search for a compact and reliable replacement for vacuum-tube amplifi-ers that led to Bell Laboratories' successful efforts to discover and develop the transistor.

Bipolar transistors were the dominant semiconductor devices of the 1950s, although an alternative device, the field-effect transistor (FET) was conceived some 25 years before the Bell Labs discoveries of 1947-1948. Although a working FET was constructed in 1951, it had no market impact until advances in process technology made production of field-effect transistors in significant volume economically practical. An FET consists of two islands of n-type silicon in a substrate of p-type silicon (or the reverse, two islands of p-type material in an n-type substrate). Unlike the bipolar transis-tor, however, the channel between the islands has no electrical connection; it is protected by a thin layer of insulating material, over which there is an electrode called a gate. Thus, only two doped regions of the transistor have direct electrical connections, but application of a charge to the insulated gate creates an electric field that influences the motion of the electrons in the substrate. The most common type of field-effect transistor uses a layer of metal for the gate, a layer of oxide (usually silicon dioxide) as the insula-tor; hence, the designation MOS (metal oxide semiconductor) for what became in the late 1970s the largest selling family of semiconductor devices.

Bipolar and MOS devices differ in their mix of functional attributes in a manner that has some economic significance. Bipolar devices tend to operate at higher speeds than MOS devices, but they consume more power. The high-speed feature of the bipolar technology makes it relatively attractive for applications involving real-time signal processing, such as radar and communications applications. The low power consumption of MOS devices, however, has made possible portable products like electronic calculators and electronic watches, which must use power sparingly to give satisfactory battery life. MOS technology also permits a denser packing of circuit elements, since lower power consumption produces less heat to be dissipated.

A transistor, while smaller and potentially more reliable than a vacuum-tube amplifier, is nevertheless simply an element in an electrical circuit. Like the other early semiconductor devices - principally diodes and rectifiers - transistors had to be connected with other discrete components to form a working circuit. Nonetheless, the invention and development of the transistor permitted substantial improvement in the performance and reliability of electronic computers. Transistors also gradually replaced vacuum-tube amplifiers in such consumer products as hearing aids, radios, and phonographic equipment, although certain applications of obvious potential - such as in television receivers and telephone equipment - were not realized for a decade or more. The truly extraordinary achievements of the microelectronics revolution required yet another major advance.

In 1952, not long after the invention of the transistor, a British physicist, G.W.A. Dummer, suggested that the next logical step beyond the transistor was "electronic equipment in a solid block with no connecting wires. The block may consist of layers of insulating, conducting, rectifying and amplifying materials, the electrical functions being connected directly by cutting out areas of the various layers" (Dummer, 1952). This extraordinary vision of an integrated electrical circuit, fabricated out of a single crystalline structure, was realized in a working model constructed just six years later by Jack Kilby of Texas Instruments. Like the first transistor built at Bell Labs by Brattain and Bardeen, Kilby's first integrated circuit would have remained an uneconomic curiosity without a major innovation in process technology. Kilby's device required that the connections between elements of the circuit be made by hand. At about the same time, however, Robert Noyce at Fairchild developed a technique that allowed metallic connections to be incorporated on a silicon surface in a batch process. Fairchild's "planar process" permitted the integrated circuit to become an economic reality.

There are two major classes of integrated circuits: digital and linear. Digital integrated circuits, which are the basis of modern computer technology, rely on the on-off switching properties of transistors and diodes to store information or to perform logical operations using binary arithmetic. Digital integrated circuits can be fabricated using either bipolar or MOSFET transistors, and there are several important families of devices that use each technology. For example, the earliest bipolar digital integrated circuits used a configuration of elements known as resistor-transistor logic (RTL). In the late 1960s, transistor-transistor logic (TTL) became the dominant type of digital integrated circuit, but emitter-coupled logic (ECL) was used for some purposes. More recently, devices employing integrated-injection logic (I^2L) have achieved significant savings in power consumption relative to TTL devices, although they operate at lower speeds. Circuits using p-MOS transistors were the first MOS devices to be widely used and remain the cheapest to fabricate. In high-performance microprocessors and semiconductor memories, n-MOS technology has largely replaced p-MOS, although the latter is still used in inexpensive equipment such as pocket calculators. Among the MOS devices, the lowest levels of power consumption can be achieved by complementary MOS (CMOS) devices, which employ both n-type and p-type field-effect transistors. Recent additions to the group of MOS technologies include silicon-on-sapphire devices (SOS), used in some military applications, and charge-coupled devices (CCD).

Linear integrated circuits are also fabricated with both bipolar and MOS technology, but these circuits are used to process analog signals. That is, they convert input signals to output signals in accordance with a predictable functional relationship, rather than simply operating as a series of on-off switches in the manner of digital circuits. Linear circuits are used, for example, as amplifiers in telephones and television receivers. They are much more expensive to design and fabricate than digital integrated circuits. Figure 2.1 illustrates the relationship of the various semiconductor devices discussed above.

Semiconductor Fabrication

There have been a series of important process innovations in semiconductor fabrication over the past two decades, but the basic process technology remains a recognizable descendant of the planar process developed at Fairchild Semiconductor in the late 1950s. In the following paragraphs, I sketch the stages of the production process in order to facilitate subsequent discussion of the economic characteristics of semiconductor fabrication.

Semiconductor fabrication begins with the manufacture of single-crystal silicon, an operation typically performed by

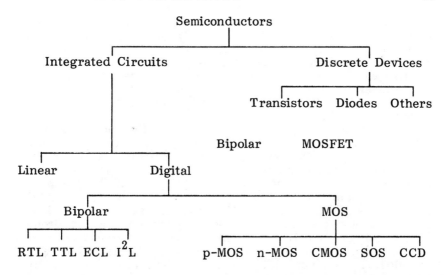

Fig. 2.1. Principal varieties of semiconductor devices.

chemical companies.   There are several techniques, but all involve the melting of polycrystalline silicon, and the controlled "growth" of a single crystal.   Dopants are introduced to the melt to impart the desired n- or p-type of semiconductivity.   The resulting material is then ground into a cylindrical rod four inches in diameter and sliced into thin "wafers" with diamond-edged cutting tools.   This is a delicate operation, since single-crystal silicon is an extremely brittle material, and the slices must be cut to a typical thickness of 0.5 mm.   The wafers are then polished to a mirror finish on one side.   The final steps in this process must be carried out in an absolutely clean environment.   For some purposes, an added layer of silicon is grown on the polished surface by the semiconductor manufacturer using a process called epitaxy.

The polished silicon wafer is heated to a temperature in the range of 1000 to 1200 degrees centigrade in an atmosphere rich in oxygen.   A layer of silicon dioxide forms on the surface of the wafer to a desired thickness determined by the temperature and duration of the process.   The oxide layer protects the doped silicon substrate and permits the selective introduction of additional dopants to specific areas of the wafer through openings etched through the oxide.   To accomplish this, the wafer is coated with a photosensitive material (photoresist).   The photoresist is exposed to ultraviolet light beamed through a master pattern, or mask, which contains an image of the desired array of openings. The photoresist is then washed with a solvent, which "develops" the image on the wafer by removing the film layer and exposing the oxide layer wherever the mask was opaque. Next, the wafer is placed in an acid solution that etches through the oxide layer to expose the silicon substrate. The remaining photoresist is then stripped away. The wafer is then ready for the introduction of dopants to the exposed areas.

The desired impurities, which alter the electrical properties of the semiconductor material, are then ordinarily introduced by a diffusion process. A batch of wafers are placed in a furnace at high temperature in a controlled atmosphere that contains the impurities. By carefully monitoring the time spent in the diffusion furnace the depth of the diffused layer can be regulated. An alternative to diffusion is a more recent technique called ion implantation, which propels a high-energy stream of dopant atoms into the exposed substrate. The preceding steps of oxidation, photolithography, and diffusion or ion implantation are then repeated a number of times using different masks in order to complete the desired circuit pattern. Finally, films of metallic and insulating materials are evaporated on the surface layer to form electrical connections within the circuit structure and to the outside.

Once the wafer-fabrication stage is complete, the individual circuits on the wafer are tested by needlelike metallic probes. Defective circuits are marked and discarded after the wafer is sliced into individual chips. Sometimes entire wafers prove to be defective. Testing at this stage permits avoidance of expensive further processing of defective parts.

The wafer is next "scribed" by a diamond cutting tool or a laser beam and sent to an assembly plant, where it is broken into individual chips, or dice. Each individual chip is then packaged into a ceramic or plastic case that protects the fragile chip and also permits its connection to the outside. This assembly stage involves two steps: die attach, where the bottom of the chip is cemented to the package, and bonding, where the circuit is welded using fine wires to metallic leads on the surface of the package. The bonding is then visually inspected, and the package sealed. Finally, the package is tested for structural integrity and the circuit is again tested for its electrical properties. Figure 2.2 illustrated the sequence of fabrication stages just described.

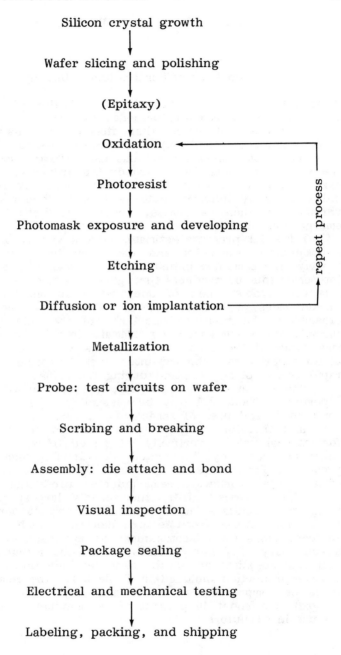

Fig. 2.2.   The stages of integrated circuit fabrication.

II. ELEMENTS OF MARKET STRUCTURE

The Structure of Demand:
End Uses of Semiconductor Devices

It is difficult to get a precise quantitative picture of the distribution of semiconductor sales by end use. There are no government statistics that itemize the distribution of semiconductor shipments by final market, except for data on government purchases for defense use. To get some sense of the various markets for semiconductor products, it is necessary to rely on a variety of estimates made by firms in the industry or by industry data services. While there is wide variance in these estimates, several qualitative conclusions emerge quite clearly.

Table 2.1 presents estimates from a variety of sources of semiconductor sales by end use over the past 20 years. Clearly, the computer industry has represented a major market for semiconductor products throughout the two decades, with a relatively stable share of one-quarter to one-third of open market sales of semiconductors. There has been a steady erosion in the relative importance of the military/aerospace market, and a corresponding dramatic increase in the relative importance of the consumer and industrial markets. It comes as no surprise that the consumer market experienced the most rapid growth of any segment during the 1970s.

There is at least one respect, however, in which the data reported in Table 2.1 may be misleading. The estimates are confined to end uses of semiconductors sold in the open market, and therefore exclude the production of semiconductors for internal use by vertically integrated firms. According to estimates made by Integrated Circuit Engineering (1980), production for internal use accounts for more than 20 percent of the total worldwide semiconductor production by United States-based firms. IBM, the world's largest producer of integrated circuits alone accounts for nearly 15 percent of the U.S. total. Apart from Western Electric, which manufactures semiconductors for telecommunications applications, and Delco, a subsidiary of General Motors, all of the remaining top ten "captive" suppliers are in the computer industry. Thus, were captive production included in Table 2.1., the relative importance of computers would rise markedly. The ICE figures suggest that about 40 percent of semiconductors are destined for use in computers.

Table 2.1. Distribution of U.S. Semiconductor Sales by End Use.

Percent of total semiconductor sales in year:

End Use	1960[a]	1965[b]	1968[a]	1972[b]	1972[c]	1974[d]	1974[e]	1979[e]
Computers	30	24	35	27	28	32	29	30
Consumer Products (calculators, watches automobiles, etc.)	5	14	10	18	22	22	24	28
Industrial Products (process controls, test equipment, office and telecommunications equipment)	15	26	20	30	26	30	33	37
Military/Aerospace	50	36	35	25	24	16	14	10

Sources: a. Texas Instruments, cited in Finan (1975).

b. William D. Witter, Inc., "Basic Report on the Semiconductor Industry for 1973/74," cited in Department of Commerce (1979).

c. J.P. Ferguson Associates, cited in Finan (1975).

d. Fairchild Camera and Instrument, cited in Department of Commerce (1979).

e. Dataquest, Inc., cited in Wilson, Ashton, and Egan (1980).

Cost and Supply: Yields, Learning, and Scale

The dramatic decreases in the real cost of semiconductor devices, and the even more astounding decreases in the cost per function (such as cents per bit of memory capacity) will be discussed in the following section on the course of technical progress.   However, to understand better the evolution of market structure as well as the trajectories of technical progress, it is important to note here several salient economic features of the fabrication technology described in the preceding section.

It should be evident that the process of semiconductor (especially integrated circuit) manufacture requires an extraordinary degree of precise environmental control.   The oxidation and diffusion processes must be carried out in sterile environments that are free of dust and impurities.   Temperatures and exposure times must be precisely controlled.   Silicon wafers, though capable of resisting high temperatures, are nevertheless extremely brittle and fragile.   And the minuteness of features etched into the surface of the wafers leaves little margin for error in the fabrication of photomasks, or in their alignment on the wafer.

An implication of these requirements for precision is that the yield of circuits that pass successfully through all stages of inspection and testing can be quite low.   Yields play a role at four distinct stages of the production process.   First, a certain percentage of wafers become broken, cracked, or otherwise damaged during the repeated cycles of oxidation, lithography, and diffusion; these must be discarded before wafer fabrication is completed.   Second, electrical testing at the wafer-probe stage leads to rejection of some fraction of the circuits.   Third, there is loss owing to improper assembly, and, finally, further losses arise when the finished, packaged chips are subjected to final electrical testing.

Table 2.2 illustrates the range of typical yields at each stage of the manufacturing process, as well as examples of yield ratios for both a mature and a new product.   It is important to note that both the lowest yields and the highest range of variation are found in the wafer-probe stage at the end of the wafer-fabrication process.   Most engineering attention to yield improvement thus focuses on the wafer-fabrication stage.   The yields on assembly and final test are typically much higher.   Since assembly and packaging operations are very labor intensive, more attention has been paid to finding low-cost labor for these latter operations than to seeking improved yields.   Much assembly work has been shifted to locations in the Far East, while wafer fabrication for the U.S. market is typically undertaken at home.

Improvements in yield are the principal source of the widely touted learning economies in semiconductor manufacture.

Table 2.2. Typical Yields in Integrated Circuit Manufacturing (in percent)

Process stage	Overall range of yields[a]		A typical mature product[b]		A typical new product[b]	
	Yield	Cumulative Yield	Yield	Cumulative Yield	Yield	Cumulative Yield
1. Wafer processing	75–95	75–95	80	80	70	70
2. Wafer probe: electrical test	5–90	3.8–85.5	40	32	20	14
3. Assembly: die attach and bond	80–95	3.0–81.2	90	28.8	85	11.9
4. Final electrical test	60–95	1.8–77.2	90	25.9	75	8.9

Sources: a. Integrated Circuit Engineering (1981).

b. J.P. Ferguson, cited in Finan (1975). The mature product is a standard TTL integrated circuit. The new product is an 1103 MOS integrated circuit.

The comparison between the new and the mature product in Table 2.2 is quite typical. With accumulated production experience for a given product, yields at all process stages tend to improve markedly, especially at the wafer-fabrication stage. Such decreases in unit cost with cumulative output are sometimes assumed to be a natural outgrowth of "experience." There is undoubtedly some relatively effortless learning by task repetition in the semiconductor industry, especially in handling and assembly operations where manual laborers learn to avoid typical mistakes or to use simple heuristics to guide performance of the task. But for the most part, "learning" is no mere artifact of "experience." Learning is a consequence of serious and sustained engineering effort to improve and perfect the production process. Moving down the learning curve is therefore not automatic; rather, cumulative output or production experience should be regarded as an input to the learning process.

In the early history of the semiconductor industry, economies of scale were believed to be insignificant. For reasons that will be more fully explained subsequently, however, the course of technical change has increased the significance of scale economies, largely through raising the minimum efficient scale of operation. The principal sources of the increase in efficient scale have been in the mounting costs of the fixed capital and R&D required to produce increasingly sophisticated and miniaturized products. These scale economies are not product specific, since process equipment, test equipment, and R&D effort can be used for multioutput production. As a consequence of learning, however, product-specific scale economies have actually been important for some time, although the point has not been widely appreciated. To the extent that learning economies take the form of improving process yields, learning permits the flow rate of output per unit of fixed capital to increase. As in chemical plants or in "stretched" versions of aircraft designs, then, increases in yield essentially extend the range of output rates over which unit costs fall. Thus, there is an interaction between learning economies and static economies of scale, or put differently, between economies to cumulative output and economies to the rate of output.

Concentration

In comparison with other industries in the U.S. manufacturing sector, the semiconductor industry has been and remains only moderately concentrated. According to Census data reported in Table 2.3, the four-firm concentration ratio for shipments of semiconductors and related devices in the United States has fluctuated between 39 and 53 percent from 1958 to 1977.

Table 2.3.   Concentration of U.S. Domestic
Semiconductor Shipments; Census Years:
1958 to 1977.

	Percent of total U.S. shipments (including value of captive production)				
	1958	1963	1967	1972	1977
Semiconductors and related devices (SIC 3674)					
4 largest companies	46	39	46	53	41
8 largest companies	64	59	62	66	60
20 largest companies	92	85	83	78	74
50 largest companies	99	97	96	91	87
Integrated circuits (SIC 36741)					
4 largest companies			67	69	46
8 largest companies			83	79	68
20 largest companies			97	91	88
50 largest companies			100	98	98
Transistors (SIC 36742)					
4 largest companies			56	70	66
8 largest companies			77	84	80
20 largest companies			95	97	94
50 largest companies			100	100	100
Diodes and Rectifiers (SIC 36743)					
4 largest companies			37	46	48
8 largest companies			61	62	68
20 largest companies			91	88	91
50 largest companies			100	99+	100

Source: Bureau of Census, Department of Commerce, Census of Manufactures, 1977 (Washington, D.C.: U.S. Government Printing Office, 1980).

There is a suggestion of an upward trend in the 1960s, with a sharp decline in concentration in the mid-1970s. But selected data from intermediate years give a less clear picture. For example, the Commerce Department (1979) reports a four-firm concentration ratio of 50 percent in 1965, which is more suggestive of a saw-tooth pattern than a trend. Perhaps more revealing than the fluctuating, yet modest, market shares of the leaders is the fact that the twenty-firm concentration ratio has fallen steadily for two decades, reflecting a continuing flow of entry and the diffusion of semiconductor technology.

When the market is more narrowly defined, at the five-digit product-class level of detail, concentration ratios are naturally higher, but in the youngest, largest, and most rapidly growing product market, integrated circuits, the trend is sharply downward over the 1970s. Older product groups have seen some increase in concentration in recent years.

Domestic concentration ratios may not be very meaningful in an industry that increasingly competes in a worldwide market. In the past decade, U.S. firms substantially expanded overseas production, while trade flows from the United States to Europe and from Japan to the United States grew rapidly. Therefore, a more realistic picture of market structure in the 1970s is probably given by Table 2.4, which presents worldwide (exclusive of the Eastern bloc) concentration ratios for 1972, 1975, and 1979. In these tabulations, captive production by firms that also sell in the open market is included, but the output of purely captive producers, such as IBM and Western Electric, is excluded. As the table indicates, worldwide concentration remained rather low throughout the decade. Moreover, important submarkets, like MOS integrated circuits, are not much more concentrated than highly aggregated markets.(1)

These moderate and relatively stable levels of concentration obscure three important underlying characteristics of the semiconductor industry's market structure. First, the industry is populated by several groups of distinctly different firms. Second, the entry of new firms has played an important role in the industry's technical development and market growth. And, third, there has been an extraordinary degree of turnover in market leadership, as technological competition has spurred the rapid expansion of successful innovators while less adaptive firms have fallen off the pace. I shall elaborate on each of these points in turn.

Varieties of Semiconductor Firms

One distinctive group of semiconductor firms consists of older established producers of electrical equipment and components. In the early 1950s these firms recognized that the invention of

Table 2.4.   Concentration of Worldwide
Semiconductor Shipments, 1972, 1975, and 1979.

	Percent of total worldwide shipments		
	1972	1975	1979
All semiconductors			
4 largest companies	32	32	30
8 largest companies	47	50	47
20 largest companies	n.a.	75	73
Integrated circuits			
4 largest companies	36	35	34
8 largest companies	49	53	52
20 largest companies	n.a.	80	78
MOS Integrated circuits			
4 largest companies	32	38	37
8 largest companies	51	60	58
20 largest companies	n.a.	n.a.	n.a.

Sources: Dataquest, Inc., Semiconductor Industry Services, Appendix B, 1980. Includes value of captive production by firms operating in open market; excludes firms with captive production only.

the transistor signalled the eventual displacement of the receiving tube as the predominant means of rectification and amplification used in electrical circuitry. The production of receiving tubes prior to 1950 and beyond was concentrated almost exclusively in eight firms, each a vertically integrated, diversified producer of electrical equipment. While the transistor was perceived as a threat to existing vacuum-tube business,(2) these firms also perceived it as an opportunity in the electrical equipment business. Thus, in contrast to producers of steam locomotives and propeller aircraft engines when challenged by a new technology, each of the leading receiving-tube firms moved quickly into transistor production in the early 1950s. By 1955, five of the eight major receiving-tube producers ranked among the top ten producers of transistors: General Electric, Philco, RCA, Sylvania, and Westinghouse. According to sources compiled by Tilton (1971), a sixth vacuum-tube firm, Raytheon, was among the top ten in 1957. Despite their early start, these vertically integrated producers of electrical equipment failed to maintain a strong position in the commercial semiconductor market. Only RCA has survived among the leaders in open market sales of semiconductors. Yet RCA does not compete directly with producers of high-volume, standardized circuits; it concentrates on more specialized circuits and on advanced technology for military application. General Electric, GTE (Sylvania), Raytheon, and Westinghouse are today insignificant factors in the commercial market, although they remain involved in the design and production of technologically sophisticated circuits for the military/aerospace market.(3)

The second important group of firms in the semiconductor industry are the so-called "captive" suppliers; these are vertically integrated firms that produce semiconductor devices exclusively (or almost exclusively) for internal use. Integrated Circuit Engineering (1981) lists nearly 50 firms with either semiconductor R&D, pilot production, or full-production capability, but fewer than 20 of these are actually engaged in full production. Most of the largest in-house producers are in the computer business; IBM, Hewlett-Packard, Honeywell, NCR, Data General, Digital Equipment, and Control Data all rank among the leading captive suppliers. While these firms do not operate as open market suppliers of semiconductor components, they do engage in open market purchases. Only IBM has been traditionally self-sufficient, but it has had recourse to the open market on occasion. Large captive production facilities are also maintained by Delco, a subsidiary of General Motors, and by Western Electric. The remaining large automobile and telecommunications equipment firms purchase semiconductors on the open market, although several, along with most aircraft and missile producers, maintain internal semiconductor laboratories.

Until 1982, Western Electric's position as a captive supplier was constrained by the terms of the 1956 consent decree, which resolved a lengthy antitrust action brought by the government. Under the decree, Western Electric, the manufacturing arm of ATT, was enjoined from the sale of semiconductors (and other products) in the commercial market, although sales for military and space applications were permitted. The effects of this and other provisions of the consent decree are discussed below in Section IV.

The third and largest group consists of all those firms that have entered into open market production of semiconductors since the early 1950s. Within this class of so-called "merchant" suppliers, there is considerable heterogeneity. First, with respect to origins, some merchant semiconductor firms were founded as new divisions of established firms in other lines of business. This subgroup includes previously large and diversified firms like Hughes Aircraft and Motorola as well as previously small specialized firms like Texas Instruments. Other firms, however, were essentially built from the ground up, organized exclusively for entry into the semiconductor business, like National Semiconductor, Intel, and dozens of others.

Second, there is considerable interfirm variation in strategic market positions among merchant semiconductor houses. Some firms, like Texas Instruments and Motorola, sell a broad line of semiconductor products, including a wide range of memory devices and microprocessors as well as older discrete devices. Others sell only a narrow range of products. Intel, for example, does not manufacture discrete devices at all; it concentrates almost exclusively on high performance n-MOS integrated circuits. A pioneer in random-access memory devices, Intel has recently deemphasized its activity in this area and concentrated on microprocessors and other logic devices. By contrast, Mostek specializes in MOS memory devices, Monolithic Memories (MMI) produces only bipolar integrated circuits, and Unitrode and Varo specialize in discrete devices.

Third, among the merchant houses, there has been significant variation in the degree and timing of forward vertical integration. Texas Instruments was an early and successful entrant into consumer electronics, but several other U.S. semiconductor houses have fared poorly (e.g., Fairchild in the digital watch business).

Finally, some industry participants believe that it is fruitful to distinguish between two very different "cultures" in which merchant semiconductor firms operate. According to this view, the "Texas" and "California" branches of the industry differ significantly in intrafirm organization and incentives, in employee mobility, in the interfirm flow of technological information, and in political orientation. Texas Instruments has successfully engendered an institutional loyalty that is rare in Silicon Valley. As a consequence, TI has been less vulnerable to attempts by competitors to hire away key employees, and it has found it easier to protect proprietary technology. By contrast, scientists and engineers working in Silicon Valley tend to identify more strongly with the professional culture of the semiconductor industry than with a particular firm. The unusually free exchange of technical information and the mobility of technical personnel in the California branch of the industry has had significant implications for the character of technical advance and a bearing on the efficacy of certain government policies, which will be discussed below.

Entry

An important feature in the development of semiconductor technology is the contribution of newly established firms. The frequency of entry and the success realized by a substantial number of new firms have been facilitated by at least three significant aspects of the industry's structural and behavioral environment. First, in the years prior to 1970, capital investment requirements were relatively low and venture capital was readily available. In the early 1950s, General Transistor became a successful manufacturer of transistors with an initial investment of $100,000. Transitron was launched for $1 million, and it maintained a position as a sales leader for more than a decade. Technological leadership came at a higher price; Texas Instruments invested about $4 million before its semiconductor operations became profitable. Entry costs rose during the 1960s, but a state-of-the-art manufacturing capability probably cost no more than $10 million by 1970.(4)

Second, although several inventions essential for economic semiconductor manufacture were protected by patents in principle, in practice the owners of these patents either pursued liberal licensing policies or neglected to prosecute infringements. The reasons for this behavior will be explored in Section IV.

Finally, entry was facilitated in the early years by the ability of a few key employees to appropriate and transfer process and product design know-how sufficient for viable operation. Indeed, the typical new firm was a "spin-off" of an established business founded by a team of key technical personnel and backed by venture capitalists or occasionally by firms seeking diversification. Thus, in the well-known genealogy of the Silicon Valley, Bell Labs begat Shockley Transistor, Shockley begat Fairchild, Fairchild begat Signetics, General Microelectronics, and Intel, among others, and each of these fourth-generation firms has numerous progeny of its own.

Since 1970 the costs of entry has risen dramatically. According to Moore (1979), the man-hour requirements of circuit design have increased more than fivefold in the last decade. The cost of photomasking equipment has increased substantially. Indeed, the cost of electron-beam writers in the era of very large-scale integration (VLSI) is expected to exceed the cost of optical printers used in current LSI technology by a factor of six or more (Robinson, 1980). Indivisibilities in design effort and best practice capital equipment have served to increase the minimum efficient scale of fabrication. These related trends imply that efficient-scale entry at or near the frontier of integrated circuit technology is many times more costly than it was a decade ago. Moreover, as the production process has grown in scale and complexity, and as circuit design has become more interactive with the parameters of particular production facilities, it has become increasingly difficult for a small number of employees to carry away sufficient know-how to establish a viable and highly competitive advanced production capability.

The evidence on new entry is consistent with the observed increases in capital requirements and minimum efficient scale. Among a sample of 90 semiconductor firms studied by Wilson, Ashton, and Egan (1980), 25 entered the industry between 1951 and 1959, a rate of 2.78 new firms per year. The entry rate accelerated in the early 1960s and again from 1968 to 1971, so that the average annual number of new firms from 1960 to 1972 was 4.69. Yet despite rapid market growth after 1975, only four new firms entered over the period 1973-1978. This precipitous decline in the rate of entry coincides with the collapse of the U.S. venture capital market, but it seems unlikely to be wholly the consequence of reduced capital availability. When conditions in the venture capital market improved in 1979, the flow of entry resumed, but the character of the entrants changed significantly. Recent entrants have not attempted to compete in the high-volume production of standardized circuits, as did a number of successful new ventures in the middle and late 1960s (notably Intel, Mostek, and AMD). Rather, recent entrants have sought to fill specialized niches in the marketplace, especially in the related areas of custom circuit design and fabrication, computer-aided design (CAD), and custom software. Semicustom design and fabrication, where silicon wafers are processed for various applications in identical fashion up to a final step of one or two custom-designed masks, has also been an attractive area of specialization for recent entrants (Integrated Circuit Engineering, 1981).

Competitive Dynamics

The reported indices of market concentration conceal, in addition to the variegated nature of the firms involved and the relative ease of entry, a remarkable rate of turnover among market leaders. As Table 2.5 indicates, only one (Texas Instruments) of the top five U.S. producers of transistors in 1955 is among the top ten producers of integrated circuits today. Five of the top ten integrated circuit producers in 1975 were not among the top ten semiconductor firms a decade earlier, and four of these firms were established after 1960. Table 2.6 summarizes the turnover in this industry by reporting the probabilities of survival among the top five and top ten firms over the specified intervals. To appreciate the extent of instability, consider that the identity of the top seven U.S. steel producers has not changed since 1955. Indeed, even the ranking within the top seven has been virtual-

Table 2.5. Leading U.S. Semiconductor Manufacturers: 1955-1980

1955 Transistors	1960 Semiconductors	1965 Semiconductors	1975 Integrated Circuits	1980 Integrated Circuits
Hughes	Texas Instruments	Texas Instruments	Texas Instruments	Texas Instruments
Transitron	Transitron	Motorola	Fairchild	National Semiconductor
Philco	Philco	Fairchild	National Semiconductor	Motorola
Sylvania	General Electric	General Instrument	Intel	Intel
Texas Instruments	RCA	General Electric	Motorola	Fairchild (Schlumberger)
General Electric	Motorola	RCA	Rockwell	Signetics (Philips)
RCA	Clevite	Sprague	General Instrument	Mostek (United Technologies)
Westinghouse	Fairchild	Philco-Ford	RCA	Advanced Micro Devices
Motorola	Hughes	Transitron	Signetics (Philips)	RCA
Clevite	Sylvania	Raytheon	American Microsystems	Harris

Sources: For 1955-1975, I.M. Macintosh, "Large Scale Integration: Intercontinental Aspects," IEEE Spectrum, June 1978, p. 54, cited in Wilson, Ashton, and Egan (1980), p. 23; for 1980, Integrated Circuit Engineering, Status 1981: A Report on the Integrated Circuit Industry, (Scottsdale, Ariz.: ICE, 1981), p. 62.

Table 2.6.   Market Leader Survival Rates of
U.S. Semiconductor Firms.

A.   Probability of Survival among Top Five Firms

From \ To	1960	1965	1975	1980
1955	.6	.2	.2	.2
1960		.4	.2	.2
1965			.6	.6
1975				1.0

B.   Probability of Survival among Top Ten Firms

From \ To	1960	1965	1975	1980
1955	.9	.6	.3	.3
1960		.7	.4	.4
1965			.5	.4
1975				.7

Source: Derived from Table 5.

ly unchanged, with only the slightest movement in rank among
the firms ranked fourth through sixth (see Oster, 1981).
     Even the data in Tables 2.5 and 2.6 conceal by aggrega-
tion much of the turbulence of "creative destruction" in the
semiconductor industry.   In a sense, each significant new
product innovation launches a technology race from which one
or two firms usually emerge with the lion's share of the
market.   At any point in time, there are dozens of product
markets, and most firms are unlikely to be among the leaders
in more than a few.   Until recently, both Intel and Mostek,
for example, had considerable success in successive generations of MOS random-access memory devices, and Intel has
been a leader in microprocessors as well.   But neither of these

firms has any significant involvement in bipolar integrated
circuit markets.   Texas Instruments is probably the only firm
that repeatedly appears among the survivors of technology
races across the complete spectrum of integrated circuit
products: bipolar and MOS, logic and memory devices.

The semiconductor industry offers one of the most striking examples available of "Schumpeterian competition" (see
Nelson and Winter, 1978; Futia, 1980; and numerous references
cited in Levin and Reiss, 1981), where the size distribution of
firms at any point in time is determined by the history of
successful and unsuccessful attempts to innovate and to imitate
the innovators.   Thus, successful innovators like Intel and
Mostek emerged rapidly from nowhere to assume a position of
market leadership.   Other would-be innovators like Sylvania
and General Electric, despite some early technical achievements, were unable to develop commercially successful new
products and declined.   A strategy of imitation - involving
high-volume low-cost production using the technical advances
of rivals - paid off early for Transitron and later for National
Semiconductor.   But successful imitation in a fast moving
technology requires a substantial investment in R&D to simply
adapt and utilize new ideas generated by others.   Thus, Transitron's failure to maintain a sizeable R&D program is held to
be responsible for its precipitous decline in market share.
[see Tilton (1971)].

III.   THE CHARACTER OF TECHNICAL PROGRESS

Measures of Technical Advance

No scalar measure of technical change can adequately capture
the phenomenal improvements in computing speed and power,
in signal processing capability, in industrial process control,
and in the range of available consumer products made possible
by innovation in the semiconductor industry.   While conventional productivity indices fail to adequately account for the
social gains attendant upon the introduction of wholly new
products or product attributes, one might nevertheless hope
that standard measures of productivity change in the semiconductor industry would help to calibrate at least in a rough
sense the pace of technical progress.   Unfortunately, it is
very likely that conventional productivity measures drastically
understate the true pace of technical advance.

First, despite frequent revisions in the product categories
used by the BLS as components of a sectoral price deflator,
the revisions typically lag by several years the introduction of
major new classes of semiconductor products.   Consequently,
available price indices fail to reflect the steep decline in

relative prices of of new products over the first years of the life cycle. Second, systematic bias arises from nonhomogeneity within individual price index components, such as MOS memory devices or TTL memory devices. As the consumption mix shifts from one kilobit (1K) to more expensive 4K and 16K devices, the average price per device falls much more slowly than the average price per unit of memory capacity. Thus, real output growth is understated by failing to capture systematic improvements in quality even within established product categories.

Despite these serious limitations, the productivity and price data are nevertheless of some interest as indicators of a very rough lower bound on the rate of overall technical advance. Total factor productivity data are not available at a disaggregated level, but the aggregated data are suggestive. Kendrick and Grossman (1980) report rates of postwar (1948-1976) productivity growth for 31 industries groups (mostly at the two-digit SIC level). Of these sectors, the two highest rates of total factor productivity growth were realized by the communications industry (SIC 48), where semiconductors are an important input, and by the electric and electronic equipment industry (SIC 36), of which the semiconductor industry is a part. The average annual rate of TFP growth was estimated to be 4.2 percent in communications and 3.7 percent in electric and electronic equipment, compared with a 2.3 percent rate of increase for the U.S. business economy as a whole.

For the semiconductor industry itself, the best available data are crude indices of labor productivity constructed from data on industrial production and employment. Webbink (1977) uses Federal Reserve data on industrial production and BLS employment data to calculate an annual index of output per worker for the semiconductor industry (SIC 3674 and 3679) from 1958 through 1974. His figures may be used to derive an average annual rate of growth of labor productivity of 5.1 percent. In contrast, the corresponding rate of growth for all manufacturing industry is 3.5 percent. From the Department of Commerce's (1979) compilation of Census data on value added and employment, one may infer an 8.0 percent annual increase in labor productivity from 1958 through 1975, but the document warns that some manufacturing stages performed overseas may be partially counted in value added, although the employment figures are purely domestic.

Finally, Finan (1981) recently completed a study of productivity for the Semiconductor Industry Association, using survey data collected from merchant semiconductor firms. His figures indicate a remarkable 20.5 percent annual increase in labor productivity from 1975 through 1979, a period in which productivity in the economy as a whole was virtually stagnant and in which productivity in the electrical equipment sector rose only 2.6 percent per annum. The utility of these most disaggregated of the available data is somewhat marred by the short sample period, which reflects a trough-to-peak phase of the business cycle. To put Finan's figure in perspective, it is well to note that the index computed by Webbink grew at an annual rate of 14.0 percent over a comparable phase of the business cycle in the early 1970s, and yet the long-term average growth rate was only 5.1 percent. Table 2.7 summarizes the various available measures of productivity growth.

Since the accuracy of productivity measurement in the semiconductor industry is highly suspect, it is useful to supplement Table 2.7 with a brief look at price movements for a few specific product categories. To begin somewhat broadly, consider Table 2.8, which reports average nominal prices per unit of transistors and integrated circuits. The record of price decline is extraordinary even without taking account of the fact that an average digital integrated circuit in 1972 had two orders of magnitude more active electronic elements than an average "unit" in 1964. Moreover, while the price of a digital circuit appears only to fall by a factor of two from 1968 to 1972, it should be noted that this period witnessed both a sixteenfold increase in the storage capacity of the best available random-access memory chips and the introduction of a revolutionary new (and initially high-priced) product, the microprocessor. Similarly, though perhaps less dramatically, the price series on transistors understates the true extent of technical change, since an average transistor by the mid-1960s operated at higher speeds, consumed less power, and was far more reliable than its predecessor of a decade before. In interpreting the transistor price data, it is useful to note that germanium was the dominant material used in the fabrication of transistors until the mid-1960s. Indeed, the quantity of germanium transistors produced annually grew steadily until 1966, the year when it was finally surpassed by silicon in physical volume. By 1972, only 5 percent of transistors produced in the United States were made of germanium.

The data in Table 2.8, which are derived from the Current Industrial Reports of the Bureau of Census, cannot be carried forward comprehensively beyond 1972 because of incomplete reporting of quantity data. Nevertheless, the piecemeal data that are available reveal continued nominal price decreases in the face of unprecedented price inflation in the economy as a whole. For example, the Current Industrial Reports combine silicon and germanium transistors after 1972, but the average unit price declines from 35 cents in 1974 to 27 cents in 1976. Similarly, the average price of a linear integrated circuit declined to 86 cents by 1976.

To round out the picture conveyed by gross statistical indicators of technical progress, some reference to the patent data seems warranted. Reclassification of patent categories impairs this effort somewhat, but even more problematic are

Table 2.7. Alternative Estimates of Productivity Growth

Source	Level of Aggregation	Measure	Time Period	Growth Rate
Kendrick and Grossman (1980)	Communications (SIC 48)	Total factor productivity	1948-1976	4.2
	Electric and electronic equipment (SIC 36)	"	"	3.7
	U.S. business economy	"	"	2.3
Webbink (1977)	Semiconductors (SIC 3674, 3679)	Industrial production per employee	1958-1974	5.1
	U.S. manufacturing industry	"	"	3.5
Department of Commerce (1979)	Semiconductors (SIC 3674)	Value added per man-hour	1958-1975	8.0
Finan (1981)	Semiconductors (Merchant semiconductor firms)	Value added per employee	1975-1979	20.5
	Electric and electronic equipment (SIC 36)	"	"	2.6
	U.S. manufacturing industry	"	"	2.1
	U.S. economy	"	"	1.0

Table 2.8. Average Price Per Unit of Transistors and
Integrated Circuits, 1954-1972
(in current dollars).

	Transistors		Integrated Circuits	
Year	Germanium	Silicon	Digital	Linear
1954	3.56	23.95	–	–
1955	2.88	20.44	–	–
1956	2.34	19.94	–	–
1957	1.85	17.81	–	–
1958	1.79	15.57	–	–
1959	1.96	14.53	–	–
1960	1.70	11.27	–	–
1961	1.14	7.48	–	–
1962	.82	4.39	–	–
1963	.69	2.65	–	–
1964	.57	1.46	17.35	30.00
1965	.50	.86	7.28	28.83
1966	.45	.64	4.34	13.39
1967	.43	.58	2.98	6.18
1968	.41	.44	2.17	3.35
1969	.37	.37	1.58	2.22
1970	.41	.38	1.42	1.86
1971	.46	.33	1.22	1.48
1972	.52	.27	1.01	1.08

Source: Electronic Market Data Book 1979, pp. 106-107.

the conceptual difficulties associated with the use of patents as
a measure of innovative activity in this industry. Quite apart
from the well-known problem that patents are highly nonho-
mogeneous in economic value, in the semiconductor industry
firms differ widely in the extent to which they seek patents on
their inventions. Many semiconductor innovations, especially
those involving the physical layout of circuits, are unpatent-

able. For many others, patents are unenforceable because of widespread cross-infringement. Thus, a number of strategic considerations influence a firm's decision to seek a particular patent; among these are the firm's patent position relative to its rivals, its employee incentive schemes, and management's aversion to the encumbrance of legal processes. The nature of patent protection will be discussed in more detail below, but it is important to note that there are substantial interfirm differences in patent strategy. Intertemporal differences in patenting behavior may be present as well. According to some industry participants, growing awareness of the inefficacy of patents as a means of protecting proprietary knowledge has led to a general reduction in their use.

For what they are worth, aggregate patent time series reveal a pattern of acceleration through the 1950s and 1960s, followed by some signs of decline in the 1970s. Table 2.9 indicates the annual total of U.S. semiconductor patents awarded to 40 leading firms over the period 1952-1968. These data were collected by Tilton (1971) from Patent Office records, and the total excludes all nonproducers of semiconductors and many small semiconductor firms. Tilton does not indicate what patent classes are included in the count, but he claims that the data cover all patents on semiconductor devices, semiconductor fabrication, manufacturing and testing equipment, and applications of semiconductors in final electronic products where their use is noted in the title of the patent.

Data on patents granted since 1968 were made available by the Office of Technology Assessment and Forecast. The first series covers what the Patent Office considers to be semiconductor inventions and is considerably broader than Tilton's definition; a far wider array of systems applications of semiconductors are counted. The second, more recent series covers a narrower patent class: semiconductor device patents (class 357) only. It is interesting to note that the former series exhibits a steady decline while the latter is reasonably level. One can only conclude that domestic patenting activity has declined sharply in semiconductor process technology and in downstream applications, though not in semiconductor devices.

Given the flaws inherent in each of the conventional economic measures of technical progress, it is instructive to look directly at the improvement in key technical parameters of semiconductor performance. A natural place to begin is with the empirical regularity observed by Gordon Moore of Intel, a phenomenon that has come to be known as Moore's Law: from the early 1960s to the mid-1970s, the density of active circuit elements per silicon chip doubled every year. This remarkable rate of miniaturization has profound implications for the speed and cost of computation. For example, computer memory capa-

Table 2.9.   U.S. Semiconductor Patents Granted to
Domestic Firms, 1952-1980

Year	(1) Semiconductor patents granted to 40 leading firms	Year	(2) Total Semiconductor- related patents	(3) Semiconductor device patents
1952	60	1968	–	141
1953	92	1969	1833	225
1954	79	1970	1705	158
1955	73	1971	2004	256
1956	186	1972	1639	158
1957	174	1973	1447	163
1958	307	1974	1242	170
1959	346	1975	1235	184
1960	322	1976	1266	190
1961	341	1977	1121	203
1962	440	1978	1003	142
1963	328	1979	804	143
1964	325	1980	–	184
1965	621			
1966	583			
1967	479			
1968	372			

Sources: (1) Tilton (1971), p. 57; (2) Office of Technology
Assessment and Forecast, U.S. Department of Commerce, special computer run, 1980; (3) Office of
Technology Assessment and Forecast, U.S. Department of Commerce, special computer run, 1981.

city, which cost as much as $1 per bit in the late 1950s, could be purchased in 1981 for $.0001 per bit. The pace of improvement has slackened somewhat in the past few years but circuit density still doubles within less than two years.

Finally, a somewhat broader picture of improvement is given by direct comparison of the technical parameters of a modern microcomputer (the Fairchild F8, introduced in 1975) with those of ENIAC, the first electronic digital computer completed in 1946. Table 2.10, drawn from Linvill and Hogan (1977), indicates that the Fairchild device, which has since been surpassed in speed, power consumption, and storage capacity, is 20 times faster and 10,000 times more reliable. The older computer consumed 56,000 times more power, was 60,000 times heavier, and required 300,000 times more space. The ENIAC was an experimental prototype; the first commercially available electronic computer, UNIVAC I, sold for prices upward of $1 million (see Katz and Phillips, 1982). The Fairchild F-8 sold for about $100 in 1977.

Table 2.10.   Comparison of ENIAC
and the F8 Microcomputer.

Characteristic	ENIAC	F8
Size	3,000 cubic feet	0.011 cubic feet
Weight	60,000 pounds	< 1 pound
Power consumption	140,000 watts	2.5 watts
Speed (clock rate)	100 kilohertz	2000 kilohertz
Read-only memory	16 K bits	16 K bits
Random access memory	1 K bits	8 K bits
Transistors or tubes	18,000 tubes	20,000 transistors
Resistors	70,000	none
Capacitors	10,000	2
Relays and switches	7,500	none
Mean time to failure	hours	years

Source: Linvill and Hogan (1977), p. 1111.

The Course of Technical Advance:
The Discrete Device Era

The history of innovation in the semiconductor industry has often been recounted (see especially Braun and Macdonald, 1978, and the references cited therein), and it seems unnecessary to reconstruct a thorough history here. Instead, the aim of this and the following subsection is to provide an interpretive characterization of the course of technical progress. In this regard, it is proposed that the history of the industry is best divided into two technological regimes: the discrete device era, inaugurated by the invention of the transistor in 1947, and the integrated circuit era, which dawned around 1960. To summarize the dominant tendencies, the invention of the transistor was followed by a sequence of key process innovations that gradually made possible the production of a high volume of reliable discrete devices at very low cost. By the time the integrated circuit was invented, the industry's basic process technology was in place, and since the early 1960s the technology has moved along a reasonably well defined natural trajectory of miniaturization.

The story of the invention of the transistor by the team of Bardeen, Brattain, and Shockley at Bell Laboratories in 1947 has been recounted in detail by Shockley (1976, and in earlier accounts cited therein) and usefully interpreted by Nelson (1962). Among the valuable lessons to be learned from these accounts is that in the generation of major inventions the forces of demand and supply work in roundabout ways.(5) There was certainly a perceived demand for invention. Indeed, Shockley (1976) recalls being most impressed by the foresight of Mervin Kelly, the research director at Bell Labs, who recognized in the late 1930s that the mechanical relays in telephone exchanges would have to be replaced eventually by electronic connections in order to accommodate the growing volume of telephone traffic. According to Shockley's own account, he perceived as early as 1938 or 1939 that Kelly's objective of electronic switching might be achieved by using phenomena in solid-state physics rather than vacuum-tube technology.

Despite the perceived need for a solid-state amplifier, the path to a working transistor was far from direct. Much of the theory of semiconductor materials had been established by the early 1930s, and in this sense the transistor was an innovation that was based on science. But applying the abstract scientific theory of semiconductors to the creation of a working amplifier required both a deepening of scientific understanding and the resolution of practical (technological) difficulties.

Specifically, Shockley's initial amplifier design, tested unsuccessfully in 1939, was based on a fairly direct application of scientific knowledge. This early design sought to operationalize the principle of the field-effect transistor which served subsequently as the basic active device in MOS integrated circuits. But it was almost a quarter of a century before theoretical and practical barriers to the construction of a viable field-effect device were overcome. Shockley's own efforts to build a field-effect amplifier were resumed after World War II, but he failed once again. His colleagues, Brattain and Bardeen, set out to understand why, and by mid-1947 Bardeen proposed that the failure of the field-effect device was a consequence of electrons being trapped in what he called surface states. Brattain and Bardeen then commenced a series of experiments to test this new theory of surface states. In the course of these experiments, almost by accident, in December 1947 Brittain and Bardeen discovered that by closely spacing two electrodes on a germanium crystal, an amplification or "transistor" effect was achieved. A satisfactory theoretical explanation of why this "point-contact" transistor worked was not immediately available, but the discovery led Shockley to recognize that the effect relied somehow on the injection of positive-charge carriers (holes) into a negatively doped (n-type) semiconductor. Taking account of these "minority carriers" led Shockley directly to the theoretical insight that permitted the conception of yet a third, previously unforeseen, amplifier design - the junction transistor, which in fact became the basis for the bipolar technology of the discrete device era.

It is clear from this sequence of events that the "supply side" - the underlying opportunities for innovation created by scientific knowledge - plays an important, but not a straightforward, role. Science-based technological opportunities are not just there to be plucked like tree-ripened fruits; scientific theory and experimental fact interact in complicated ways. The field-effect transistor, the device predicted by prior theoretical understanding (the apparent fruit to be plucked), failed to work. The first working device, the point-contact transistor, had no prior basis in theory, yet empirical observation of this device led to deeper theoretical understanding and to a new device, the junction transistor, based on the enriched theory.

Translating Shockley's theoretical conception to a commercial product required essential process innovation. Management at Bell Labs and Shockley himself initially thought that ordinary polycrystalline germanium or silicon would prove a satisfactory material from which to construct junction transistors, but Gordon Teal insisted that imperfections in the crystalline structure would vitiate the goal of producing low-cost, reliable and uniform parts. He worked for two years to develop a commercially feasible method of pulling single crystals of germanium (see Teal, 1976). By 1952 the commercial production of junction transistors was underway at Western Electric.

The earliest germanium junction transistors were expensive to produce. They contained significant levels of impurities, impairing reliable performance. Operation was restricted to relatively low temperatures and to low electrical frequencies. Moreover, germanium was a relatively rare material. In the form of germanium dioxide, it cost about $300 per kilogram in 1952; it was worth more than gold after purification (Braun and Macdonald, 1978). Thus, in the early 1950s, the R&D agenda for improvement of the transistor was reasonably clear. Effort was devoted to seeking greater materials purity and hence greater device reliability, to developing devices with a greater range of feasible operating conditions, and to exploring the possibility of using semiconductor materials other than germanium.

Success was achieved in each of these areas over the course of the first decade. Further materials research at Bell Labs led to the zone-refining techniques, which permitted the production of germanium several orders of magnitude more pure than obtained with Teal's original methods. To expand the range of operating conditions, in 1952 engineers at General Electric developed a method of alloying indium to a thin wafer of germanium. If the germanium layer were sufficiently thin, the alloy junction transistor could operate at much higher frequencies than the original Western Electric devices. Further development at Philco in 1953 of a jet-etching method made practical the achievement of the desired thinness. With the new technique, Philco manufactured a surface barrier transistor of very high quality, but the thinness of the germanium layer rendered it rather delicate.

Meanwhile, engineers at Bell Labs and at General Electric pursued yet another tack in order to achieve more reliable devices with the desired frequency performance. The diffusion process permitted impurities in vaporized form to penetrate the semiconductor material in precisely controlled quantities. The effectiveness of the diffusion process was substantially enhanced by the selective use of oxide masks that regulated the diffusion of dopants to desired areas of the semiconductor material. A significant feature of the diffusion process was that semiconductors could be processed in batches rather than only individually. Thus, the superior precision of the diffusion process improved device reliability and frequency response, and the transition to batch production process greatly reduced cost.

Despite major improvements in the frequency range and reliability of transistors, the use of germanium still greatly restricted the application of transistors in high-temperature environments. When Gordon Teal moved in 1953 to assume the directorship of the Central Research Laboratory at Texas Instruments, he immediately launched a program to improve techniques for pulling single crystals of silicon. Texas

Instruments pursued the dual objectives of becoming the first producer of a silicon transistor and the innovator in the commercial production of high-purity silicon for use by the entire industry. Both objectives were realized quickly: the first by 1954, the latter by 1956.

Silicon became the dominant semiconductor material only gradually; production of silicon devices did not overtake production of germanium devices until 1966. The military provided a major source of demand for silicon, because of its need for devices capable of operation at high temperatures. For commercial purposes, however, germanium had superior frequency response characteristics at normal temperature, and it was initially cheaper to produce (despite the higher cost of the raw material), because the technical difficulties of purifying silicon were only gradually overcome. For the long run, however, silicon offered the advantage that it was more suitable for the formation of oxide layers and for the deposition of metallic films. This rendered silicon the material of choice in the subsequent development of integrated circuits.

The drive toward cheaper, more reliable devices capable of operating at high frequencies and high temperatures culminated in the crowning achievement of the discrete device era - the development of the planar process at Fairchild Semiconductor in 1958. The immediate consequence of this innovation was to render possible the mass production of yet cheaper and more reliable transistors, although their technical specifications were somewhat inferior to current best practice. The ultimate consequence, however, was a process technology ideally suited to manufacture increasingly miniaturized integrated circuits.

The planar process was built upon the diffusion and oxide-masking innovations of Bell Labs. It is in essence the cycle of repeated steps of wafer oxidation, coating with photoresist, photomask exposure and development, etching, and diffusion that is described in Section I above (see Figure 2.2). The earlier methods involving oxidation and diffusion had been used to build the "mesa" transistor, a device so named because it was literally a raised layered structure from which the sides had been etched away. Mesa transistors performed well and they were durable, but the electrical connections to each layer of doped material had to be made laboriously by hand. The planar process, in contrast, produced a device that was essentially flat. This permitted electrical connections to be made between doped regions by the evaporation of metallic film on appropriate portions of the wafer. This feature, significant in reducing the production cost of discrete devices, was indispensable to the successful batch production of integrated circuits.

The Course of Technical Advance:
The Integrated Circuit Era

Not long after the invention of the transistor, G.W.A. Dummer
of the British Royal Radar Establishment enunciated his remarkable vision of electrical circuits embedded in a solid block.
He pursued this idea through the auspices of the Royal Radar
Establishment. Under contract to the RRE, the Plessey
Company in Britain explored the concept of a solid-state
circuit, and they went so far as to construct a cardboard
model of an integrated circuit in 1956.

Meanwhile, unaware of details of the work in Britain, the
U.S. Air Force displayed an intense interest in the miniaturization of electronic components throughout the 1950s. Working
without direct government support, but well aware of the
military interest in miniaturization, Jack S. Kilby at Texas
Instruments had the critical insight in the summer of 1958 that
the passive components of an electronic circuit (resistors and
capacitors) could be fabricated out of semiconductor material
just as were the active devices (transistors and diodes).
Since all circuit components could be made of the same type of
material, he reasoned that they could also be made literally out
of the same piece of material and interconnected to form a
circuit (Kilby, 1976). By September 1958, Kilby constructed
the first working integrated circuits. These were fabricated
by hand, using germanium. A patent was sought in February
1959, and the invention was announced one month later.

Kilby's device was little more than a laboratory curiosity,
but Robert Noyce at Fairchild immediately grasped the potential
of the planar process for translating the idea into a commercially viable product. Noyce filed a patent on the planar
integrated circuit in July 1959. A lengthy patent dispute
ensued, which was eventually resolved in favor of Noyce.

By the early 1960s, the integrated circuit era was underway, with the new devices finding their first application in
areas where economical use of space was most important –
namely, missile guidance systems and hearing aids. As the
variety of applications grew, and as integrated circuits very
gradually displaced discrete devices, the course of further
technical development was quite clear. There was none of the
profound interaction between scientific theory and technological
practice that had characterized the first years of the development of the transistor. The integrated circuit concept did not
rest on any novel application of scientific theory; it was an
engineering achievement. Moreover, the fundamental process
technology was in place. Hence, a simple engineering heuris-
tic dominated the course of technology over the next two
decades: make it smaller.

The advantages of miniaturization were of particular
importance to the government and to manufacturers of compu-

ters. The military and space agencies were especially interested in economizing on size, weight, and power consumption.
And both the government and computer manufacturers were
concerned about reliability. By 1960 discrete components had
relatively low failure rates, but complex circuitry introduced
two problems. First, if many components were needed in a
system, failure rates could be unacceptably high even if
individual components had long expected lives. Second, a
principal source of equipment failure was in the connections
between components. Reliable integrated circuits offered a
potential solution to both problems.

Moreover, there were obvious economic advantages to
increasing the density of circuitry on a chip. Most of the cost
of fabrication is incurred in wafer processing and in subsequent assembly and packaging operations. Many costs depend
on the number of wafers processed, and many, especially in
the final stages, depend directly on the number of chips.
Making one chip do the work of four may increase the cost per
chip, but almost certainly by less than a factor of four.

Thus, progress in the integrated circuit era has moved
along a natural trajectory toward increasing the density of
circuit elements per chip. By the end of the 1960s the industry moved to what has been called Medium Scale Integration
(MSI), which is conventionally defined as involving (for logic
devices) 10 to 100 digital logic gates. By the mid-1970s the
industry was producing devices with Large Scale Integration
(LSI), which contained 100 to 1000 gates or the equivalent.
In the early 1980s the industry will move to Very Large Scale
Integration (VLSI), where circuit complexity exceeds 1000
gates. In terms of the more widely publicized random-access
memory devices, the LSI era began with the 16K RAM and the
VLSI era will begin with the production of 256K RAMs.

Each step along the miniaturization trajectory required a
family of related technological advances. Scaling down individual circuit elements required the etching of finer lines in
the silicon substrate. By 1980 best practice commercial
devices had line widths in the two-micron range. Shrinking
feature sizes required in turn lithographic equipment of higher
resolution. Technology is currently approaching the limits of
conventional optical lithography, and electron-beam and x-ray
lithography are expected to dominate in the VLSI era. Moreover, finer line widths necessitated the use of increasingly
purer silicon, since even minute impurities can impede the
proper flow of current through a connection two microns in
diameter. Similarly, the techniques of doping via diffusion or
ion implantation became ever more precise. Perhaps an even
more demanding requirement of miniaturization was the need
for innovation in testing equipment suitable to assure quality
control. Finally, improvements in computer software for the
design, analysis, and evaluation of increasingly complex
circuitry was required.

Emphasis on the dominant trajectory of miniaturization should not obscure several notable advances in the design and development of new varieties of semiconductor devices. Perhaps the most important new device innovation since 1960 was the development of the MOS transistor, the practical realization of Shockley's conception of a field-effect amplifier. The decisive technological work on MOS devices was accomplished at RCA and Fairchild in the early 1960s, but General Microelectronics and General Instrument were the first to make a serious effort to market MOS integrated circuits in 1965.

As noted above in Section I, MOS devices are slower than bipolar devices, but they consume less power and dissipate less heat. Consequently, MOS technology permits a denser packing of circuit elements, and MOS circuits have therefore tended to lead the way down the miniaturization trajectory. Through the 1970s successful RAM devices at each level of complexity (1K, 4K, 16K, 64K) were fabricated first with MOS, later with bipolar, technology.

The other notable achievement of the integrated circuit era has been the wide variety of new devices brought to market since the mid-1960s. Random-access memories have been mentioned previously; they are a fundamental building block of computer technology. Perhaps even more significant was the development of the microprocessor, the computer on a chip, initially introduced by Intel in 1971. More recently, there have been advances in the design of logic circuits that promise to make them nearly as flexible as memory devices. Chips consisting of arrays of digital logic gates, on which final electrical connections can be made to the specification of the users, have introduced a whole new class of so-called "semicustom" devices which is likely to enhance the range of applications possible in the coming VLSI era.

The trend to miniaturization of integrated circuits has had profound consequences for the market structure of the semiconductor industry. Miniaturization has significantly raised both the capital requirements for semiconductor production and the minimum efficient scale of operation. Moreover, miniaturization has also pushed the industry in the direction of increased vertical integration by blurring the lines between electronic components and systems. As more and more functions are built onto a single chip, system design is no longer a matter of configuring standardized components. Chip and system design have become increasingly interdependent. Thus, producers of downstream electronic products have greater incentive to acquire the capability for in-house design and production of customized circuits. And merchant suppliers of integrated circuits have greater incentive to design products around their innovative circuitry. Further implications of miniaturization for the market structure of the semiconductor industry and the likely feedback effects on future innovation are discussed in Levin (1982).

Sources of Innovation

Without question the lion's share of innovation in semiconductor technology has been the work of private firms. The government has played an important role in funding research and development, and its role as a purchaser of semiconductor devices has been even more important, but government research labs have done relatively little in the way of fundamental innovation. No truly major innovations have emerged from government labs, and government agencies hold only a modest share of total U.S. semiconductor patents. Table 2.11 lists those government agencies that rank among the leading organizational patent holders. As is clear, governmental patenting is concentrated among five agencies: the military service branches, NASA, and the AEC/DOE (Atomic Energy Commission and its successor Department of Energy). The government share of semiconductor device patents is rather negligible; it is only 2.1 percent of the total over the 1968-1980 period. In the broader category of semiconductor-related patents (including process patents and systems applications), government activity is more substantial, comprising 4.7 percent of total patents issued between 1969 and 1979.

Universities have contributed substantially to the advance of semiconductor technology, but not much in the form of patented inventions. Only two universities, MIT and Caltech, rank among the top 100 institutional holders of patents. MIT ranks 50th in device patents and 52nd in total semiconductor-related patents over the periods covered in Table 2.11, while Caltech ranks 90th in device patents.

Patent counts understate the importance of the university role. In the late 1940s, researchers at Purdue came remarkably close to inventing the point-contact transistor before Brittain and Bardeen. In the 1950s and 1960s the aggressive development policy of Stanford University, together with the strength of the physics and electrical engineering faculties at Stanford and the nearby University of California at Berkeley, was responsible for the location of numerous merchant semiconductor firms in the Santa Clara Valley. University-industrial ties in the Bay Area and in the Boston area were especially close in these early years, and many new ventures involved university faculty in important consulting or managerial roles. As the industry developed, however, university research became increasingly removed from the practical needs of commercial technology. While the industry concentrated on perfecting mass-production methods for germanium and later for silicon devices, universities turned their attention to exploration of the more exotic semiconductor materials: the III-V and II-VI compounds (formed from elements in the designated columns of the periodic table) and organic substances. By the 1970s basic research in solid-state physics most relevant to the in-

Table 2.11. U.S. Semiconductor Patent Grants to
Government Agencies, 1968-1980.

Agency	(1) Total semiconductor-related patents 1969-1979		(2) Device patents only 1968-1980	
	Rank	No. of Petents	Rank	No. of Patents
Navy	11	426	34	19
Army	15	225	22	31
NASA	21	170	45	14
Air Force	30	118	41	15
AEC/DOE	56	58	65	7
Others	–	20	–	14
Total government patents		1017		100
Government patents as percentage of total patents		4.7		2.1

Sources: (1) Office of Technology Assessment and Forecast,
U.S. Department of Commerce, special computer run,
1980; (2) Office of Technology Assessment and Forecast, U.S. Department of Commerce, special computer run, 1981.

dustry was being done primarily at Bell Labs and at IBM.
Nevertheless, there have been a few technology areas in which
university research has continued to lead the industry, most
notably in the development of computer software to aid in the
design, simulation, and analysis of complex circuitry.

Within the private sector, the predominant locus of
innovative activity has been firms that produce semiconductor
devices for the market or for internal consumption. The
contribution of specialized equipment suppliers should not be
neglected, especially in the areas of test and lithographic
equipment, but the fundamental process innovations have come
from semiconductor producers rather than capital-goods suppliers. For example, all the major process innovations identified by Tilton (1971) and nearly 80 percent of those listed by

Wilson, Ashton, and Egan (1980) were the consequence of work done by semiconductor-manufacturing firms. All exceptions on the list compiled by Wilson, Ashton, and Egan are recent innovations in the area of lithography. More generally, Von Hippel (1977) examined key innovations at each stage of the semiconductor-fabrication process and found that even with respect to new capital equipment, users rather than suppliers most often dominated the innovation process. Nevertheless, in the lithography area, firms like Perkin-Elmer have made substantial contributions, and major improvements in the manufacture of high-purity silicon wafers have been recently achieved by Japanese suppliers.

Within the semiconductor industry itself, significant innovation has come from firms of all types and sizes. The preceding narrative has noted in particular the major contributions of Bell Laboratories and some of the receiving-tube firms in the early years, and the contributions of Texas Instruments, Fairchild, and Intel in subsequent years. But new products and incremental process improvements have come from all quarters of the industry. Table 2.12 reports the number of patents granted since 1952 to leading semiconductor firms. As before, the counts in each column are based on a somewhat different aggregation of patent classifications.

While the use of patent data in this industry is subject to the important caveats noted above, it is nevertheless interesting to look at the broad patterns suggested by the data. Table 2.13 presents a useful summary.

The data support the impression that Bell Labs and the old receiving-tube suppliers were the technological leaders in the early years. Together, Bell and the eight receiving-tube firms had 56 percent of the patents granted to Tilton's sample of 40 leading U.S. semiconductor producers through 1968. The share of these nine firms declined later, to 41 percent of the patents granted to semiconductor firms from 1969 through 1979. The patent share of both open-market and captive suppliers increased accordingly. In this later period open-market suppliers had a relatively high share of patents on semiconductor devices, while captive suppliers including Bell apparently concentrated more on process and applications patents. Receiving-tube firms had a surprisingly high concentration of patents in the device area, which may have been a consequence (especially for RCA and Westinghouse) of a focus on technologically sophisticated devices for the military market.

Given the weakness of the patent data, somewhat greater light is cast on the sources of progress by looking at the distribution of major innovations among firms. Tilton (1971), in consultation with expert scientists and engineers, compiled a list of major process and product innovations. Those that occurred in the discrete device era are tabulated in Table

Table 2.12.   U.S. Semiconductor Patent Grants by
Firm and Type of Firm, 1952-1980

	(1) Process, Device and Selected System Patents	(2) Total Semi- conductor- related Patents	(3) Device Patents Only
	1952-1968	1969-1979	1968-1980
Receiving-tube firms			
RCA	668	1093	259
General Electric	580	865	208
Westinghouse	410	538	142
Sylvania[a]	158	145	16
Philco-Ford[b]	130	67	10
Raytheon	72	67	27
Others	35	–	–
group total	2,053	2,775	662
Major captive suppliers			
Bell Laboratories[c]	835	1,153	138
IBM	521	1,496	251
Honeywell	160	332	27
Sperry Rand	139	152	16
General Motors–Delco	133	132	31
Burroughs	–	124	10
Hewlett Packard	–	118	19
NCR	–	77	6
Tektronix	–	70	6
Others	–	273[e]	23[f]
group total	1,788	3,927	527
Open-market suppliers			
Texas Instruments	286	632	159
Motorola	190	623	85
Hughes	160	243	44
ITT	111	161	23
Bendix	77	114	3
Fairchild	52	180	55
Sprague	52	93	23
Rockwell	–	203	13
Signetics	–	119	49
National Semiconductor	–	110	36

continued

Table 2.12.   U.S. Semiconductor Patent Grants by
Firm and Type of Firm, 1952-1980 (continued)

	(1) Process, Device and Selected System Patents	(2) Total Semi- conductor- related Patents	(3) Device Patents Only
	1952-1968	1969-1979	1968-1980
Open-market suppliers (continued)			
TRW	-	71	15
Intel	-	69	15
Harris	-	52	16
General Instrument	-	50	8
Others	359	22[g]	17[h]
group total	1,287	2,742	561
Identified semiconductor firms	5,128[d]	9,444	1,750
Other firms	-	4,028[i]	473[j]
Total semiconductor patents	-	13,472	2,223

a. Includes patents awarded to GTE Microelectronics.
b. Includes patents awarded to Ford Motor Co.
c. Includes patents awarded to Western Electric.
d. Total includes patents of 40 large firms only.
e. Includes all manufacturers of electronic equipment identi-
    fied by Integrated Circuit Engineering (1981) as possess-
    ing semiconductor R&D or production capability with 20 or
    more semiconductor related patents.
f. Includes all manufacturers of electronic equipment identi-
    fied by Integrated Circuit Engineering (1981) as possess-
    ing semiconductor R&D or production capability with three
    or more device patents.
g. Includes all open-market suppliers with 20 or more semi-
    conductor-related patents.
h. Includes all open-market suppliers with three or more
    device patents.
i. Includes both all other types of firms (e.g., process-
    equipment suppliers, consulting firms, and all semiconduc-
    tor firms with less than 20 semiconductor-related patents).
j. Includes both all other types of firms and all semiconduc-
    tor firms with less than three patents.

Sources: (1) Tilton (1971, p. 57; (2) Office of Technology
          Assessment and Forecast, U.S. Department of Com-
          merce, special computer run, 1980; (3) Office of
          Technology Assessment and Forecast, U.S. Depart-
          ment of Commerce, special computer run, 1981.

Table 2.13. Distribution of Semiconductor Patents
by Type of Firm, 1952-1980
(Percentage of Patents Granted to
Identified U.S. Semiconductor Firms).

Type of Firm	Percentage of Process, Device, and Selected Systems patents	Percentage of Total Semiconductor-Related Patents	Percentage of Semiconductor Device Patents
	1952-1968	1969-1979	1968-1980
Receiving-tube firms	40	29	38
Captive suppliers	35	42	30
Bell Laboratories	(16)	(12)	(8)
Other captive suppliers	(19)	(30)	(22)
Open-market suppliers	25	29	32
Total	100	100	100

Source: Derived from data in Table 2.12.

2.14.   There is some inevitable arbitrariness in any such list,
but Tilton claims that his represents a consensus of many
opinions.

A more inclusive list of innovations since 1960 is provided
by Wilson, Ashton, and Egan (1980), who relied on consulting
engineers at Ferguson Associates for the selection of innovations.   The Wilson, Ashton, and Egan list, summarized in
Table 2.15, is divided into three categories: new device structures, new processes, and new product families.   Their criterion of what constitutes a "major" innovation is clearly less
stringent than Tilton's.   They attribute 42 major innovations
and new product families to U.S. firms over the period 1960-
1977.   (Three others are attributed to foreign firms or are
unattributed.)   Tilton, on the other hand, identified only 12
major innovations in the 1950-1960 period, two of which are
attributed to foreign firms.

Table 2.14. Distribution of Major Innovations by
Firm and Type of Firm; Discrete Device Era: 1950-1960.

Type and name of firm	All innovations[*]		Product Innovations		Process Innovations	
	Number	Percent	Number	Percent	Number	Percent
Receiving-tube firms	3.0	30	3.0	33	2.0	29
General Electric	(1.5)	(15)	(1.5)	(17)	(1.0)	(14)
Philco	(1.0)	(10)	(1.0)	(11)	(1.0)	(14)
RCA	(0.5)	(5)	(0.5)	(6)	(0.0)	(0)
Captive suppliers	4.5	45	3.5	39	4.0	57
Bell Laboratories	(4.5)	(45)	(3.5)	(39)	(4.0)	(57)
Open-market suppliers	2.5	25	2.5	28	1.0	14
Texas Instruments	(1.5)	(15)	(1.5)	(17)	(0.0)	(0)
Fairchild	(1.0)	(10)	(1.0)	(11)	(1.0)	(14)
Total	10.0	100	9.0	100	7.0	100

---

*Six innovations are listed by Tilton as both process and associated product innovations. These are counted only once in the column reporting "all innovations."

Source: Derived from data on major innovations presented in Tilton (1971), p. 16-17. Only innovations through 1960 are included in the tabulation, and foreign innovations are omitted. Where multiple innovators are noted, credit for the innovation is allocated equally.

Table 2.15. Distribution of Major Innovations by
Firm and Type of Firm; Integrated Circuit Era:
1960–1977.

Type and name of firm	Total Innovations		Device Structures		New Products		New Processes	
	Number	Percent	Number	Percent	Number	Percent	Number	Percent
Receiving-tube firms	7.0	16	4.5	32	1.5	14	1.0	6
RCA	(6.0)	(14)	(4.0)	(29)	(1.0)	(9)	(1.0)	(6)
GTE Sylvania	(1.0)	(2)	(0.5)	(4)	(0.5)	(5)	(0.0)	(0)
Captive suppliers	5.5	13	1.0	7	0.0	0	4.5	26
Bell Laboratories	(3.5)	(8)	(0.0)	(0)	(0.0)	(0)	(3.5)	(21)
IBM	(2.0)	(5)	(1.0)	(7)	(0.0)	(0)	(1.0)	(6)
Open-market suppliers	25.5	61	8.5	61	9.5	86	7.5	44
Fairchild	(7.0)	(16)	(3.0)	(21)	(2.0)	(18)	(2.0)	(12)
Intel	(4.3)	(10)	(1.0)	(7)	(2.3)	(21)	(1.0)	(6)
Texas Instruments	(3.8)	(9)	(0.5)	(4)	(1.8)	(16)	(1.5)	(9)
Mostek	(2.8)	(7)	(1.0)	(7)	(0.8)	(7)	(1.0)	(6)
Others*	(7.5)	(18)	(3.0)	(21)	(2.5)	(23)	(2.0)	(12)
Equipment suppliers	4.0	10	0.0	0	0.0	0	4.0	24
Total	42.0	100	14.0	100	11.0	100	17.0	100

*Includes five firms with one innovation each (General Instruments, Harris, Motorola, National Semiconductor, and Signetics) and five firms that shared credit for one innovation each (Advanced Memory Devices, American Microsystems, General Microelectronics, Siliconix, and Standard Microsystems).

Source: Derived from data in Wilson, Ashton, and Egan (1980), pp. 40–41. Foreign innovations are omitted. When multiple innovators are noted, credit for the innovation is allocated equally.

Table 2.14 supports even more strongly than the patent data the dominance of Bell Labs and the receiving-tube firms in the technology of the discrete device era. Together they generated seven of the ten major innovations, with Bell alone accounting for four and sharing credit for a fifth. Even more strikingly, only one major process innovation came from outside this group, namely the planar process developed at Fairchild.

The role of new firms in the technology of the integrated circuit era is dramatically illustrated in Table 2.15, which indicates that 61 percent of total major innovations and 86 percent of major new-product families were introduced by merchant semiconductor houses. Of the specific firms indicated, only Motorola, with one innovation to its credit, was involved in the electronics field before the invention of the transistor. Many - including Intel, Mostek, Harris, Signetics, AMD, AMI, General Microelectronics, Siliconix, and Standard Microsystems - were established after the invention of the integrated circuit. The decline of the receiving-tube producers is evident, despite the importance of RCA in the early development of MOS device structures. RCA is credited with only one major innovation since 1970 - the process technology for manufacture of silicon-on-sapphire (SOS) circuits, high-performance devices tailored to military application. Not surprisingly, Bell Labs retains an important position as a major innovator in process technology, but its absence from the commercial marketplace has undoubtedly diminished its role in product innovation.

It is of interest to consider why technologically sophisticated, vertically integrated electronics firms like IBM, RCA, GE, and Westinghouse have impressive patent statistics in recent years, but have contributed only a modest share of major innovations. One possibility arises from the fact that many sorts of product innovations, like circuit design concepts, are not patentable. Yet innovative circuit design is a critical element in many new products. Thus, open-market suppliers may concentrate on inherently less patentable areas of innovation. Relatedly, the large integrated firms tend to focus R&D on longer-run research. If the results of such research are somewhat more basic, more remote from commercial application, there may be more inclination to rely on patents and to depend less on secrecy. In addition, the structure of incentives in large, bureaucratic organizations may favor the use of patents to identify the contributions of particular scientists and engineers. Finally, in large, older firms institutional mechanisms for processing patent applications are more likely to be routinized. Open-market semiconductor houses tend to be less bureaucratic and to view patents and the attendant legal processes as an encumbrance. Thus, for all these reasons, the patent statistics are likely to exaggerate the importance of large, vertically integrated firms relative to merchant semiconductor houses.

The record indicates that small firms and new entrants have had a substantial impact on advancing mainstream semiconductor technology along its dominant miniaturization trajectory. Fairchild, a new entrant, moved to a position of sales leadership on the strength of its major process innovations of the late 1950s. A decade later Intel and then Mostek emerged as leading firms with innovative product designs exploiting the opportunities for miniaturization latent in MOS technology. Today, however, it is much more difficult to imagine a grassroots entrant moving directly to market leadership in high-volume product areas like semiconductor memory or logic devices. Large, established firms have considerable cumulative R&D experience involved with solving the technological problems of miniaturization. Moreover, the capital and R&D investment requirements for efficient-scale operation now constitute formidable barriers to establishing a position in a major product market via innovation. Several large Japanese electronics firms have recently joined the group of firms currently working on the technological frontier, but they are firms with years of cumulative experience aided by a major program of government support.

It is therefore likely that the next several generations of general-purpose memory and logic devices will be introduced by large established firms. Such devices are the types most likely to realize the remaining latent economies of miniaturization, which most experts expect to persist for some years to come. Innovation (and even imitation) along this trajectory will be costly in virtually every area of technology. And, as in the past, innovation will require related advances in lithography, materials quality, circuit design, packaging, software, and testing. Only large established firms are likely to have the human, organizational, and financial resources necessary to pursue these related developments simultaneously.

Certain areas of opportunity nevertheless remain open to smaller firms and new entrants. Many of these opportunities arise as a consequence of innovative microelectronic applications proposed by small-and medium-scale downstream systems producers that lack independent semiconductor fabrication capability. While many downstream firms have made innovative use of standardized circuits produced in large volume by leading merchant semiconductor firms, others have increased the demand for custom-designed circuits for specialized applications. Virtually all new semiconductor firms established in the wave of new entry since 1979 have specialized in one or more of the related areas of custom or semicustom design, custom or semicustom fabrication, computer-aided design (CAD), and custom software.

Small firms may prove to have a comparative and perhaps absolute advantage in custom and semicustom work. Many custom demands can be served cost-effectively by technology that is not on the frontier of the miniaturization trajectory. Consequently, custom design and fabrication does not require investment in human capital and in state-of-the-art process equipment on the scale of a full-line merchant semiconductor firm.   Similarly, many industry experts believe that the most fruitful applications of CAD tools will be well within the miniaturization frontier.   Nonetheless, innovations in CAD and in custom design and fabrication may have high payoff in terms of enhanced productivity in downstream industries, even if they do not significantly advance best practice semiconductor technology.

IV.   THE IMPACT OF PUBLIC POLICY ON INNOVATION

It is widely agreed that certain policies of the U.S. government have contributed significantly to the extraordinary record of innovation in the semiconductor industry.   The principal task of this section is to recount in brief the history of these policies, and to identify the reasons for this success.   At a time when "industrial policy" is being touted as a remedy for slow productivity growth and unfavorable trade balance in a number of industries, it is particularly important to understand the conditions under which such policies can succeed. The semiconductor success story affords an opportunity to extract some lessons for the conduct of policy to promote technological progress.

To this end attention will be focused on those policies that have had a considerable impact: public procurement of semiconductor devices and public support for research, development, and production capability.   These are not the only policies relevant to innovation in the semiconductor industry, but they are undoubtedly the most important.   To give a more complete picture of the role of government, several other areas of policy will be considered in somewhat less detail.   In particular, antitrust, patent, education, and tax policies will be considered.   Of course, neither tax policy nor support of scientific and technical education has been focused specifically on the semiconductor industry, but they will be treated here for two reasons.   First, some believe that their impact is important, and second, significant policy change in both areas is presently being sought by groups within the industry (see Semiconductor Industry Association, 1981).

Public Procurement of Semiconductor Devices

Of the various policy instruments that have directly or indirectly influenced the evolution of semiconductor technology, none has been of such fundamental importance as public procurement of electronic components and systems for purposes of national defense and space exploration. Although there is no evidence that latent military demand played a significant role in inducing the invention of the transistor, there can be little doubt that the presence of a large potential military market increased the rate and influenced the direction of technical change in the 1950s and 1960s. In particular, two of the most important innovations of the period - the silicon transistor and the integrated circuit - were both developed with the military clearly envisioned as the first large customer. As will become evident in the following exposition, not only the size but also the particular character of military demand was of great importance for the rapid evolution of semiconductor technology. And it will thereafter become clear that even policies designed to operate directly on the "supply side" of the semiconductor market - such as R&D support and subsidy of productive capacity - worked as well as they did precisely because of the military's role as a large buyer with well-defined needs.

Work at Bell Labs toward the invention of the transistor proceeded independently of government funding. The Bell System itself constituted a sufficiently large potential market for solid-state devices to warrant private investment in development work following the discoveries of 1947-1948. Yet Bell was immediately cognizant of the potential interest of the military in the transistor. Indeed, there was substantial concern in early 1948 that disclosure of the transistor to the military prior to public announcement might lead to restriction of its use or to its classification for national defense purposes. Thus, Bell did not disclose the invention to the military until one week prior to public announcement.

The military services were deeply interested in potential application of the transistor from the very beginning. The use of electronic equipment for military purposes had grown dramatically during and immediately after World War II, and the proliferation of vacuum tubes in circuitry of increasing complexity had begun to cause serious problems of equipment unreliability. In 1952 a study revealed that 60 percent of electronic equipment in the Navy's fleet was not operating satisfactorily, and one-half of all equipment failures were attributed to problems with receiving tubes (Speakman, 1952). The Navy was thus prepared to pay steep prices for devices that offered a significant improvement in reliability. Meanwhile, the Air Force was concerned not only with the reliability of its airborne electronic equipment, but with its size and weight as well. It reacted enthusiastically to the early transistor developments. In 1952 the Air Force estimated that 40 percent of its electronics could be handled by transistors, with a saving of 20 percent in size, 25 percent in weight, and, optimistically, 40 percent fewer failures (Speakman,

1952). Within a year, after the military had gained some experience with the first commercially produced transistors, it was recognized that reliability at this level was some years away (see Braun and Macdonald, 1978).

Nevertheless, the military stood ready as a willing buyer for new devices that promised improvements along several key performance dimensions. The military needs were quite specific, and they established clear targets for innovative effort. Smaller components were needed where space was at a premium, as in airborne systems or in more prosaic field equipment such as portable radios. Weight savings were sought under similar circumstances, and they could be achieved by two means. Substitution of semiconductors for tubes promised some direct weight savings, but perhaps more important was the lower power consumption of semiconductor devices. This permitted the use of smaller and lighter power sources. The Air Force was especially interested in devices capable of withstanding adverse operating conditions, such as high temperatures and high levels of radiation. Such devices were needed for the development of reliable missile guidance systems. Finally, the need for low failure rates was pervasive, and, as noted previously, the need for individual component reliability grew with the complexity of circuitry.

To satisfy these needs, the military was willing to support R&D, to subsidize engineering effort required to install production capacity, and to pay premium prices for the procurement of new devices. Data reported to the Defense and Business Services Administration of the Department of Commerce (1960) indicate that the average unit price for devices sold to the military was roughly twice that received from private sector customers in the middle and late 1950s. In part, the price disparity reflects the military's role as a first buyer of most new devices. Military consumption was heavily weighted toward younger products with prices unaffected by cumulative economies of experience. The price premium also reflects the government demand for devices of the highest quality. Indeed, there is some evidence that batches of devices produced for the military, but failing to test to exacting military specifications, were sold to private-sector customers at discount prices (Golding, 1971).

Aggregated data on sales to the private and public sector are available since 1955. These are reported in Table 2.16. If there is any bias in these data, it probably runs in the direction of understatement of the government share.(6) Not revealed in the table is the importance of government purchases at the very outset of commercial transistor production. In 1952, approximately 90,000 point-contact transistors were manufactured, mostly at Western Electric (Braun and Macdonald, 1978). All of the Western Electric's sales and virtually all of the rest went to the military (Kraus, 1973).

Table 2.16. Government Purchases of
Semiconductor Devices, 1955-1977.

Year	Total Semiconductor Shipments (millions of dollars)	Shipments to Federal Government* (millions of dollars)	Government Share of Total Shipments (percent)
1955	40	15	38
1956	90	32	36
1957	151	54	36
1958	210	81	39
1959	396	180	45
1960	542	258	48
1961	565	222	39
1962	575	223	39
1963	610	211	35
1964	676	192	28
1965	884	247	28
1966	1123	298	27
1967	1107	303	27
1968	1159	294	25
1969	1457	247	17
1970	1337	275	21
1971	1519	193	13
1972	1912	228	12
1973	3458	201	6
1974	3916	344	9
1975	3001	239	8
1976	4968	480	10
1977	4583	536	12

*Includes devices produced for Department of Defense, Atomic
Energy Commission, Central Intelligency Agency, Federal
Aviation Agency, and National Aeronautics and Space Admin-
istration equipment.

Source: 1952-59 data from U.S. Department of Commerce,
Business and Defense Services Administration, Electronic Components: Production and Related Data,
1952-59, Washington, D.C. 1960.

1960-68 data from BDSA, "Consolidated Tabulation:
Shipments of Selected Electronic Components," mimeo,
Washington, D.C., annually.

1969-77 data from U.S. Department of Commerce,
Bureau of Census, Current Industrial Reports, Series
MA-175, "Shipments of Defense-Oriented Industries,"
Washington, D.C., annually.

The earliest semiconductor devices were used by the military for experimental purposes, but they were soon incorporated into communications equipment, such as radios. A detailed breakdown is not available, but apparently the Army Signal Corps led service units in the purchase of semiconductors during the early and mid-1950s. A significant jump in demand occurred in 1958, when the Air Force made the commitment to rely on semiconductors for the electronics in its major missile program. The Minuteman Missile Reliability Program involved 13 component suppliers. Among them, Motorola received a $1.7 million contract for supply of germanium transistors. Fairchild, a new entrant, received a $1.5 million contract for silicon-diffused transistors, and the cash flow from this large order had a significant impact on the ability of Fairchild to fund development work on the planar process.

The articulated needs of the military provided strong inducement for two of the most important innovations in the evolution of semiconductor technology: the silicon transistor and the integrated circuit. By his own account, Gordon Teal (1976) turned his attention to the growth of single-crystal silicon in 1951 because silicon offered the potential for fabrication of transistors capable of meeting high-temperature performance specifications for military equipment. A conscious aim of the management of Texas Instruments when it hired Teal in late 1952 was to be the first to make a silicon transistor available to the military. As a consequence of its efforts, and the fortuitous choice of pursuing a grown junction approach while competitors experimented with an alloy junction design, Texas Instruments enjoyed a three-year monopoly on the silicon transistor. The breakthrough in silicon secured for TI its position as the largest merchant supplier of semiconductor devices.

The inducement that military demand provided to the invention of the integrated circuit was somewhat less direct. In the case of the silicon transistor, the entire industry knew that the prize was available to the successful innovator. The military's need for equipment capable of operation in high-temperature environments was well specified, and the thermal properties of silicon were well known. In the case of the integrated circuit, military demand was not so sharply specified. Instead, it was clear that the military had an interest in smaller, lighter, and more reliable electronic equipment, but there were several possible means of achieving these goals. Each service branch had initiated its own R&D program to further progress toward miniaturization of electronic components, and each program took a strikingly different approach to the problem. In fact, none of the approaches was strictly consistent with the integrated circuit concept that emerged from research at TI and Fairchild. Yet the very diversity of R&D approaches taken by the military services suggested that any significant advance toward miniaturized electronics would find a major customer. Thus, latent demand for smaller and more reliable devices created an atmosphere conducive to innovation despite the absence of a specific prior commitment to the integrated circuit. Kilby's (1976) own account of his invention makes clear that TI had the military clearly, indeed exclusively, in mind during the course of early R&D work on the integrated circuit.

Following Kilby's initial work, the government moved quickly to support the development and pilot production of the integrated circuit in a manner to be described below. But by 1962 there was still considerable debate among potential military and civilian users about whether the integrated circuit could be made sufficiently reliable to gain wide acceptance. No major commitments had yet been made by private-sector customers. Indeed, IBM, which in 1960 was the largest single private-sector customer of every major semiconductor house (Linvill and Hogan 1977), took a very cautious approach to the new technology. After careful study, IBM opted against the use of monolithic integrated circuits in its new 360 series of computers (Kraus, 1973).

Two key procurement decisions of government agencies were responsible for moving the integrated circuit into pro- duction on a significant scale. In 1962 NASA announced its intention to use integrated circuits for its prototype Apollo spacecraft guidance computer. Shortly thereafter, the Air Force declared that its improved version of the Minuteman ICBM would make extensive use of integrated circuits in its guidance package. In retrospect, these were good decisions. But they involved considerable risk. The Air Force and its prime contractor, Autonetics, carefully considered three alternative strategies to achieve size and weight reductions in the Minuteman guidance package. One involved repackaging discrete components. The second entailed use of thin-film hybrid circuits, similar to the approach taken by IBM in its 360 computer. The third was to use integrated circuits (U.S. Air Force, 1965). Of the three, the integrated circuit route was understood to promise the greatest potential for size and weight reductions, but the technological distance to be trav- eled in order to achieve necessary levels of reliability was the farthest. In short, the Air Force chose the high-payoff, high-risk approach over more conservative options.

Minuteman II development contracts were let in late 1962 to Texas Instruments and Westinghouse. Initial shipments took place that year, and a round of procurement contracts were awarded during 1963 to TI, Westinghouse, and RCA. The principal Apollo contract went to Fairchild. The two programs together accounted for virtually all sales of integrated circuits through 1963 and most of 1964. Table 2.17 illustrates the importance of government procurements in the early years of

Table 2.17. Government Purchases of
Integrated Circuits, 1962-1968.

Year	Total Integrated Circuit Shipments (millions of dollars)	Shipments to Federal Government[a] (millions of dollars)	Government Share of Total Shipments (percent)
1962	4[b]	4[b]	100[b]
1963	16	15[b]	94[b]
1964	41	35[b]	85[b]
1965	79	57	72
1966	148	78	53
1967	228	98	43
1968	312	115	37

a. Includes circuits produced for Department of Defense, Atomic Energy Commission, Central Intelligence Agency, Federal Aviation Agency, and National Aeronautics and Space Administration.

b. Estimated by Tilton (1971).

Sources: Tilton (1971), p. 91. Total shipments data originally drawn from Electronic Industries Association, Electronic Industries Yearbook, 1969, Washington, 1969. Government share calculated by Tilton from data in BDSA, "Consolidated Tabulation: Shipments of Selected Electronic Components."

the integrated circuit era. As procurements for the Minuteman and space programs expanded over the mid-1960s, substantial contracts went also to Motorola, Signetics, General Microelectronics, and Siliconix, facilitating their entry into integrated-circuit production.

Military and space procurement programs succeeded in accelerating the development of integrated circuit technology for several reasons. First, the willingness of the military and NASA to pay high initial prices provided a clear incentive for producers to enter the field, justifying substantial investment in R&D and productive capacity. Second, the large volume of orders conferred cumulative production experience on the contractors, facilitating the learning required to reduce costs

over time. Together, the increased competition from new
entry and cumulative production experience drove prices down
(see Table 2.8 above), rendering integrated circuits viable in
an increasing variety of commercial applications. Third,
progress payments throughout the contract period eased potential cash-flow problems, reducing the economic risk of technical
failure (Golding, 1971). Fourth, the military provided an
unusually high level of user feedback, enabling suppliers to
correct those defects in design and production techniques that
were responsible for equipment failure. Finally, the pace of
innovation was undoubtedly accelerated by the exacting demands of the Air Force. Like the predecessor Minuteman I
Reliability Program, Minuteman II procurement contracts specified device performance targets beyond what was immediately
feasible, and they required that goals be achieved within a
very short time. For example, one TI contract called for the
design, fabrication, and delivery of 18 new types of integrated
circuits within a six-month period (Platzek and Kilby, 1964,
cited in Golding, 1971).

It has been argued that the government's presence as a
large potential customer with clearly specified needs provided a
substantial inducement to innovation in the semiconductor
industry. Latent government demand accelerated the initial
development of two key innovations - the silicon transistor and
the integrated circuit - and the scale and character of procurement programs served to stimulate further technical advance in the wake of these major innovations. In addition to
these more or less direct impacts of government procurement,
several indirect impacts on technical change should be noted.

For instance, in both the transistor and integrated circuit
eras, there were spillovers leading to improvement of the
products sold to nonmilitary customers. In some cases, devices designed for or purchased by the military were transferred directly and successfully to the civilian market. For
example, Fairchild's family of Micrologic circuits in the early
1960s was designed initially for general-purpose computers.
Yet large-scale purchases of these circuits by NASA in 1963
contributed to substantial learning economies. The following
year, Fairchild announced substantial price reductions in the
commercial market, which led to wide use of these circuits in
the computer industry (Golding, 1971). More typically, spillover is less direct. Texas Instruments did not achieve substantial commercial-market penetration with the circuits it
designed for the Minuteman program, but its production experience with devices made to stringent standards of reliability
facilitated the design of numerous commercially successful
integrated circuits.

Another indirect but nevertheless important influence of
government procurement policy in the 1950s and 1960s was
transmitted through the practice known as "second sourcing."

In assembling complex weapons systems, prime military contractors have typically sought to avoid reliance on a single
source of important components. Using a second source has
the obvious advantage of hedging against the risk that technological or production schedule difficulties experienced by a
single subcontractor will slow progress on an entire weapons
system. Second sourcing of electronic components was actively
encouraged by the Defense Department, and the practice served as a means of transferring technological capability both
directly and indirectly. In most cases, second sources were
simply supplied with product specifications, but sometimes
substantial transfers of know-how were involved.

The encouragement of second sourcing by the military
helped to diffuse advanced technology more widely through the
industry, and also facilitated the entry of new firms. While
most new firms entered with some differentiated products of
their own, second-source contracts were often a means of
securing cash-flow and production experience sufficient to
mount a successful marketing effort with proprietary products.
Conversely, innovative entrants sometimes found it advantageous to seek out larger, established firms to second-source
their new products in order to assure customers (including the
military) who were wary of reliance on an inexperienced supplier (Finan, 1975).

Whether through purchase of second-source or proprietary products, the government encouraged the diffusion of
technological capability during the 1950s and 1960s by its
liberal inclination toward new firms. For example, Transitron
began its rapid rise to market leadership when the military
authorized use of its gold-bonded diode prior to any significant commercial sales of the device. (Tilton, 1971). Moreover, according to data collected by the Department of Defense
and analyzed by Tilton (1971), new firms accounted for a
substantially larger share of the military market than of the
commercial market in 1959. The role of the Minuteman program
in facilitating the entry of Signetic, Siliconix, and General
Microelectronics provides another illustration of official
encouragement of the diffusion of technological capability.

During the late 1960s and 1970s, as the market for integrated circuits grew, the relative importance of military
demand as a stimulus to innovation declined. As production
technology matured and followed its trajectory toward miniaturization, products and device performance improved incre-
mentally albeit rapidly. Large price premiums for innovations
along this trajectory no longer seem warranted. Since four
16K RAMs can, in most cases, do the work of a 64K RAM,
there is a fairly tight limit to the price premium that a military
buyer, not to mention a commercial one, would willingly pay.
The example doubtless oversimplifies, but the point remains
that there is now a large commercial market providing adequate
incentive for incremental innovation.

In the past the exacting performance standards of the military had a significant impact on the improvement of device reliability and on increasing the range of feasible operating conditions. These achievements had valuable spillover effects in the commercial market, in areas such as mainframe computer and automotive applications. In recent years, however, some industry participants perceive a wider chasm between the device performance improvements desired by the military and commercial users. Desired military specifications for package strength and radiation hardness are far in excess of the standards required for most civilian uses. Thus, the scope for technological spillover from the military to the commercial market has probably narrowed. Indeed, most spillover today probably flows in the reverse direction. Recent advances in circuit density and the associated improvements in memory and microprocessor capacity have been aimed first at civilian applications. The high costs of designing and testing new devices to meet military specifications means that the circuits now purchased by the military are several years behind the best practice technology in terms of line width and circuit function density.

Government Support for Research, Development, and Production Capability

Since the end of World War II, the U.S. government has devoted substantial funds to semiconductor research and development. The bulk of these funds, especially in the early years, came from the military services, but support has come also from NASA, the National Bureau of Standards, and the National Science Foundation. From the late 1940s through the mid-1960s military-sponsored R&D was focused relatively sharply on the objectives of developing smaller, lighter, and more reliable electronic components with performance characteristics suitable for advanced weapons systems. Although there was some support from relatively basic research in the semiconductor area, the principal thrust of military R&D was to provide a "supply push" to complement the "demand pull" of military and space procurement programs. Along with this R&D effort, in the 1950s the military advanced funds of comparable scale to subsidize construction of a national capability to produce reliable transistors in large volume. These "production refinement" contracts added an additional, quite significant, push on the supply side.

During World War II the government funded significant research on the properties and applications of semiconductor materials. Silicon rectifiers, directly descended from the "cat's whisker" of amateur crystal radio, had proven effective as detectors in wartime radar. As a consequence, all semiconductor research had been placed under the direction of the Radiation Laboratory at MIT, which was responsible for coordinating the national research effort in radar technology. Through the radar program, significant work on germanium was funded at Purdue University, leading to the near-discovery of the transistor effect, and work on the properties of silicon was supported at Bell Labs. After the war, however, the research of Bardeen, Brattain, and Shockley was supported by private funds.

Once informed of the initial Bell discoveries in July 1948, the government responded quickly. Funds were immediately authorized to expedite transistor development and production, and Bell was awarded an R&D contract in 1949. By the time Bell reached the point of readiness to produce experimental devices for the military in 1951, the Department of Defense had given the Army Signal Corps responsibility for funding pilot production capability. The Signal Corps proceeded to finance construction of production lines at Western Electric, GE, Raytheon, RCA, and Sylvania. All were in operation by mid-1952 (Kraus, 1973). The cost of these production engineering measures (PEMs) was nearly $13 million (Golding, 1971).

The production of germanium transistors was barely underway when the Air Force launched a major R&D program aimed at development of a silicon transistor. Over a period of several years the Air Force spent $5 million toward this end (Golding, 1971). Meanwhile, the Signal Corps had an ongoing R&D effort of its own to complement the PEMs. A portion of this effort was directed toward the diffusion of technical information. Projects sponsored at Bell Labs covered a range of subjects from basic research to device design and performance evaluation, and Signal Corps contracts called for wide dissemination of technical reports on this work. The reports included considerable information on research supported by Bell's own funds. Thus, major industrial laboratories were provided with access to recent advances in semiconductor technology. Efforts to spread knowledge were complemented by government sponsorship of conferences and symposia.

After research at Bell Labs led to development of the diffusion method of fabrication, batch processing of transistors and other discrete devices became feasible. At this point, the Department of Defense moved quickly to develop large-scale industrial capacity for semiconductor manufacture. In 1956, the Signal Corps committed $14 million to production refinement contracts in the transistor area. The contracts called for federal support of all engineering design and development effort, while the firms involved paid for capital equipment and plant space. Twelve firms received contracts that called for the installation of capacity to produce some 30 different types of silicon and germanium transistors. Production lines were to be capable of producing 3,000 transistors per month of each type, thus augmenting annual transistor production capacity by over one million units. Additional funds followed the initial dose, including engineering support for capacity to produce diodes and rectifiers. In all, it is estimated that the Signal Corps provided about $50 million for production engineering between 1952 and 1964 (Braun and Macdonald, 1978).

Table 2.18 reports published figures on government funding of both R&D and production engineering for the years 1955-1961. Unfortunately, detailed data are not available for other years. It is important to note that these figures include only direct funding for semiconductor R&D, but a significant amount of government-sponsored R&D was financed indirectly through prime contractors for weapons systems. A special survey conducted by the Department of Defense in 1960 indicated that total semiconductor R&D funded directly and indirectly was $13.9 million in 1958 and $16.2 million in 1959 (Department of Commerce, 1961). This is more than double the figures reported for directly financed R&D in those years. The 1960 survey also revealed that government funds supported approximately one-quarter of all R&D effort in the field.

The most striking feature of this account of government R&D and production engineering support during the 1950s is the impression it conveys of the government's flexibility and responsiveness to technological developments. Although no major technological breakthroughs were directly funded by the government, the Department of Defense stood ready at every critical junction to seize opportunities to develop the technology and production capability. It provided R&D support to Bell in response to the announcement of the transistor. It subsidized pilot production as soon as it was feasible. Once the rudimentary problems of manufacturing germanium devices were solved, the government moved to support R&D on silicon, where the production problems were tougher. And as soon as mass production of both silicon and germanium devices appeared feasible, funds were immediately provided for substantial increments to production capacity.

The record strongly suggests that the government's capability to respond to developments quickly and flexibly was closely linked to its role as a potential user of the new technology. Early funding to Bell Labs and to others for pilot production was undertaken principally to provide the military with devices that it could use and test in its communications equipment and weaponry. Funding for silicon R&D was motivated by the clear knowledge that aircraft and missile applications required silicon's high-temperature characteristics. Development of mass-production capability was not motivated by some abstract desire to promote the industry, but by concrete needs. Production lines were funded for specific types

Table 2.18. Direct U.S. Government Funding for Semiconductor Research, Development, and Production Refinement, 1955–1961 (in millions of dollars).

	1955	1956	1957	1958	1959	1960	1961
Research and Development	3.2	4.1	3.8	4.0	6.3	6.8	11.0
Production refinement:							
Transistors	2.7	14.0	–	1.9	1.0	–	1.7
Diodes and rectifiers	2.2	0.8	0.5	0.2	–	1.1	0.8
Total	8.1	18.9	4.3	6.1	7.3	7.9	13.5

Source: U.S. Department of Commerce, Business and Defense Services Administration, Semiconductors: U.S. Production and Trade (Washington, D.C.: Government Printing Office, 1961).

of semiconductor devices, and no sooner was the capacity in place than the Air Force followed with high-volume orders for its Minuteman I missile program. As a sophisticated consumer of devices, the government was aware of technological constraints, and its knowledge permitted a flexible response to new opportunities. Thus, the effectiveness of government's "supply-side" push was critically dependent on its clearly specified demands as a user of semiconductor technology.

The role of government R&D in the development of the integrated circuit is a bit more subtle. It was noted earlier that, unlike the case of the silicon transistor, the government had no clearly articulated demand for an integrated circuit as such. Instead, there was a clear requirement for more compact and more reliable electronic systems and subsystems in military equipment. Braun and Macdonald (1978) argue that in fact these two demands were conflated; it was simply assumed that miniaturization would lead to more reliable devices. Of course, there was no necessary connection here, although in retrospect the conjecture was borne out. In any event, by the early 1950s there was a clearly perceived demand for miniaturization, and each military service moved over the course of the decade to define its own approach to the problem.

The first major effort toward miniaturized electronic circuits was funded by the Navy working in cooperation with the National Bureau of Standards. The project, known as Tinkertoy, had dual objectives: to reduce the size of electronic circuits and to automate the process of circuit assembly. Between 1950 and 1953, about $5 million was spent, both within NBS and with outside contractors, and a viable assembly method was developed. The concept involved forming one to four passive components on each of several ceramic wafers. The wafers were stacked and the components interconnected by vertical riser wires. At the top of the wafer stack were active circuit elements in the form of miniaturized vacuum tubes. The project was a technical success, but the contemporaneous development of viable methods for producing transistors rendered Tinkertoy obsolete by the time its goal was achieved.

Several years later, the Tinkertoy approach was revived by the Army Signal Corps in its Micromodule program. This time around transistors were used as the active components, but the passive components were again embedded in stacked ceramic wafers. Between 1957 and 1963 the Signal Corps spent $26 million on the program. The bulk of the funds went to RCA, but small contracts were awarded to other component manufacturers to encourage them to repackage their parts into wafers, compatible with Micromodule assemblies. RCA was expected to develop production capability for one million units per year by 1964, but by 1963 it was clear that the integrated circuit concept dominated the Micromodule alternative. Thus, like its predecessor, Tinkertoy, the Micromodule program was abandoned as obsolete.

An alternative approach to miniaturization was to construct circuits on a single ceramic wafer, rather than by forming connections between tiers of wafers. The construction of such two-dimensional (2-D) circuits involved the use of passive components and electrical connections formed by the evaporation of thin metallic films on the ceramic substrate. Discrete active components made of semiconductor material were embedded in the substrate. These "thin film" or "2-D" circuits were favored by the Army's Diamond Ordnance Fuze Laboratories (DOFL), which funded a significant program of in-house and contract research in the area from 1957 through 1959. The Navy saw promise in this approach, and in 1958 it commenced an R&D program aimed toward the fabrication of active devices using thin-film techniques.

While the Army Signal Corps was supporting Micromodule development and DOFL and the Navy were pushing thin-film circuits, the Air Force conceived an entirely different and more radical approach to the problem of miniaturization. The Air Force approach, which it called "molecular electronics," envisioned the development of solid-state devices that performed the functions of electronic circuits but did not correspond part for part to conventional circuits. Kilby (1976) notes that a quartz crystal was used as an illustrative example of a molecular electronics device, since it performs the functions of inductance and capacitance without a part-for-part correspondence to conventional circuit elements. It was expected that semiconductor materials would be used in the construction of molecular electronics devices, but it was never very clear just how the Air Force proposed to move beyond the idea of a circuit to these radically different types of electronic devices. The Air force approach stirred great controversy within the military establishment. Only after two years of negotiation did the Air Force finally award its first R&D contract under the program. The $2 million per year award to Westinghouse was granted over the strenuous objections of the other services.

Kilby's initial work toward building an integrated circuit out of a single piece of semiconductor material was undertaken without direct R&D support from any of these government programs. Indeed, his personal account suggests that he was not eager to pursue any of the approaches favored by the services. He describes his feelings in the summer of 1958 thus: "I felt it likely that I would be assigned to work on a proposal for the Micro-Module program when vacation was over unless I came up with a good idea very quickly. In my discouraged mood . . ." (Kilby, 1976, p. 650).

The integrated circuit did not fit neatly into any of the ongoing military R&D programs aimed toward miniaturization. Since Kilby's device was fabricated out of a single block of semiconductor material, it was not a Micromodule. Nor was it made of thin film deposited on a ceramic substrate. And it was emphatically a circuit, and hence not a molecular electronics device. Nevertheless, beginning in October 1958, five months prior to public announcement of the breakthrough, Texas Instruments informed each of the services about the new concept. Only the Navy showed little interest. The Signal Corps began at once to define a contract to support further development, but they hoped that TI would be able to demonstrate that the new device was fully compatible with the Micromodule approach. Initial reaction within the Air Force was mixed. According to Kilby (1976), strong adherents of the molecular electronics concept felt the TI approach did not qualify for support under the ample budget available for work in molecular electronics. It was a circuit, and the aim was to transcend the circuit. Moreover, Kilby's device used resistors, which wasted power. Nevertheless, a group led by R.D. Alberts of the Wright Air Development Center was persuaded that the integrated circuit concept, building as it did on existing semiconductor technology, provided a more orderly and natural transition to miniaturized electronics than did the ambitious molecular electronics idea.

By Kilby's account, "Alberts' group then provided the first in a series of contracts which proved invaluable in sustaining the project during the critical years" (Kilby, 1976, p. 651). This first contract, in the amount of $1.15 million, was awarded in June 1959, only three months after TI announced its invention. Like much of the earlier military support for transistor R&D, even this first contract was targeted toward objectives with a specificity that only a knowledgeable potential consumer could have provided. In particular, the contract called for the development of integrated circuits capable of performing several specific functions, and it required devices fabricated out of silicon, rather than the germanium used in TI's first demonstration models.

According to the official account (U.S. Air Force, 1965), R&D progress at TI was so rapid that by late 1960 a major production refinement contract was awarded in the amount of $2.1 million. The new contract called for establishment of a pilot production facility capable of turning out 500 integrated circuits per day. Simultaneously, the Air Force awarded another contract calling for the immediate construction of a small digital computer using integrated circuits. The success of this latter project led directly to yet another contract in which TI was to build a piece of equipment involving direct substitution of integrated circuits for discrete components. As a consequence, two identical computers were built, one with 587 integrated circuits and another with 9,000 discrete components. Air Force-sponsored public demonstrations of these two machines in October 1961 were influential in persuading the other service branches and the uncommitted semiconductor firms that the future belonged to the integrated circuit.

THE SEMICONDUCTOR INDUSTRY   73

Meanwhile, at Westinghouse, work on molecular electronics was redefined in the direction of integrated circuits. Specifically, two Air Force contracts totalling $4.3 million was awarded to Westinghouse in 1960 and 1961 for development of equipment using integrated circuits. Shortly thereafter, the Air Force began to expand its program of support for integrated circuit development beyond TI and Westinghouse. Numerous modest R&D contracts were awarded by the Air Force in the period 1961-1964, and a major program was launched at Motorola to support the development of hybrid linear integrated circuits, which combined aspects of thin-film and semiconductor technology. Fairchild shunned government involvement in its R&D, despite its readiness to sell circuits to NASA.

Success of the Air Force R&D programs encouraged other government agencies to get involved in the integrated circuit area. In 1961 NASA partially funded the development of the Series 51 computer at Texas Instruments, and by 1964 the Air Force share of total government R&D funds for integrated circuits fell below 50 percent, despite steady growth in the level of funding (Golding, 1971). In all, during the period 1959-1964, it is estimated that the government allocated $32 million to integrated circuit R&D, with the Air Force responsible for 70 percent of the total (Kleiman, 1966).

In reviewing this brief history of government involvement with the development of the integrated circuit, it is once again clear that success depended on the flexibility and responsiveness of government R&D support, which was in turn made possible by its role as a potential consumer of new technology. No major innovations emerged directly from government-funded R&D; indeed, several large-scale programs ended in failure. But the government's overriding objective - reliable, miniaturized electronic devices and systems - was kept in sufficiently clear focus that when a breakthrough occurred in the private sector it found enthusiastic support and substantial aid. There is little doubt that Air Force funding of development work at Texas Instruments accelerated the pace at which the integrated circuit became commercially viable, just as did its large-scale procurements commencing in 1962.

Paradoxically, none of the specific R&D programs of the military services aimed toward miniaturization resulted in success. Tinkertoy and Micromodule were obsolete by the time their technical objectives were realized. Thin-film technology, while subsequently developed for some special applications, did not prove the key to the miniaturization problem. And molecular electronics was a concept far too advanced for its time. Indeed, 25 years later it still does not appear readily feasible. At the time, each of these programs was criticized as being too rigid in its conception, and of course none succeeded. Yet throughout the period the overall military objective was quite

clear. The demand for miniaturized and reliable devices was
well understood, and R&D funding even if ultimately misdi-
rected served to create an atmosphere of urgency, underscor-
ing the military's latent demand. The impression that demand
for smaller and more reliable devices mattered more than
achievement of the technical goals of specific R&D programs
was of course confirmed by the enthusiastic support provided
for the integrated circuit.

Since the mid-1960s the relative importance of government-supported R&D has declined considerably.   Absolute
levels of funding increased somewhat during the 1960s, but fell
steadily through the 1970s in real terms until the recent
increase spurred by inauguration of the VHSIC Program, which
will be discussed below. The military gave a significant boost
to the early work on MOS technology at RCA, although as
noted earlier the commercialization of MOS devices was undertaken by other firms not funded by the government.   RCA's
work on CMOS and later on silicon-on-sapphire technology was
also partially supported by the military.   Silicon-on-sapphire
devices are particularly invulnerable to transient doses of
gamma radiation, and this feature makes them attractive in
certain military/aerospace applications despite their high cost.

On the whole, however, government funding of semiconductor R&D in recent years has been spread rather thinly over
a wide range of technological areas and a large number of
firms, and it has not played a major role in technical advance.
To be sure, military standards for resistance to extreme
conditions of temperature and radiation have remained higher
than for nearly all commercial applications, and military
requirements for high-speed performance have also tended to
exceed those of the commercial market. Thus, the military has
continued to support work toward those ends, but during the
1970s technological spillover from commercial R&D to military
application was probably greater than flows from military R&D
to commercial application.   This, of course, has reversed the
pattern that prevailed in the 1950s and early 1960s.

One area in which government R&D support has remained
central rather than peripheral is in relatively basic research in
more exotic semiconductor technologies that may prove valuable
over the long run. In particular, research on gallium arsenide
and other III-V compound semiconductors has been supported
primarily by the military and the National Science Foundation.
Gallium arsenide has received substantial military support as a
consequence of its theoretical capability to operate at very
high speeds.   Indeed, it is estimated that, prior to the
initiation of the VHSIC program, nearly one-half of the 1978
and 1979 funds allocated by the Defense Department to integrated circuit R&D supported work on materials other than
silicon.(7)

Before leaving the topic of government support for semi-conductor R&D, some mention should be made of the work on measurement technology at the National Bureau of Standards. For more than two decades the NBS has carried out a modest but continuing effort involving the characterization of materials and devices, the development of test structures and methods, and the evaluation and control of production processes. Virtually all the projects initiated by the NBS have been widely disseminated. The costs of R&D in areas such as device characterization and standards for measurement, testing, and process control have been modest, but a careful study completed by Charles River Associates (1981) concluded that the social rates of return from three selected NBS projects ranged from 60 to over 100 percent.

Antitrust Policy

In January 1949 the Department of Justice filed an antitrust suit against Western Electric and its parent company, American Telephone and Telegraph.(8) Among the government's principal complaints was a 1932 agreement by AT&T, General Electric, RCA, and Westinghouse. The agreement established a patent pool among the four firms, which might not in itself have caused concern. But the agreement also provided that AT&T would not enter the broadcasting business, in return for which the three electrical equipment firms would refrain from establishing a system of wire lines for transmission of network broadcasts. As relief, the Department of Justice proposed that AT&T divest itself of Western Electric.

After seven years of litigation, the suit was settled with the issuance of a consent decree in January 1956.(9) Under the terms of the settlement, AT&T agreed to grant to any applicant royalty-free licenses on any of its patents issued before the date of the decree. All future Bell patents were to be made available at reasonable royalties, provided that the licensee grant licenses at reasonable royalties on any of its patents sought by the Bell System. In addition, AT&T and its subsidiaries were barred from "engaging in any business other than the furnishing of common carrier communications services." Among the several exceptions to this last provision was permission to furnish services to any agency of the federal government. Effectively, then, the ban prohibited Western Electric from selling semiconductors (or computers) in the commercial market, although sales to the government were permitted. In return for these concessions, the government abandoned (at least for the next 17 years) its effort to break up the Bell System.

Interestingly, the consent decree did little more than ratify what was already the corporate policy and practice of

AT&T. The original patents covering the transistor and its use in circuits had been made available to all comers in 1952 for royalties ranging from 0 to 5 percent of sales. In 1953 the maximum royalty rate on all semiconductor patents was lowered to two percent of sales. Moreover, before 1956 Western Electric had made no attempt to penetrate the commercial market for semiconductor devices. A modest share of its production was used in telephone equipment and in experimental applications; the remainder was sold to the military (Braun and Macdonald, 1978).

The naive inference to be drawn from this recitation of facts is that antitrust policy had no effect on AT&T's behavior, and hence no impact on the structural and technological evolution of the semiconductor industry. But the issue requires a more subtle treatment. Clearly, in the years following the invention of the transistor and prior to the consent decree, AT&T operated under severe constraint. Any attempt to dominate the youthful semiconductor industry would have surely jeopardized its antitrust case. Nevertheless, it is at least arguable that Bell would have behaved no differently even in the absence of antitrust prosecution. Thus, there are two interesting questions to be addressed. First, did antitrust policy, as manifested both by the latent threat of divestiture and by the subsequent consent decree, have a significant effect on the behavior of AT&T? And if so, would the structure of the semiconductor industry and the course of its technological evolution have been substantially altered?

There are good reasons to believe that Bell would have disseminated the capability to manufacture transistors even in the absence of antitrust constraint. Widespread licensing of Bell patents had been a corporate policy of long standing. As noted, Bell pooled patents with the large electrical equipment manufacturers from the 1930s onward, and it made licenses available to others outside the pool on reasonable terms. Shortly before the Justice Department filed its antitrust complaint, an article appeared in the Bell Telephone Magazine explaining and documenting Bell's willingness "to make available on reasonable terms to all who desire them non-exclusive licenses under its patents for any use" (McHugh, 1949). Since the telephone-operating companies represented an enormous protected market for electrical equipment, AT&T reasoned that it ordinarily had more to gain than lose from the wide availability of its technology. Others might build on Bell's advances in ways that would ultimately lower the cost and improve the quality of telephone services. Thus, Bell saw advantages in both the products and the knowledge spillovers that might return to it as a result of licensing its patents. Consequently, Bell did not seek to exploit its patents to maximize royalties, but instead it attempted to maximize access to the technology of its licensees. Typically, Bell sought (and has continued to seek) cross-licensing agreements, in which Bell grants nonexclusive licenses in return for the right to use the patents held by the licensee on a royalty-free nonexclusive basis (Golding, 1971).

Apart from these general reasons favoring a liberal licensing policy, there were specific reasons to license the transistor. First of all, it was inevitable that an attempt to exploit transistor technology exclusively would invite a lawsuit from RCA, which held the rights to the semiconductor patents arising from the wartime work at Purdue. Although it was unlikely that Bell's patents would be invalidated entirely, the central Shockley patent on the point-contact transistor contained some extremely broad claims that might well have been narrowed considerably by the courts. Second, the transistor was probably just too big and too important to be effectively exploited within one organization. A Bell vice president observed: "We realized that if this thing [the transistor] was as big as we thought, we couldn't keep it to ourselves and we couldn't make all the technical contributions. It was to our interest to spread it around. If you cast your bread on the water, sometimes it comes back angel food cake" (attributed to Jack Morton, cited in Tilton, 1971). The credibility of this remark is supported by the fact that Bell went well beyond the terms of the consent decree in its efforts to disseminate transistor technology. In addition to licensing its patents liberally, Bell held a series of landmark symposia on semiconductor technology at which it conveyed substantial information and know-how to its licensees. Although these symposia were held before the antitrust settlement, Bell continued to transfer know-how along with its patents after 1956, by encouraging licensees to visit its facilities and confer with its scientists and engineers. This policy of transferring know-how along with patent licenses was a new one at Bell, further supporting the view that antitrust policy was not decisive in the decision to spread transistor technology. According to the same Bell executive, "There was nothing new about licensing our patents to anyone who wanted them. But it was a departure to tell our licensees everything we knew" (attributed to Jack Morton, cited in Tilton, 1971).

Even if Bell had chosen to monopolize knowledge of semiconductor technology, it is difficult to imagine how it could have managed to develop and supply transistors suited to the full range of potential applications: radios, hearing aids, computers, telephone equipment, and military systems of all kinds. It is just possible to imagine that AT&T would have dominated the computer industry had it chosen to take that course. But it is virtually inconceivable that it would have become the monopoly supplier of electronic equipment, or even of semiconductor components, to the Department of Defense. This brings us to the final reason why antitrust policy cannot be held responsible for Bell's decision to disseminate transistor technology.   Even if Bell were otherwise predisposed toward retaining its monopoly, it is a virtual certainty that the Department of Defense would not have permitted it.   The national security interest in the availability of the technology to a broad range of weapons systems and communications-equipment suppliers would have compelled licensing by executive pressure or, if necessary, by legislation.

It is most likely, then, that AT&T would have spread knowledge of semiconductor technology even in the absence of antitrust constraint.   But it is rather more difficult to assess whether Bell would have entered the commercial market for semiconductors or computers had no antitrust suit been filed or threatened.   Western Electric's prior restriction of its manufacturing activity to the telecommunications business cannot be determinative here, because it had never before been presented with so promising an opportunity as the transistor.   The most reasonable conjecture is that Bell would have inevitably become involved to some extent in commercial sale of at least semiconductor components.   Since Western Electric in fact and under any imaginable scenario was in the business of selling transistors to producers of weapons systems, it is only natural to assume that it might have developed some nonmilitary business with the same customers.   It is much more uncertain whether Western Electric would have aimed to dominate the commercial semiconductor market, or whether and when it would have even entered the computer business.

If one assumes that antitrust prohibition of entry into the commercial semiconductor market was in fact a binding constraint on AT&T's behavior, it is reasonable to ask whether the consent decree had any substantial impact on technological change or market structure.   One can only speculate, but the answer is probably no.   Of course, it is likely that Western Electric would have gained and retained a position as one of the industry's leading firms, but once the basic patents were available all around, it is unlikely that Western Electric could have dominated the industry or much ·altered the pace of technical advance.   Entry costs were low, and key employees were mobile.   Indeed, it is unlikely that Western Electric's presence in the commercial market would have had any effect on Teal's decision to move to Texas Instruments or on Shockley's decision to go to Palo Alto.   Thus, TI would likely still have been first with a silicon transistor, and Fairchild first with the planar process.

If antitrust constraint was decisive in keeping AT&T out of the computer industry, the effects were rather more likely to be substantial.   As Katz and Phillips (1982) demonstrate, IBM's dominance of the mainframe computer business emerged partly as a consequence of good strategic decisions, but also partly as a consequence of the failure of others to grasp what in retrospect appear to have been attractive opportunities. The presence of an additional player with the technological acumen of AT&T might well have altered the outcome.

Patents

In theory, patents provide a stimulus to innovative activity by granting prospective innovators exclusive rights to exploit their inventions. Society accepts the losses entailed by grant of temporary monopoly power in return for the induced increase in inventive activity.(10)   In many industries, patents play the role described by theory.   For example, patent protection provides a powerful incentive to develop new pharmaceuticals, despite the possibility of inventing close chemical substitutes for patented drugs (see Grabowski and Vernon, 1982).   In a survey of U.K. firms, Taylor and Silberston (1973) found that 64 percent of pharmaceutical R&D expenditures were dependent on patent protection, and, in firms producing other finished and speciality chemicals, 25 percent of R&D expenditures were so dependent.

For a variety of reasons, including both historical accident and the inherent nature of the technology, patents provide very little protection to innovators in the semiconductor industry, and in consequence they provide little incentive to innovative activity.   For example, seven of the eight electronics firms queried by Taylor and Silberston (1973) replied that none of their R&D expenditure was dependent on the existence of patent protection, and the eighth indicated that less than 5 percent was so dependent.

To understand the reasons for the inefficacy of semiconductor patents, it is necessary to distinguish between those areas of semiconductor technology in which inventions are patentable and those in which they are not.   New production processes - such as methods of growing single crystals, oxide masking, and epitaxy - can be and have been patented, though the breadth of claims is often open to dispute.   New pieces of process equipment - such as test devices, optical printers, and electron-beam lithographic equipment - are also clearly patentable, although patents may cover only those specific features of the equipment that are genuinely novel. In addition, new semiconductor-device structures - such as the junction transistor, the field-effect transistor, and the C-MOS transistor - have been patented.

Circuit designs are also patentable insofar as they can be specified in a patent claim.   But this is where matters become murky.   Clearly it is possible to patent a new logical configuration of circuit elements that perform a specific function, such as a particular design for a memory cell or a particular design for a logic gate.   To the extent that it can be described in

writing, a novel configuration of memory cells or logic gates in
an integrated circuit could also be patented. But what appears impossible to patent is the actual physical layout of an
integrated circuit of even moderate complexity. The principal
reason for this is the difficulty of rendering a full verbal
description of the circuit layout. While pictorial representa-
tions may be used in patent applications to illustrate or clarify
the language of a claim, a drawing cannot itself constitute a
valid patent claim. There are conflicting precedents on the
precise extent to which claims may refer to diagrams to convey
their meaning. In any event, it has generally been assumed
within the semiconductor industry that large-scale integrated
circuits cannot be patented since their layouts cannot be
described in the form of a valid patent claim.(11)

In the patentable areas of semiconductor technology, the
original Bell Labs patents were indispensable to any firm
seeking to enter the business. Had Bell chosen to exploit
them exclusively and had it successfully defended the validity
of its claims against the challenge of RCA, there is little doubt
that from a legal standpoint it could have excluded all competition in the manufacture and use of transistors. For
reasons described above, Bell chose to make its patents available to all comers, and the pattern it initiated of nonexclusive
licensing became the norm of industry behavior. Under the
terms of the consent decree, the price of licensing Bell patents
issued after January 1956 included availability of the licensee's
own patents at reasonable royalties. Bell patents on oxide
masking, diffusion, and epitaxy were all issued after this
date, and these techniques were indispensable for economical
manufacture of transistors. Thus, any firm contemplating the
exploitation of its own patents had to share them with Bell.

In the integrated circuit era, the key Fairchild and Texas
Instruments patents occupied a position similar to the earlier
Bell patents. From the time of filing, the two firms disputed
the claims of each other's patent applications. In a 1966
settlement, each dropped its opposition and agreed not to
dispute its rival's patents for a period of ten years. The
agreement left third parties with the option of licensing
integrated circuit patents from either Fairchild or TI. The
former firm was by far the more aggressive in seeking licensees, and indeed royalties from its patents on the planar
process were a significant fraction of Fairchild's net income in
the late 1960s and early 1970s (Finan, 1975).

Interestingly, if only one firm had held a reasonable claim
to patents on integrated circuit technology, it might have
attempted to monopolize it. Of course, Bell would have had to
be licensed in order for the integrated circuit monopolist to
produce semiconductors at all, but Bell was enjoined from
competing for commercial sales. Fortuitously, the coincident
claims of TI and Fairchild assured that integrated circuit

technology would be widely disseminated, since an agreement between the two to refrain from licensing to others would surely have been challenged under the antitrust laws.

Once the critical patents were freely available, the cumulative nature of technical progress in the industry guaranteed that patents would either be widely cross-licensed or simply ignored. Von Hippel (1982) provides a useful hypothetical illustration. Suppose Firm X invents and patents an improvement on the planar process. Legally, Fairchild may prevent Firm X from using its own invention unless it takes a license from Fairchild on the original process. But whether or not Firm X undertakes production, it may in turn prevent Fairchild from using its improvement on the planar process without a license. An obvious solution to the stalemate is to cross-license.

In fact, however, many semiconductor firms simply neglect to arrange to license all the patents that might conceivably be used to block its activity. Typically, most merchant firms regularly review cross-licensing arrangements with technologically progressive captive suppliers like IBM and AT&T, but they often infringe on one another's patents. A patent holder typically ignores such infringement, since it is likely to be guilty of reciprocal offense. When an infringement claim does occasionally arise, the accused firm usually responds by proclaiming its innocence and sending the accuser "a pound or two" of possibly germane patents which it believes the accuser to be infringing (von Hippel, 1982). The result is almost inevitably a cross-license rather than litigation.

The preceding example illustrates clearly why many semiconductor firms go to the time and trouble to file for patents despite their widespread infringement and limited enforceability. From the standpoint of a single firm, an arsenal of patents is a useful deterrent against the infringement claim of others, justifying investment of $2,000 to $10,000 per filing. From the standpoint of society, however, patenting activity in this particular industry, given the cumulative and interdependent nature of technical advance, is probably, like the arms race, a negative-sum game.

Imitation by infringement is thus a common practice in the patentable areas of semiconductor technology, but in the area of circuit design, imitation is literally a line of business. Firms devote considerable R&D resources to "reverse engineering" of new circuits fabricated by competitors, and there are even a handful of firms that specialize in the provision of reverse engineering services. So rampant is this activity that the term "second sourcing," which originally referred to technology-transfer arrangements demanded by large buyers, has become in industry parlance a euphemism for outright copying. Both large and small firms engage in unauthorized second sourcing. For example, a copy of Mostek's best-selling

16K random-access memory chip could be bought in 1980 from Texas Instruments, Fairchild, or ITT, or, indeed, the photo-masks used to fabricate the circuit could be purchased for $100,000 from Mosaid, a small Canadian company specializing in reverse engineering (Business Week, April 21, 1980).

The absence of patent coverage for circuit layouts has induced some firms to seek protection for proprietary designs by "potting" their circuits, encapsulating them in epoxy resin (Shapley, 1978). To decode the circuitry inside, an imitator must find a solvent for the coating that leaves circuit compon-ents intact. According to industry sources, however, such deceptive packaging techniques rarely withstand the reverse engineering efforts of a determined chemist.

For many years, reverse engineering of circuit designs was considered by the industry more a nuisance than a serious problem. Lead time was often sufficient to develop production experience with a new product, and, through learning, the innovator often gained the advantage in cost and quality control relative to imitators. Recently, however, there has been greater concern about the lack of protection for circuit layouts, because of the substantial investment required for VLSI designs. For example, the American Electronics Associa-tion conducted an unsuccessful lobbying campaign in 1979-1980 to extend copyright protection to integrated circuit layouts. An alternative approach to protecting circuit designs, advo-cated by Levine (1981), is to invoke the common-law doctrine of unfair competition, as exemplified in the case of Interna-tional News Service vs. Associated Press (39 S.Ct. 68, 1918) and recently revived in the area of "pirated" tape recordings.

These attempts to increase the appropriability of circuit designs, however, are unlikely to succeed in practice even if they do in law. So long as semiconductor technology continues to advance cumulatively, with circuit design innovations incorporated into successively improved devices, cross-imitation will be indispensable to continued coexistence of rapid tech-nical progress and market competition. Thus, it is a rea-sonable conjecture that legislation or litigation to enhance the legal appropriability of circuit designs would produce a situation similar to that which prevails in the presently patentable areas of semiconductor technology. Specifically, one would expect widespread cross-imitation in forms both legal (licenses) and illegal (infringement). A concerted attempt to enforce appropriability strictly in a regime where technical opportunities are abundant and where advance is cumulative and interdependent is likely both to reduce competition and to impede technical progress.

Public Support for Higher Education

Perpetuating the advance of a technology that rests on a well-developed scientific base requires a steady influx of highly trained research scientists and engineers. Even incremental improvements in process technology as sophisticated as semiconductor fabrication requires scientific and engineering training that would have been unimaginable two decades ago. Advances in circuit design and software, areas that are as much art as science, require a deep understanding of principles in addition to the spark of imagination.

Scientists and electrical engineers entering the semiconductor industry have always needed a substantial dose of on-the-job training, but formal higher education to the bachelor's level and, for many jobs, beyond is an indispensible prerequisite. Through the 1950s and 1960s, American universities by all accounts provided these educational services at a high standard of excellence, and a handful of U.S. institutions were acknowledged worldwide as the prominent centers for training in electrical engineering and in associated areas of solid-state physics.

It is difficult to be certain that federal government support for higher education played an essential role in these developments since the manpower demands generated by the young and growing electronics industries provided substantial impetus to development of an educational infrastructure. But the evidence suggests that the government's supply-side push was an important complement to industry's demand pull. From the early 1950s onward, the Department of Defense supported doctoral candidates in solid-state physics in addition to funding faculty research. Defense funds totalling one to two million dollars annually supported over 100 doctoral candidates during the 1950s (Linvill, 1962). After Sputnik, student support increased dramatically at all degree levels, and funding grew steadily throughout the 1960s. Detailed data on total government outlays for education in solid-state physics and electrical engineering are not available, but the data in Table 2.19 on National Science Foundation obligations for student support are representative of the pattern that prevailed in these specific fields. As is clear, funding rose rapidly in the decade after Sputnik; then support fell steadily after 1966, declining severely during the Nixon and Ford administrations. In the late 1970s, funding has remained roughly constant in nominal terms, masking continuation of a significant decline in real terms. In constant dollars, federal support for science students had declined by 1980 to less than 20 percent of its 1966 levels.

The data on degrees conferred strongly suggest that government funding matters a great deal. Given that the market for electronic equipment generally (as well as that for

Table 2.19. Estimated NSF Obligations for
Student Support, 1952-1980
(in millions of dollars).

Year	Funds	Year	Funds
1952	1.54	1967	47.81
1953	1.38	1968	48.41
1954	1.80	1969	42.66
1955	1.85	1970	45.67
1956	2.36	1971	37.55
1957	3.00	1972	24.11
1958	4.22	1973	21.16
1959	14.71	1974	18.13
1960	15.94	1975	19.00
1961	15.23	1976	22.50
1962	20.01	1977	23.03
1963	26.65	1978	21.45
1964	33.37	1979	22.40
1965	43.35	1980	21.18
1966	52.21		

Sources: National Science Foundation, Science Education
Databook, Washington: NSF, 1980, p. 19.

semiconductor components) grew as rapidly in the 1970s as in
the 1960s, one would expect to find continued increase in the
supply of new physicists and electrical engineers. But such is
not the case. The number of doctoral degrees awarded annually in physics grew fourfold from 1950 to 1970, then declined
35 percent over the next decade. Master's degrees in physics
doubled and then also fell 35 percent over the same period
(National Science Foundation, 1980). In engineering overall,
degrees awarded at all levels peaked in 1972, and declined
slightly thereafter at the bachelor's and master's levels. At
the Ph.D. level, degrees awarded fell 30 percent in just five
years (National Science Foundation, 1980). In electrical
engineering specifically, the peak also came in 1972, and the
number of total graduates declined sharply for five years,

though some evidence of resurgence appeared at the end of the decade.

The striking contrast between recent developments in engineering education in the United States and Japan is a matter of great concern to the industry. Table 2.20 shows the annual number of electrical engineering graduates at all levels in the United States and Japan. Over the decade, the United States experienced no growth at all, while in Japan the number of graduates nearly doubled. In per capita terms, the two countries were on equal footing in the mid-1960s, with about 80 electrical engineering graduates per million population in each country. By 1977, the United States had 66 graduates per million; the Japanese had 185.

The Semiconductor Industry Association (1981) reported that there were 20,000 unfilled skilled positions in the industry in 1979, and it projected that the manpower situation will deteriorate throughout the 1980s unless there is a substantial increase in engineering graduates. Given the experience of the 1970s, when demand growth alone proved inadequate to meet the industry's needs, the prognosis is not particularly good without increased government support.

Table 2.20. Electrical Engineering Graduates at all Degree Levels, 1969-1979, United States and Japan.

Year	U.S.	Japan
1969	16,282	11,848
1970	16,844	13,889
1971	17,403	15,165
1972	17,632	16,052
1973	16,815	17,345
1974	15,749	17,419
1975	14,537	18,040
1976	14,380	18,258
1977	14,085	19,257
1978	14,701	20,126
1979	16,093	21,435

Source: Semiconductor Industry Association, The International Microelectronic Challenge (Cupertino, Calif.: Semiconductor Industry Association, 1981), p. 29.

Tax Policy

Although changes in federal income-tax policies for both
individual and corporations have economywide repercussions,
two specific changes in tax policy are believed by many industry participants to have had an especially significant impact on
the vitality of the semiconductor industry. These particular
policies are the gradual increases in capital gains taxation from
1969 through 1976 and the elimination of the qualified stock
option in 1976. The alleged impact of these policies on
technical advance is perforce indirect, working through more
direct effects on the rate of entry of new firms and the
mobility of key employees.

The Tax Reform Act of 1969 increased from 25 percent to
35 percent the standard rate of taxation on capital gains for
individuals in the highest income-tax bracket. Additional
provisions for a minimum tax on overall income and for reductions in maximum tax benefits on personal service income
combined to add more than 14 percentage points to the standard rate for some individuals. Thus, the maximum effective
tax on capital gains increased from 25 to over 49 percent in a
period of three years.

The effect of these provisions on the semiconductor
industry was especially severe because of its dependence on
venture capital to finance the entry of new firms. Since the
returns to venture capital come largely in the form of capital
gains, tax increases on this type of income reduced the relative attractiveness of financing risky new businesses. As
noted in Section II, there was a sharp decline in the rate of
entry in the years following these tax increases. Of course,
the reduced rate of entry cannot be wholly attributed to the
tax changes and attendant conditions in the venture-capital
market, since the cost of entry increased rapidly over the
same period. Nevertheless, the evidence for a significant
impact is fairly convincing because venture capital began to
flow again when the maximum capital gains tax was reduced to
28 percent by the Tax Reform Act of 1978. As a consequence,
there was a sharp increase in the number of new semiconductor firms established in 1979. It is too early to assess the
impact of the further reduction in the effective maximum rate
of capital gains taxation embedded in the Economic Recovery
Tax Act of 1981.

Qualified stock options were viewed by semiconductor
firms as an especially inexpensive form of executive compensation prior to their elimination in 1976. Under previous law,
qualified options were not taxed when granted or when exercised; they were taxed only when the stock was sold, and
then only at the capital gains rate. Wilson, Ashton, and Egan
(1980) found that industry executives believed that the elimination of qualified stock options reduced the mobility of key

executives and engineers, impeding the flow of information among established firms as well as reducing the incentives for key personnel to move to new firms.

The link between these changes in tax policy and technological performance is only indirect. Since recent entrants to the industry were important sources of product innovation in the 1960s and early 1970s, it is distinctly possible that by restricting the flow of entry in the mid-1970s tax policy had an adverse impact on technical progress. Moreover, to the extent that interfirm mobility has been reduced, new knowledge may diffuse more slowly. It was argued in Section III, however, that the nature of past technical change has altered the role that new firms are likely to play. Quite apart from tax consideration, the cost and scale required for entry near the frontier of mainstream integrated circuit technology probably preclude entrants from taking a leading role in progress along the miniaturization trajectory. New firms are more likely to contribute to technical advance in specific areas like CAD, custom and semicustom design, and fabrication. To the extent that tax policies have constrained entry, it is likely that any adverse technological impact has been limited to areas such as these.

A somewhat more direct impact of government policy on the technological effort of small firms has been observed in the area of financial reporting, which is regulated by the Securities and Exchange Commission. In 1974 the Financial Accounting Standards Board promulgated a rule (FASB No. 2) requiring all companies issuing certified financial statements to report R&D as expenses in the year incurred. This standard, as adopted by the SEC in 1975, eliminated the option of capitalizing R&D for financial reporting purposes, though firms were allowed to retain the option of capitalization for tax purposes. The new standard has the effect of depressing reported current earnings of small new entrants in high-technology industries where large R&D outlays are often necessary to establish a market position. If capital markets were fully efficient, one would expect the new standard to produce no change in firm behavior or in asset valuation. Nevertheless, it is widely believed that reported current earnings are given undue weight in business decision making. In a careful study of the effects of FASB No. 2, Horwitz and Kolodny (1981) found that the R&D expenditure of firms previously capitalizing their R&D declined significantly relative to those of a control group previously expensing R&D.

V. CURRENT POLICY ISSUES IN THE
LIGHT OF PAST EXPERIENCE

International Competitiveness

Without question the issue that has dominated public-policy discussion at the outset of the 1980s is the actual or impending loss of U.S. leadership in semiconductor technology. Few would seriously maintain that the U.S. industry has performed poorly. Indeed, in the 1970s the rate of growth of semiconductor sales by U.S. firms exceeded the sales growth of the previous decade. As described in Section III, technical progress along the miniaturization trajectory has continued at a swift pace. Nevertheless, by the end of the 1970s, the leading Japanese firms appeared to have developed technological capability on a par with the largest and most advanced U.S. firms.

The data tell the story. According to the Department of Commerce, Japanese exports of total semiconductors to the United States surpassed the U.S. exports to Japan in 1977. Integrated circuit imports from Japan surpassed exports to Japan in 1978, and U.S. imports of MOS integrated circuits from Japan first exceeded flows in the reverse direction in 1979 (Department of Commerce, cited in Semiconductor Industry Association, 1981). These data reflect more than a shortening of the imitation lag. In the high-volume area of advanced random-access memories, the Japanese have quite literally taken the lead. By 1979, three Japanese firms (Nippon Electric, Hitachi, and Fujitsu) ranked among the top five in worldwide sales of 16K RAMs, accounting for about one-third of the total market (Dataquest, cited in Business Week, December 3, 1979). In 1981, Hitachi grabbed a 40 percent share of the world market in 64K RAMs, leaving Mostek and Intel, the previous U.S. leaders in random-access memory sales, far behind. Motorola managed to stay in the race with 20 percent of the market, a share matched by Fujitsu. In all, Japanese firms accounted for 69.5 percent of worldwide sales of 64K RAMs, according to Dataquest's preliminary estimates for 1981 (Dataquest, cited in Fortune, December 14, 1981).(12) At the same time, U.S. firms continued to maintain both technological and market advantages in sophisticated logic devices and microprocessors.

The Japanese drive to technological parity was facilitated by a substantial program of government coordination and R&D support. The five-year VLSI program, which began in 1974, had much in common with successful U.S. military efforts of the past. The program had clear technical objectives, specified in terms of device complexity and speeds to be attained by 1980. Research and development funds from the government were more than matched by private expenditures of the five large firms involved. The government facilitated the dissemination of information but the firms retained patents and proprietary rights to products developed with VLSI funding. Most important, by simultaneously making a commitment to the development of a strong national computer industry, the Japanese created an environment in which there was perceived to be a large secure market for the products developed through semiconductor R&D. Rather than committing itself to be the primary user of new devices in the manner of the U.S. military, the Japanese government achieved some of the same effects by committing itself to massive support of the primary users.

The Japanese challenge has elicited proposals for an enormous variety of policy responses. Among the policy options under current discussion are restrictions on direct foreign investment in the United States, relaxation of restrictions on access to Japanese markets via trade and direct investment, provision of capital to U.S. firms on terms comparable to those enjoyed by Japanese firms, tax credits for R&D, increased support for scientific and engineering education, and restrictions on technology transfer abroad. No attempt will be made here to sift through the merits and defects of each of these proposals, but the past experience of government involvement in the semiconductor industry does shed light on at least two of the current proposals: increased support for higher education and tax credits for R&D.

Among the inferences to be drawn from the historical record is that government funding influences the supply of trained manpower. In the face of dramatic decreases in federal support, even the powerful influence of rapid growth on the demand side was insufficient to generate increases in the supply of physicists and engineers in the 1970s. If expansion of the supply of trained personnel is a desirable objective, significant increases in government support would seem an appropriate instrument, since the record suggests that in this area the market is likely to provide too little, too late.

On the other hand, there is little in the historical record to suggest that tax credits for R&D, or some other form of undirected R&D subsidy, is likely to be productive of more, or more successful, innovative effort. To the contrary, the success of past government policies in eliciting valuable R&D effort has depended on several features previously enunciated: the government's specification of clear technological objectives without rigidly specifying the means to achieve them, its role as a user of the technology, and the assurance of substantial demand for the products generated by R&D. Past military R&D and procurements programs contained each of these features, and, as noted, the Japanese VLSI program resembles the military model in some respects. But undirected R&D

subsidies contain none of the ingredients common to previously successful programs. No doubt, R&D tax credits would elicit some increase in real resources devoted to innovative effort, but the impact per dollar of tax expenditure is likely to be modest for several reasons. First, unless the tax credits are restricted to incremental R&D, many federal dollars would be devoted to pure transfer payments. Second, even tax credits on incremental R&D would contain substantial elements of pure transfers, since most firms in this growing industry would steadily increase R&D without subsidy. Third, firms will find means to redefine considerable engineering effort as R&D. Finally, the tax system already favors R&D over capital investment, and many in the industry would argue that it is not limitations of knowledge, but the massive capital investment required, that presently constrains more rapid technical progress.

The VHSIC Program

Amidst public concern for the future of the semiconductor industry, the U.S. government launched in 1980 a major program of R&D support for military applications of advanced technology. The impetus for initiating the VHSIC (Very High Speed Integrated Circuit) program was quite independent of civilian concerns about the industry's future or its standing relative to Japanese competition. Rather, planning for the program began in 1978 immediately after military intelligence reports revealed that the U.S. advantage in the electronics embodied in fielded weapons systems had been significantly eroded. The principal objective of the VHSIC program is to establish the capability for fielding weapons systems utilizing high-speed integrated circuits of submicron feature size by the end of the decade. Technically, one of the program's central goals is defined as an increase of two orders of magnitude in a critical parameter which is the product of speed (clock rate) and circuit density (gates per cm^2).

The design of the VHSIC program appears quite sensible in the light of past experience. As in successful past government efforts, the technological goals are quite clearly specified, and there is an understood commitment that technological success in the design of VHSIC chips will be followed by procurements. Indeed, each contractor is required to work toward the development of chips for two specific systems applications. At the same time, the program does not insist on a single technological approach, as the military did in the abortive Micromodule and molecular electronics programs. The only rigid requirement for VHSIC funding is that R&D effort be confined to silicon materials. This decision was dictated by the program's relatively short time horizon; most experts feel that the more exotic technologies, from gallium arsenide semiconductors to Josephson junction superconductors, will not be sufficiently well developed for military application by 1987.

The premature obsolescence of the Tinkertory and Micromodule efforts suggests that confining R&D to silicon might be a mistake, given the plausible alternatives on the horizon. Yet it is only VHSIC program funds that are so restricted, and at the outset of VHSIC, all ongoing R&D programs in the three service branches were ordered to cut back their funding of silicon technology. The net result is that military funding of silicon R&D has increased by less than the approximately $35 million annually expended under VHSIC, while funding of more advanced technologies has increased substantially, perhaps by as much as $10 million annually.

Technologically, the goals of the VHSIC program are highly compatible with the continued pursuit of miniaturization in the commercial segment of the semiconductor business. The military has certain specialized needs, such as the ability of circuitry to perform under extreme conditions of temperature and radiation. But much R&D funded by VHSIC, such as support for advanced lithographic techniques to facilitate realization of submicron feature sizes and support for improved CAD, software, and testing methods, should have significant spillovers to commercial application. In turn, the independent pursuit of similar technological objectives for commercial purposes should facilitate the achievements of VHSIC goals. Indeed, the planned Department of Defense expenditure of approximately $200 million over seven years is far less than industry will spend on its own, but there is an emerging consensus that the added stimulus provided by VHSIC funds will move forward the realization of submicron circuits by two or three years.(13)

When the VHSIC program was first announced, it was enthusiastically received by most major suppliers of military electronics systems, but several leading merchant semiconductor firms expressed serious reservations and some chose to abstain from bidding on VHSIC contracts. A major concern was that the VHSIC program would divert scarce R&D resources, in particular critical personnel, from pursuit of commercial objectives. It was feared that VHSIC would handicap U.S. firms in competition with the Japanese for leadership in VLSI technology. These fears seems to have been misplaced, as industry participants have come to recognize the substantial complementarity between VHSIC and commercial VLSI objectives. On the other hand, it would be a mistake to view the VHSIC program as a direct response to the Japanese government's support of the semiconductor industry. While it now appears that VHSIC will provide an indirect boost to U.S. firms in technological competition with the Japanese, merchant semiconductor firms still seek policy assistance more directly related to meeting the Japanese challenge, as noted above.

Early critics of VHSIC also questioned the program's emphasis on supporting large-scale, vertically integrated research efforts. As envisioned, the program involved vertically integrated firms or teams of firms, and each proposal was expected to tackle a range of issues from circuit-fabrication technology and process equipment to insertion of circuits into weapons systems. Critics feared that the emphasis on large firms and vertically integrated teams would hasten concentration of the semiconductor industry and reinforce the trend toward vertical integration, allegedly threatening the vitality of a highly competitive and dynamic merchant semiconductor industry. Congress initially delayed funding of the program until it received assurances that the program would not have an anticompetitive impact on the industry.

As it has developed, the major portion of R&D support will be allocated to vertically integrated contractor teams responsible for developing the technology necessary at all levels to use submicron integrated circuits in operational weapons systems. Initial nine-month Phase 0 contracts were awarded to nine such teams in 1980, and in May 1981 six of the teams were selected as contractors for Phase I of the program, which will extend into 1983. It is unlikely that confining VHSIC support to six teams (five of which involve merchant semiconductor firms; one contract was won by IBM) will increase concentration in the industry, especially since several nonparticipating firms will be pursuing VLSI technology with private resources on a significant scale. But the initial congressional worries about market concentration did encourage the DOD to develop a program design that preserved niches of opportunity for small, nonintegrated firms as well as university research laboratories.

Paralleling the mainstream Phase I and II efforts will be a series of much smaller contracts to be awarded on a continuing basis throughout the duration of the program. These smaller Phase III contracts will focus on narrow technical problems, where significant contributions, complementary to the Phase I and II objectives can be expected from firms outside the mainstream program. It is expected that Phase III contracts will be concentrated in areas such as lithography, CAD, software, and testing. In concept, Phase III represents a reasonable safeguard against the somewhat remote possibility that the VHSIC program will unduly accelerate the industry toward maturity and stagnation. In any case, it appears to be an example of organizational design well suited to maximizing technical advance. On the one hand, major support will be given to not one but several large-scale, vertically integrated efforts. On the other hand, substantial funds, one-third of the total budget, will be reserved for smaller-scale projects complementary to the program's overall objectives. In principle, such an organizational design can be used to generate innovation from both large and small firms in the areas where each has a comparative advantage.

Given this rather creative institutional design, the results of the first round of Phase III contract awards were somewhat discouraging. The first Phase III contracts were let several months before the due date for Phase I proposals, and consequently, virtually half (77 of 157) of the proposals submitted came from the large firms involved in the mainstream Phase 0 program. Evidently, Phase 0 winners saw in Phase III an opportunity to impress the DOD with good work prior to the major funding decisions on Phase I proposals. Of the 157 proposals received, only 4 came from qualified small businesses and only 8 came from nonintegrated semiconductor firms. Only 1 of these 12 was among 53 funded proposals, while 24 contracts were awarded to Phase 0 participants. Somewhat more encouraging was the award of 11 Phase III contracts to 5 different universities.

It is evident that if the VHSIC program is to benefit from innovative ideas from a variety of sources, more attention must be paid to encouraging the submission of proposals from small and nonintegrated firms. Managers of the program are aware of this problem, and they have taken steps to simplify drastically the format of the second-round, Phase III request for proposals. Indeed, there is a growing recognition throughout the DOD that opportunities for small-firm participation in R&D support programs have been diminished by the escalating complexity of the contracting process. In a promising development, the DOD initiated in April 1981 a new Defense Small Business Advanced Technology Program. Proposals were solicited by a lucid 23-page document, a striking contrast to the 100 pages of boilerplate contained in the first-round request for VHSIC Phase III proposals. If the VHSIC program follows this lead, prospects will be enhanced for the preservation of a dynamically competitive semiconductor industry structure with variegated sources of innovation.

SUMMARY AND CONCLUSIONS

Seen against the perspective of the varied government activities influencing technical progress throughout the U.S. economy, the record of public policy in the semiconductor industry appears to have been uncommonly successful. Yet even in this sector some policies have failed and others have achieved unintended consequences. Thus, at least two costly military R&D programs were obsolete before completion, and another was hopelessly ahead of its time. Patents, intended to protect the appropriability of returns from invention and thus to provide incentive to innovate, turn out to have had little of their intended effect in a technology where advance is rapid and cumulative. Indeed, but for two accidents of history – the invention of the transistor by AT&T rather than by a firm less willing to disseminate its knowledge and the simultaneous occurrence of complementary product and process innovations at TI and Fairchild in the late 1950s – the patent system might have retarded the advance of semiconductor technology.

Nonetheless, the achievements of government procurement and R&D policies were striking, and reflection upon them yields insights of considerable importance for future policy design. First among these is the importance of clearly perceived technological objectives, beyond the immediately feasible but not hopelessly far beyond. To suggest that the government limit its ambition is only to take seriously the experience of electronics and all other technologies. One has only to pursue the collection of vignettes in Jewkes, Sawyers, and Stillerman (1958) to become convinced of the pervasive role of serendipity in the realization of truly radical inventions. Major innovations are simply too much to expect from the government R&D. The government could not have revolutionized electronics (i.e., invented the transistor) nor, for that matter, is it likely to cure cancer.

What the government at its best can do is move flexibly and responsively to support promising new technologies early in their life cycle. In the early years of semiconductor technology, the government stood ready with funds for R&D and production engineering to support each significant new technical development. Such flexibility and responsiveness might have been impossible were the government not itself a major consumer, indeed the major consumer, of the new technology. It was thus a position to evaluate technological alternatives intelligently, and to provide valuable user feedback to suppliers. Finally, the importance of the government's role as a high-volume purchaser of quality goods at premium prices cannot be exaggerated. The presence of government demand reduced the risks of investment in new technology, and the government's willingness to purchase large volumes at premium prices permitted the accumulation of production experience necessary for the realization of dynamic economies and the penetration of the commercial markets.

All this by no means suggests that the appropriate response to the current threat to U.S. leadership in semiconductor technology is a return to the policies of the 1950s and early 1960s. The policies reviewed here, in which the government takes the role of informed first user and largest buyer of new technology, were appropriate at an early stage of technological evolution. Today, the technology has matured considerably; advances are incremental and indeed predictable. Commercial users have clear needs for technological improvements, and some commercial users, such as computer manufacturers, represent a much larger potential market than the government. The government is no longer able to direct technical advance in accord with its own needs as a consumer. The pattern of progress is now fairly well understood; it is mostly a question of influencing its pace. Thus, the policy problems of a more mature technology require a different set of instruments, but the lessons learned in the semiconductor industry's youth may yet find fruitful application elsewhere, perhaps in the superconductor technology of the next generation, in the production of solar energy, or in biotechnology.

NOTES

1. It is interesting to note that in 1979 three of the top eight firms in worldwide semiconductor sales were Japanese and one was European. Surprisingly, four of the top eight were Japanese in 1972. In integrated circuits, only two of the top eight firms throughout the 1970s were Japanese, and none was European. In the most rapidly growing market segment, MOS integrated circuits, three Japanese firms were among the top eight in 1979, an increase over the two represented in 1972.

2. Though conceived as a replacement for the vacuum tube, the transistor found many of its earliest uses in applications for which the vacuum tube was poorly suited (e.g., hearing aids, airborne computers, and mobile communications systems). Sales of receiving tubes continued to grow throughout the 1950s, driven largely by rapid expansion of the markets for television receivers and phonographic equipment.

3. Two explanations are commonly offered to explain the relative demise of the vertically integrated electrical equipment firms in the semiconductor business. First, bureaucratic rigidity in these large organizations is felt to have impeded flexible and rapid response to changing technological and market conditions. Second, the electrical giants failed to perceive soon enough that the most promising areas for early application of semiconductor technology were outside their traditional lines of business. For a detailed discussion of these points, see Golding, 1971.

4. For further examples and discussion of entry costs, see Golding, 1971, pp. 108-120; Tilton 1971, pp. 87-89; and Wilson, Ashton, and Egan, 1980, p. 167.

5. This point is emphasized by Mowery, 1980.

6. In the surveys conducted by the Business and Defense Services Administration, semiconductor producers were asked to report total shipments destined for end use by the government. Thus, sales to private firms of components to be em-

ployed in military equipment were intended to be recorded as shipments to the government. But to the extent that semiconductor producers were not aware of the uses to which private-sector customers put their components, the data are unreliable and probably understate shipments destined for use by the government.

7. This estimate was made available to the author in a private communication from the Department of Defense.

8. U.S. v. Western Electric Company, Inc. and American Telephone and Telegraph Company, Civil Action 17-49, U.S. District Court of New Jersey, January 14, 1949.

9. U.S. v. Western Electric Company, Inc. and American Telephone and Telegraph Company, Civil Action 17-49, Final Judgement, January 24, 1956.

10. For a lucid treatment of the economic theory of patents, see Nordhaus, 1969.

11. For an interesting discussion of the relevant case law and one lawyer's view that circuit layouts could be patented, see Levine, 1981.

12. The Japanese ascended to domination of the 64K RAM market with unexpected rapidity. A Business Week article of October 6, 1980 cited a forecast by an industry expert that 64K RAMs would sell for $17 in 1982. By late 1981, Japanese firms were selling reliable 64K devices at $8, and they anticipated prices in the $4 to $6 range in 1982 (Fortune, December 14, 1981).

13. For an interesting and detailed survey of progress made during the first nine months of VHSIC funding, see Aviation Week and Space Technology, February 16, 1981, pp. 48-85.

REFERENCES

Asher, N.J. and L.D. Strom. The Role of the Department of Defense in the Development of Integrated Circuits. IDA Paper P-1271. Arlington, Va.: Institute for Defense Analysis, May 1977.

Aviation Week and Space Technology. "Technical Survey: Very High Speed Integrated Circuits." February 16, 1981, pp. 48-85.

Braun, E. and S. Macdonald. Revolution in Miniature. Cambridge: Cambridge University Press, 1978.

Business Week. "Japan is Here to Stay." December 3, 1979, pp. 81-86.

_____. "How 'Silicon Spies' Get Away with Copying." April 21, 1980, pp. 181-88.

_____. "The Chip Makers' Glamorous New Generation." October 6, 1980, pp. 117-22.

Charles River Associates. Productivity Impacts of Government R&D Laboratories: The National Bureau of Standards' Semiconductor Technology Program. 2 volumes. Boston: Charles River Associates, 1981.

Dataquest. Semiconductor Industry Services, Appendix B. Cupertino, Calif.: Dataquest, 1980.

Dummer, G.W.A. "Electric Components in Great Britain," in Progress in Quality Electronics Components, Proceedings of a Symposium of the IRE-AIEE-RTMA, Washington, D.C. May 1952.

Electronic Industries Association. Electronic Market Data Book 1979. Washington: Electronic Industries Association, 1979.

Finan, W.F. "The International Transfer of Semiconductor Technology Through U.S.-Based Firms." NBER Working Paper No. 118. Washington: National Bureau of Economic Research, December 1975.

_____. "The Semiconductor Industry's Record on Productivity," in American Prosperity and Productivity: Three Essays on the Semiconductor Industry. Cupertino, Calif.: Semiconductor Industry Association, 1981.

Fortune. "Japan's Ominous Chip Victory." December 14, 1981, pp. 52-57.

Futia C. "Schumpeterian Competition." Quarterly Journal of Economics, 94, June 1980, pp. 675-95.

Golding, A.M. "The Semiconductor Industry in Britain and the United States: A Case Study in Innovation, Growth, and Diffusion of Technology." D. Phil. Dissertation, University of Sussex, 1971.

Grabowski, H.G. and J. Vernon. "Government Policy and Innovation in the Pharmaceutical Industry," in R.R. Nelson, ed., Government and Technical Progress: A Cross-Industry Analysis. New York: Pergamon Press, forthcoming, 1982.

Horwitz, B. and R. Kolodny. "The FASB, the SEC, and R&D." Bell Journal of Economics, 12, Spring 1981, pp. 249-262.

Integrated Circuit Engineering. Status 1980: A Report on the Integrated Circuit Industry. Scottsdale, Arizona: ICE, 1981.

_____. Status 1981: A Report on the Integrated Circuit Industry. Scottsdale, Arizona: ICE, 1981.

Katz, B. and A. Phillips. "Government, Technological Opportunities, and the Structuring of the Computer Industry: 1946-61," in R.R. Nelson, ed., Government and Technical Progress: A Cross-Industry Analysis. New York: Pergamon Press, forthcoming 1982.

Kendrick, J.W. and E.S. Grossman. Productivity in the United States: Trends and Cycles. Baltimore: Johns Hopkins University Press, 1980.

Kilby, J.S. "The Invention of the Integrated Circuit." IEEE Transactions on Electron Devices, ED-23, July 1976, pp. 648-54.

Kleiman, H.S. "The Integrated Circuit: A Case Study of Product Innovation in the Electronics Industry." D.B.A. Dissertation, George Washington University, 1966.

Kraus, J. "An Economic Study of the U.S. Semiconductor Industry." Ph.D. Dissertation, New School of Social Research, 1973.

Levin, R.C. "Innovation in the Semiconductor Industry: Is a Slowdown Imminent?" in H.I. Fusfeld and R.N. Langlois, ed., Understanding R&D Productivity. New York: Pergamon Press, 1982.

_____, and P.C. Reiss. "Tests of a Schumpeterian Model of R&D and Market Structure," in Zvi Griliches, ed., R&D, Patents and Productivity. Chicago: University of Chicago Press, forthcoming 1982.

Levine, H. "Very Large Scale Protection for VLSI?," in N.G. Einspruch, ed. Microstructure Science and Engineering/ VLSI, Volume I. New York: Academic Press, 1981.

Linvill, J.G. "How Can Solid-State Electronics Mature Without Losing its Youth?" Solid State Journal, 2, April 1962, pp. 13-17.

Linvill, J.G. and C.L. Hogan. "Intellectual and Economic Fuel for the Electronics Revolution." Science, 195, March 18, 1977, pp. 1107-13.

McHugh, K.S. "Bell System Patents and Patent Licensing." Bell Telephone Magazine, January 1949, pp. 1-4.

Moore, G.E. "VLSI: Some Fundamental Challenges." IEEE Spectrum, 16, April 1979, pp. 30-37.

Mowery, D. "The Semiconductor Industry." Unpublished paper, Stanford University, 1980.

National Science Foundation. Science Education Databook. Washington: National Science Foundation, 1980.

Nelson, R.R. "The Link Between Science and Invention: The Case of the Transistor," in R.R. Nelson, ed., The Rate and Direction of Inventive Activity. Princeton: Princeton University Press, 1962.

Nelson, R.R. and S.G. Winter. "Forces Generating and Limiting Concentration under Schumpeterian Competition." Bell Journal of Economics, 9, Autumn 1978, pp. 524-48.

Nordhaus, W.D. Invention, Growth and Welfare. Cambridge, Mass.: MIT Press, 1969.

Oster, S.M. "Firm Strategies in an Adverse Environment: the Steel Industry." Working paper, Yale University, October 1981.

Platzek, R.C. and J.S. Kilby. "Minuteman Integrated Circuits: A Study in Combined Operations." IEEE Proceedings, 52, December 1964, pp. 1669-78.

Robinson, A.L. "Giant Corporations from Tiny Chips Grow." Science, 208, May 2, 1980, pp. 480-84.

Semiconductor Industry Association. The International Microelectronic Challenge. Cupertino, Calif.: Semiconductor Industry Association, May 1981.

Shapley, D. "Electronics Industry Takes to 'Potting' its Products for Market." Science, 202, November 1978, pp. 848-49.

Shockley, W. "The Path to the Conception of the Junction Transistor." IEEE Transactions on Electron Devices, ED-23, July 1976, pp. 597-620.

Speakman, E.A. "Reliability of Military Electronics," in Progress in Quality Electronics Components, Proceedings of a Symposium of the IRE-AIEE-RTMA. Washington, May 1952.

Taylor, C.T. and Z.A. Silberston. The Economic Impact of the Patent System: A Study of the British Experience. Cambridge: Cambridge University Press, 1973.

Teal, G.B. "Single Crystals of Germanium and Silicon - Basic to the Transistor and Integrated Circuit." IEEE Transactions on Electron Devices, ED-23, July 1976, pp. 621-39.

Tilton, J. International Diffusion of Technology: The Case of Semi-Conductors. Washington: Brookings Institution, 1971.

U.S. Air Force. Integrated Circuits Come of Age. Washington: U.S. Air Force, 1965.

U.S. Department of Commerce, Business and Defense Services Administration. Electronic Components: Production and Related Data, 1952-1959. Washington: Department of Commerce, 1960.

U.S. Department of Commerce. Semiconductors: U.S. Production and Trade. Washington: U.S. Government Printing Office, 1961.

U.S. Department of Commerce. Report on the Semiconductor Industry. Washington: U.S. Government Printing Office, 1979.

Utterback, J.M. and A. Murray. Influence of Defense Procurement and Sponsorship of Research and Development on the Civilian Electronics Industry. Cambridge: MIT Center for Policy Alternatives, 1977.

von Hippel, E. "The Dominant Role of the User in Semiconductor and Electronic Subassembly Process Innovation." IEEE Transactions on Engineering Management, EM-24, May 1977, pp. 60-71.

_____. "Appropriability of Innovation Benefit as a Predictor of the Functional Locus of Innovation." Research Policy, forthcoming, 1982.

Webbink, D.A. The Semiconductor Industry: A Survey of Structure, Conduct and Performance. Staff Report to the Federal Trade Commission. Washington: Federal Trade Commission, 1977.

Wilson, R.W., P.K. Ashton, and T.P. Egan. Innovation, Competition, and Government Policy in the Semiconductor Industry. Lexington, Mass.: Lexington Books, 1980.

3
The Commercial Aircraft Industry

David C. Mowery
Nathan Rosenberg

Judged against almost any criterion of performance - growth in output, exports, productivity, or innovation - the civilian aircraft industry must be considered a star performer in the American economy. American commercial aircraft dominate airline fleets the world over, and the air transportation industry, a primary beneficiary of technical progress in commercial aircraft, has compiled an impressive record of productivity growth since 1929.(1) Along with this perform-ance record, however, the aircraft industry presents important anomalies in structure and conduct to the student of industrial organization and technical change. Fierce price competition coexists with high levels of producer concentration and sig-nificant product differentiation. The industry has received substantial government support for research on commercial aircraft through the National Advisory Committee on Aeronau-tics (NACA, 1915-1958), and the National Aeronautics and Space Administration (NASA, 1958-present). In addition, the military sector of the industry has benefited from federal procurement and research support. In short, federal support of technological advance in the aircraft industry has been substantially greater than such support in any other industry during the period from 1925 to 1975.

In this chapter, we will examine the innovation process within the commercial aircraft industry, focusing particularly upon the role of government policy in affecting the pace and character of innovation within the industry, as well as the structural context within which such innovation has occurred. We will argue that government policy has influenced innovation in the aircraft industry through its impact upon the demand for aircraft, in both the military and civilian spheres, as well as through direct support of research. The peculiar structur-al combination of high levels of producer concentration and intense price and quality competition among producers also reflects the influence of government policy, in the provision of both a market and research and development funding for military aircraft. This government role also has encouraged the development of a vertically disintegrated industry structure, and an important role for subcontractors in the production of aircraft. One important result is a great reliance in the aircraft industry upon the contractual provision of complex technologies, to a much greater extent than is observed in other high technology industries. The importance of subcontracting in the commercial aircraft industry also reflects the extremely high costs and uncertain demand faced by innovators in this area.

The discussion opens with a summary examination of important aspects of the process and product technologies that underlie the commercial aircraft industry. We next consider briefly the structure and historical evolution of the industry and aircraft technology. The role of government-sponsored research, in both the military and civilian sectors of the industry, is covered in the subsequent section of the chapter. The general character and impact of government regulatory, research, and procurement policies is discussed next, followed by a conclusion exploring the relevance of the aircraft industry's experience for other sectors of the economy.

ASPECTS OF PROCESS AND PRODUCT INNOVATION IN THE COMMERCIAL AIRCRAFT INDUSTRY

The commercial aircraft industry has reaped considerable benefits as a technological "borrower," in at least two specific ways. Many of the significant innovations in commercial aircraft design, including many incorporated in the DC-3 (the first great commercial success in the industry), were originally developed by manufacturers of airframes and engines for military applications; such a list would include the air-cooled engine that powered the DC-3, as well as the high-bypass turbofans associated with the L-1011, DC-10, and B-747. "Borrowing" goes beyond applications to commercial designs of components developed for military purposes, as we argue below. Important benefits are reaped by airframe and engine manufacturers who are able to share development, or less often, production tooling, costs between military and civilian designs that are less closely related. Borrowing of another sort also has played a key role in the development of commercial aircraft technology. Aircraft have benefited to an unusual extent from technological developments in other industries. Noteworthy examples are the metallurgical and materials industries, from which have come a wide range of new alloys and composite materials, as well as the chemicals and petroleum industries, where important developments in fuels were achieved before World War II, and electronics, which has provided since 1940 a steady stream of crucially important innovations, ranging from radar to airline reservation and navigational computers. The aircraft industry is unusual in the extent to which it has benefited from the interindustry flow of innovations that typifies the modern economy.

The ability of the commercial aircraft industry to benefit from technical developments in so wide a range of seemingly unrelated industries reflects another important aspect of the commercial aircraft industry, namely, the high degree of systemic complexity embodied in its products. The finished commercial aircraft is composed of a wide range of components for propulsion, navigation, and the like, that are individually extremely complex. The interaction of these individually complex systems is crucial to the performance of an aircraft design, yet extremely difficult to predict from design and engineering data, even with computer-aided design techniques. Uncertainty about aircraft performance is also exacerbated by the still modest state of scientific theory concerning the behavior of such key components as materials. A substantial element of technological uncertainty thus exists in the design and production of a new aircraft. Performance, in many cases, cannot be predicted definitively before the initial flight. The major aircraft manufacturers frequently have pursued production and design strategies (e.g., subcontracting) aimed at insulating themselves from the adverse consequences of such uncertainty.

A final aspect of considerable significance in the commercial aircraft industry concerns the need to achieve large production runs for a given aircraft in order to take advantage of learning curves and to defray high development expenses. Economies of scale and learning curves (the latter phenomenon having been first observed in the production of airframes) play a major role in affecting production costs and the overall profitability of a given aircraft. High development costs, which have become important with the advent of the jet engine and reflect the systemic complexity of aircraft technology, render the greatest possible production of a given aircraft design very important. This has in turn endowed with great importance the "family concept" in aircraft design, in which a given aircraft, such as the Boeing 727, spawns a succession of modified designs, frequently through stretching the fuselage. Modern aircraft are designed so as to develop and exploit technological trajectories.

THE DEVELOPMENT OF INDUSTRY STRUCTURE, 1925-1975

The development of the commercial aircraft industry's structure may be divided into four periods of unequal length, each of which saw a different pattern of development: 1920-1934, 1934-1940, 1940-1945, and 1945-1975. Prior to the First World War, the American aircraft industry scarcely existed; total aircraft output (i.e., private, military and commercial aircraft) numbered 49 in 1914. However, military production of foreign designs and the American Liberty engine increased dramatically during the war, to an annual rate of 21,000 aircraft by late 1918 (see Rae, 1968, pp. 1-2). Over the entire 1920-1975 period, the industry has grown substantially and become much more concentrated. At present, only three producers of airframes and two domestic engine manufacturers are of major importance in the commercial market.

1923-1934

The 1920-1934 period was one during which military and commercial aircraft production were gradually distinguished from one another. Peacetime military procurement also came to play a role in airframe and (particularly) engine development during this period. In the immediate aftermath of World War I, the market for aircraft collapsed with the cessation of military demand, and a surfeit of war surplus aircraft became available for purchase. Aircraft production declined from 14,000 in 1918 to 263 in 1922, according to Holley (1964), but slowly revived, particularly after the military announced plans in 1926 to maintain a total aircraft fleet of 26,000 by 1931, and as the Kelly Air Mail Act of 1925 transferred responsibility for transportation of air mail from the Post Office to private contractors. Also of importance during the 1920s were the increasing level and quality of the research being carried out by the National Advisory Committee on Aeronautics, established in 1915. Military support of aircraft engine development during this period culminated in the foundation of the Pratt and Whitney aircraft engine firm in 1925, on the strength of interest from the Navy in the Pratt and Whitney Wasp.

The revival of the aircraft industry gave rise to a series of mergers in the late 1920s that produced, for the first and only time in the history of the industry, several vertically integrated firms, combining air transport, airframe manufacture, and engine production. United Aircraft, founded in 1929, was comprised of Boeing Aircraft, Boeing Air Transport, Pratt and Whitney, Chance Vought Aircraft, the Hamilton Standard Propeller Corporation, and Stearman Aircraft. North American Aviation, incorporated in 1928, included Curtiss

Aeroplane, Wright Aeronautical, and had large minority stock-
holdings in Transcontinental Air Transport and Western Air
Express (subsequently combined to form TWA). Other major
consolidations of the late 1920s included the Aviation Corporation and the Detroit Aircraft Corporation.

The onset of the Depression placed all manufacturers
under considerable stress, but the air mail scandals of 1933
and the Air Mail Act of 1934 were the crucial events in the
dissolution of these consolidated firms.(2) Under the terms of
the 1934 act, air transportation and aircraft manufacture had
to be separated; United Aircraft divested itself of Boeing and
United Airlines, North American divested what were to become
the Eastern and TWA airlines, and the Aviation Corporation
"spun off" American Airlines. The 1934 act also abandoned
the goals of previous airmail legislation (the McNary-Watres Act
of 1930), which had included a substantial element of subsidy
for the development of air transportation, in specifying that
minimum cost was to be the sole criterion for awards of mail
contracts.

The data in Table 3.1 show the dominance of the aircraft
market during the 1920s and early 1930s by government procurement (the figures are distorted slightly by the fact that
Curtiss-Wright and United were the only firms producing engines). In both the military and commercial sectors, moreover, a small number of firms were dominant. The share of
total sales of the two largest firms, Curtiss-Wright and United,
was in excess of 70 percent in the military, and over 90 percent in the commercial, markets.

1934-1940

During the 1930s, four airframe producers and two engine
manufacturers comprised the bulk of the civilian aircraft
industry. Boeing, Douglas, Lockheed, and Curtiss-Wright all
were active producers of commercial airframes, while Curtiss-
Wright and Pratt and Whitney were the major engine producers. This period also saw the production of the first
monocoque airframe passenger transports, the Boeing 247 and
the DC-2 and DC-3. The last-named aircraft dominated the
commercial aircraft market through the remainder of the decade, based primarily upon its efficient operating characteristics
for passenger transport. The data in Table 3.2, from Phillips
(1971), demonstrate the complete dominance by Douglas of the
commercial market. The 1930s also were the period during
which passenger, rather than mail, carriage became the central
activity of commercial air carriers.

The other major airframe manufacturers survived primarily
on military contracts during the 1930s; by the 1938-1939 period, of course, the military market was expanding rapidly.
Throughout this period, military production remained of great

Table 3.1. Aircraft and Engine Sales, 1927–1933

Companies	Government Sales	Percentage of Total Government Sales	Commercial Sales	Percentage of Total Commercial Sales	Total Sales	Percentage of Total Sales
United	$ 50,184,443	39.7	$28,056,208	48.0	$ 78,240,651	42.8
Curtiss-Wright	44,755,590	35.4	26,813,517	45.9	71,569,107	38.7
Douglas	14,437,623	11.4	1,412,790	2.4	15,850,413	8.6
Glenn Martin	9,895,605	7.8	none	–	9,895,605	5.4
Consolidated	4,307,632	3.4	1,118,231	1.9	5,425,863	2.9
Great Lakes	2,451,993	1.9	905,719	1.5	3,357,712	1.8
Grumman	452,195	0.4	153,492	0.3	605,687	0.3
Totals	$126,485,081	100	$58,459,957	100	$184,945,038	100

Source: Rae, 1968, p. 43.

Table 3.2. Estimated Deliveries of Newly Produced Aircraft
to Domestic Trunk Carriers, 1936–1941

Deliveries of Particular Types

Year	Total Deliveries of New Aircraft	DC-3	L-10	L-12	L-14	L-18	Beechcraft 18	B-307
1936	42	29	10	3	–	–	–	–
1937	54	47	–	–	6	–	1	–
1938	24	21	–	–	3	–	–	–
1939	41	40	1	–	–	–	–	–
1940	112	95	–	–	–	12	–	5
1941	36	35	–	–	–	1	–	–
Total	309	267	11	3	9	13	1	5

Source: Phillips, 1971, p. 94.

107

importance to the major commercial manufacturers. Despite Douglas Aircraft's dominance in civilian air transport markets, the greater unit value of military aircraft enabled other producers to avoid financial disaster. Holley (1964) noted that in 1937, 2,281 civilian aircraft were sold for a total cost of $19 million, while sales of military aircraft, totalling 949 units, were valued at $37 million.(3) The data in Table 3.3 display the shares of military contracts in total sales for the major airframe manufacturers for the 1931-1937 period.(4) The role of subcontractors also remained rather minor during this period, as commercial producers strove to utilize the substantial excess capacity more fully in their own factories.

Table 3.3.   Share of Military Contracts in Total Sales, 1931-1937.

Manufacturer	Percentage of Total Sales
Boeing	59
Chance Vought	75
Consolidated	79
Curtiss	76
Douglas	91
Martin	100
Grumman	75

Source: Holley, 1964, p. 22.

1940-1945

During the wartime period, there effectively was no commercial aircraft industry. All airframe and engine producers, as well as such nonaircraft firms as the Fisher Body division of General Motors and Ford Motor, worked feverishly to produce military designs. Several aspects of this period merit comment. The heavy demand for aircraft spurred the growth of firms such as Convair (formerly Consolidated Vultee), Bell Aircraft, and the Martin Corporation, raising formerly minor commercial producers to the status of viable competitors. Proprietary control of military aircraft designs was also reduced during this period, as cross-licensing of designs for maximum production was commonplace. Substantial "in-kind" technology transfer took place. In the rush to increase production, subcontracting came to play a crucial role in the aircraft industry. The large size of production runs also forced much greater attention to production engineering, including maximum exploitation of scale economies and learning curves. The in-house research and engineering capabilities of the major producers were expanded greatly as well. Finally, and of great importance for the postwar period, the development of the first American jet engine, based upon the British design developed by Whittle, was assigned by the Army Air Force to General Electric, on the basis of the firm's past experience in turbine design.

1946-1975

The postwar period was one in which the technology of the jet engine came to dominate commercial aircraft, causing substantial shifts in the relative importance of firms in both the airframe and aircraft engine sectors of the industry. An important consequence of the adoption of jet engine and electronics technologies in the modern commercial airliner was a spectacular rise in the magnitude of development costs in the production of a new commercial aircraft design. Miller and Sawers (1968, p. 267) note that:

> in the 1920's the cost of engineering development for an airplane was counted in tens of thousands of dollars - $25,000 for the Lockheed Vega and £5,000 for the prototype of the Hawker Hart; in the 1930's it ran into hundreds of thousands - about $150,000 for the DC-2 and $3,300,000 for the DC-3; as the 1940's began it reached the millions - $3,300,000 for the DC-4; by the end of the war it was in the tens of millions - $14,000,000 for the DC-6 and $29,000,000 for the two prototypes of the B-47; in the 1950's it ran into hundreds of millions - $112,000,000 for the DC-8 and $468,000,000 for the McDonnell Phantom; and in the 1960's it reached thousands of millions with the XB-70, which cost $1,500,000,000 for two prototypes.

Figure 3.1 provides a graphic illustration of this trend. McDonnell Douglas faced $625 million of deferred development costs on the DC-10, ten years after the aircraft's introduction in 1969. More recently, the development costs for the Boeing

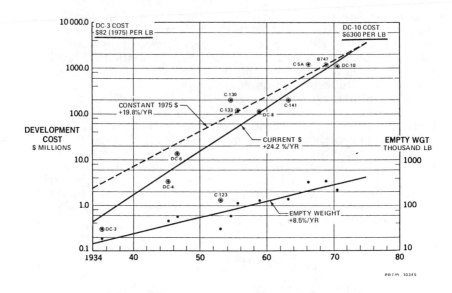

Fig. 3.1. Transport aircraft development cost trends.

Source: DOD–NASA–DOT study "RADCAP" August 1972 and McDonnell Douglas Corporation, Douglas Aircraft Company, Office of Planning; from Gellman and Price (1978), p. 12.

767 have been estimated to be in excess of $1 billion. The rapid growth of these costs in effect means that an increasing proportion of the costs of introducing a new aircraft are incurred during the phase of greatest uncertainty concerning market prospects and technical feasibility.

The jet engine was originally developed for American military applications by General Electric during and after World War II. General Electric, Westinghouse, and the Allison division of General Motors all had substantial development programs underway in 1945; they were joined by Pratt and Whitney shortly thereafter. By the 1960s, however, only General Electric and Pratt and Whitney remained as major factors in the commercial jet engine market (which by the early 1960s essentially defined the commercial engine market).

Table 3.4 gives the shares of the commercial market held by the major airframe producers during the postwar period up to 1965. Douglas, Lockheed, Convair, and Martin dominated the commercial market during the heyday of the four-engine

Table 3.4.   Shares of Commercial Aircraft Deliveries,
1947-1965.

Year	Douglas	Boeing	Lockheed	Convair	Martin
1947	74.3%	0.0%	17.1%	0.0%	8.6%
1948	21.7	0.0	5.2	60.0	13.0
1949	1.7	17.2	32.8	48.3	0.0
1950	9.8	0.0	60.8	11.8	17.6
1951	47.6	0.0	23.8	0.0	28.6
1952	14.9	0.0	16.2	16.2	52.7
1953	29.1	0.0	9.4	58.3	3.1
1954	66.7	0.0	9.3	24.0	0.0
1955	38.2	0.0	43.6	3.6	0.0
1956	40.0	0.0	8.0	15.2	0.0
1957	67.0	0.0	18.4	12.8	0.0
1958	59.8	0.0	14.6	0.0	0.0
1959	10.8	30.1	57.8	0.0	0.0
1960	39.2	35.1	11.3	14.4	0.0
1961	9.6	45.6	12.0	19.2	0.0
1962	12.1	51.5	0.0	31.8	0.0
1963	8.7	69.6	0.0	21.7	0.0
1964	6.7	91.6	0.0	1.7	0.0
1965	12.6	78.5	0.0	0.0	0.0
Total	30.8	21.7	17.4	17.2	6.9

Source: Phillips, 1971, pp. 110-111.

propellor transport.   After 1958, however, when Boeing introduced the 707 and Douglas followed with the DC-8, Lockheed,
Martin, and Convair all went into eclipse, out of which only
Lockheed would emerge in the early 1970s as a competitor in
the wide-body designs with the L-1011.   Boeing has come to
dominate the commercial market over the last 10 to 15 years as
thoroughly as Douglas dominated the commercial market of the
1930s.   Table 3.5 contains data on the relative importance of
Boeing, Lockheed, McDonnell Douglas, and General Dynamics
in the U.S. commercial jet aircraft market through 1973; by
1973, over 63 percent of commercial jet aircraft in service with
American airlines had been produced by Boeing.
     The commercial air transport industry was regulated by
the Civil Aeronautics Board during the entire postwar period;
this customer industry increased slightly in concentration,
while price competition in transport was largely absent.   Price
competition among the airframe producers remained intense,

Table 3.5.   Jet Aircraft in Service on United States Airlines. *

Year	Boeing	BAC	General Dynamics	McDonnell Douglas	SUD	Lockheed	Other	Total
1958	6							6
1959	66			18				84
1960	113		14	75				202
1961	170		39	93	17			319
1962	216		60	100	20			396
1963	237		65	104	20			426
1964	357		67	114	20			558
1965	476	17	65	134	20			712
1966	645	54	63	196	20			978
1967	661	54	63	205	20		3	1003
1968	883	57	59	321	20	1	3	1340
1969	1146	60	52	503	20			1781
1970	1331	60	47	610	20			2068
1971	1408	59	46	622			1	2136
1972	1395	62	49	619		1	6	2132
1973	1341	58	49	650		18	2	2118

*The above figures are as of December 31 each year except 1973, when the effective date is August 31. Others includes Dassault and Hamburger Flugzeugbau.

Source: "Aircraft in Operation by Certified Route Air Carriers," U.S. Department of Transportation, Federal Aviation Administration, FAA Statistical Handbook of Civil Aviation, various years; from Carroll, 1975, p. 147.

however, despite increased producer concentration. The failure of the Convair 880, and the subsequent problems of the
Douglas DC-9, were both due in part to aggressive efforts by
their producers to underprice the competition. More recently,
the introduction of wide-body transports was marked by fierce
competition between McDonnell Douglas and Lockheed in both
price and delivery date. The market for American commercial
aircraft became an international market during the postwar
period, aided by substantial government assistance in the
finance of purchases by foreign concerns. This represented a
great change from the situation of the 1930s, when barriers to
trade in aircraft were substantial. According to Carroll
(1975), as of early 1969, "In the total world [jet] aircraft
fleet, 2747 of the total 3494 (or 78.6%) are United States made"
(p. 153).

The intense competition among major airframe producers
during the postwar period has been responsible for several
near-failures of major firms. The Douglas Aircraft Corporation
approached bankruptcy as a result of poor financial management and overly energetic sales efforts for the DC-9 in 1966.
Despite an order backlog of $2.3 billion, Douglas was forced to
merge with McDonnell Aircraft in 1967, with the acquiescence
of the Department of Justice, and the aid of a federally
guaranteed loan of $75 million. The Douglas firm, producing
largely civilian aircraft, complemented the primarily military
product line of McDonnell, and McDonnell Douglas moved
quickly to begin work on the DC-10 wide-body transport.
Sales competition between the McDonnell Douglas DC-10 and the
Lockheed L-1011, as well as the bankruptcy of Rolls-Royce and
the C-5A debacle, left the Lockheed Aircraft Corporation
financially ravaged. Collapse of Lockheed was averted in 1971
only by a federal loan guarantee of $250 million. To an
unprecedented extent in the 1960s and 1970s, then, the federal
government was directly involved in determining the structure
of the commercial aircraft industry.

Market Structure and Conduct in the Jet Age

As was noted above, the coexistence of high levels of producer
concentration and fierce price competition make the commercial
aircraft industry an unusual one within manufacturing. This
aspect of the industry reflects several unique structural
features, which receive greater attention below. As Carroll
(1972, 1975) and others have noted, the relationship of aircraft producers and airline consumer through 1978, that is,
prior to deregulation of air transportation, closely approximated that of bilateral oligopoly. The market for commercial
aircraft was dominated by large orders from a small number
(approximately four) of major trunk carriers. Domination of
the market by this group was in part a result of airline regu-

lation by the Civil Aeronautics Board (see below). Within this environment of bilateral oligopoly, airlines tended to have the upper hand in purchase negotiations, playing competing suppliers off against one another, as in the case of Douglas, Boeing, and Convair in the early 1960s:

> once the decision has been made to purchase, it becomes desirable to place orders for sizable fleets. Further, the fairly concentrated nature of the air carrier industry insures that the orders made by individual airlines are large relative to the total market. From this, and the situation of the sellers, a large airline derives considerable market power from its purchasing decision [Carroll, 1975, p. 158].

The willingness of aircraft producers to undertake the expensive and risky tasks of development of new aircraft designs for which an insufficient market might exist reflects the importance of early delivery of new designs to airlines under the CAB regime. The advantages to airline customers of multiple suppliers of new aircraft designs also led them to encourage competition in aircraft production; thus, Juan Trippe of Pan American placed the first orders for commercial jet aircraft with both Boeing and Douglas. Nonetheless, the desire of producers to enter into such ruinous competition, as well as its recurrence, are not easily explained without consideration of the role of government military procurement (as well as the federal government's evident reluctance, in time of financial crisis, to allow a producer to go bankrupt.) As Table 3.6 shows, military sales have remained very important during the postwar period for all of the major commercial airframe producers, and have provided a steady source of profits with which to support commercial gambles. Carroll (1975, p. 162) points out that

> large government and space involvement, provides the safety net that catches a plummeting airframe company. Large backlogs of government contracts furnish rather steady income during periods when commercial activities make sales and earnings volatile. Government-sponsored research provides the bulk of airframe technology. Finally, the government simply will not allow a major defense contractor to fail completely, whatever its commercial sins.

The current structure of the commercial aircraft industry places considerable reliance upon contractual relationships in the design, production, and procurement of complex capital goods; the industry exhibits a low degree of vertical integration. The nature, causes, and consequences of this market

Table 3.6. Commercial Sales as a
Percentage of Total Sales.*

Year	General Dynamics	McDonnell-Douglas	Boeing	Lockheed
1957		31.5	2.1	
1958		21.2	4.0	12.5
1959		11.9	25.4	
1960		46.7	31.1	
1961	10.8	37.5	22.7	
1962	13.3	22.8	23.8	3.0
1963	19.4	22.4	14.7	
1964	20.5	23.5	35.6	
1965	21.9	33.2	49.6	
1966	22.0	46.4	52.3	5.0
1967		32.2	57.1	
1968		46.5	69.2	
1969		46.0	64.3	6.0
1970	1.3	29.6	78.4	4.0
1971	1.7	28.5	76.7	3.0
1972	4.3	40.9		

*A blank indicates data were not available.

Source: Company Annual Reports, various years, Moody's
Industrials and Moody's Handbook of Common Stocks,
various years; from Carroll, 1975, p. 148.

structure are of some interest. Production of new aircraft
requires extensive negotiations between the airframe and
engine producers. Performance specifications and guarantees
for engines are absolutely crucial to the success of a given
airframe design. However these may be highly unrealistic at
the time a contract is signed. Both the Boeing 747, using
Pratt and Whitney engines, and the Lockheed L-1011, relying
upon Rolls-Royce for engines, encountered severe difficulties
in meeting original performance specifications. For the 747,
the initial range and weight goals had to be abandoned, while
Lockheed nearly collapsed following the failure of the Rolls-
Royce firm, a failure due in large part to design problems with
the L-1011 engines. One result of the increasing complexity of
aircraft engine technology has been the acquisition by such
major airframe producers as Boeing or Northrop (an important

military producer and civilian subcontractor)' of a substantial
in-house expertise in engine design, engineering, and performance evaluation, duplicating that of the engine manufacturers.

The subcontracting of production of new aircraft designs
has also grown substantially in importance in recent years,
owing to mushrooming development costs and the increasing
complexity of aircraft components. Rae (1968, p. 83) states
that in the 1930s subcontracting "constituted less than 10
percent of the industry's operations." By the mid-1950s,
however, 30 to 40 percent of the assembly work for the turboprop Lockheed Electra was subcontracted. With the introduction of the Boeing 747, six major subcontractors accounted for
70 percent of the assembly of the aircraft, according to
Hochmuth (1974); a major subcontractor for the fuselage assembly was Northrop. Subcontractors for both the 747 and
the upcoming 767 are also required by Boeing to share a substantial portion of the development costs. Thus, subcontracting has increasingly come to fulfill a risk-sharing role in the
aircraft industry.(5)

An additional reason for the growth in subcontracting in
both the military and commercial aircraft sectors is the
increasing complexity of such aircraft components as avionics.
Major airframe producers simply do not have the requisite
in-house competence to develop and produce these complex
systems themselves. Occasionally, the decision is made to
proceed with in-house development of a given component, as a
means of acquiring such expertise; this is far more common
with military development contracts than in the commercial
sector.

The final nexus of contractual relations in the aircraft
industry is that between producers and airlines. As was
noted above, competition among producers is intense in
the areas of price, delivery date, and performance specifications. The importance of a large initial order for a new
aircraft design has increased considerably, reflecting the
concomitant growth of development costs. Producers must
have a guarantee that at least a substantial portion of these
development costs will be recouped prior to undertaking prototype development. The airlines placing these initial orders
thus are in a position of considerable power to dictate the
performance characteristics of a given aircraft; the negotiation
and specification of these performance criteria necessitate a
large in-house technical and engineering staff in each of the
major commercial carriers. Further, since there is considerable variation in the route structure of each carrier, the
performance characteristics viewed by each as most desirable
often vary substantially. The airline placing a large initial
order thus may be in a position to influence the characteristics
of a new generation of aircraft. As a result of this "user-

active" pattern of new product development, the contrasting financial health and route structures of airline firms exert a major influence on the direction of technical change in the commercial aircraft industry. As is the case elsewhere in this contractual system, the ability of producers to meet performance specifications is rarely certain at the time such commitments are made.

Given the complexity of the technologies involved, as well as the severe uncertainties that are inherent in the production and procurement of such a technologically sophisticated capital good, there exist considerable transactions costs within this industry structure. Extensive parallel engineering staffs are maintained by airframe manufacturers, engine producers, and airline purchasers. Certain segments of the commercial aircraft market, notably short-haul aircraft, do not appear to have been well-served by the innovation process. Considerable resources are invested in negotiation and (not infrequently) litigation. Finally, the incentives for misrepresentation of performance characteristics - Arrow's "moral hazard" - and competition in price and delivery dates may have deleterious effects upon product safety. The crash of the DC-10 near Paris in 1974 involved an improperly designed cargo door, whose dangers were not disclosed rapidly to airlines and which was modified by McDonnell Douglas only after considerable delay (and not at all for certain aircraft, such as the one that crashed in 1974). The faulty door was particularly dangerous because of the design of the hydraulic system of the DC-10; unlike the L-1011 or B-747 which had four, only three hydraulic systems were built into the plane by McDonnell Douglas, and were located close together, making the aircraft susceptible to a severe loss of control in the event of an accident. One account (Eddy, Potter, and Page, 1976) notes that the incentives faced by McDonnell Douglas to produce a wide-body transport prior to Lockheed were partly responsible for these design defects. The 1979 Chicago crash of a DC-10 demonstrated the vulnerability of the aircraft's hydraulic system, and revealed additional difficulties of moral hazard and communications concerning engine maintenance.(6) The market interface in many of the transactions involving commercial aircraft production and procurement occasionally may result in severe impediments to the free flow of information and/or full revelation of details of design and performance. Offsetting these potential costs of market-mediated fabrication and procurement processes, of course, are the substantial benefits of competition among airframe and engine producers. It is extremely unlikely that a greater degree of vertical integration in the commercial air transportation sector would have produced as rapid a pace of innovation, service quality improvement, and productivity growth. The pace of innovation, however, probably has been affected more heavily by military research and procurement policies, as well as regulation by the Civil Aeronautics Board of commercial air transportation, than by industry structure.

III. THE RECORD OF TECHNICAL PROGRESS

The innovative performance of the commercial aircraft industry may be captured in part by two measures: available seats multiplied by cruising speed (AS x V_c), and costs per available seat mile.(7) While these two measures do not translate straightforwardly into an index of total factor productivity such as that provided by Kendrick (1961, 1973) on a highly aggregated basis for air transportation, they have the advantage of being available separately for various aircraft designs. Over time, with the introduction of successive "generations" of aircraft, as AS x V_c has risen, and costs per available seat mile have fallen. Figure 3.2 and 3.3 display the evolution of these two aircraft performance measures during the 1920-1975 period. Examining direct operating costs per seat mile, the quantum drop represented by the DC-3 stands out quite clearly; as Phillips noted, the seat mile costs of the DC-3 aircraft were "so much lower than those of alternate aircraft that even with a relatively low load factor its passenger mile costs were often lower than those for other planes" (1971, p. 94). Another major drop in seat-mile costs came with the introduction of the wide-bodied transports, incorporating large, high-bypass-ratio jet engines. The evolution of cruising speed and capacity (AS x V_c) shows the large jump that came with the introduction of four-engine transports immediately after World War II, as well as the improvement that was registered with the first jet engine transports. It is interesting to note that, alone of the successive generations of aircraft, the wide-body transports incorporate major increases in available seat velocity, and significant declines in direct operating costs per seat mile. As Rosenberg et al. (1978) noted, since the appearance of the monocoque airframe design in 1933, costs per seat mile have declined tenfold, while passenger capacity and speed have risen by a factor of 20.

An additional important feature of technical progress in aircraft is overlooked in these tables, which present operating costs as of the year of introduction of a given design. An important element of technical change and performance improvement in this industry operates during the life of a given airframe design, in the "beta phase" of the innovation process (see Enos, 1962, for further discussion). For the Boeing 247, the first monocoque passenger airframe design, seat-mile operating costs declined from 7¢ in 1933 to 5¢ in 1940 (Figure

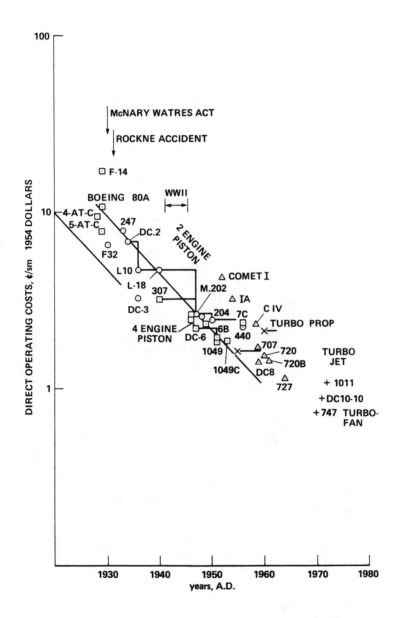

Fig. 3.2. Direct operating costs of multiengine American transports, first year of operation (1954 dollars).

Source: Rosenberg, Thompson, and Belsley, 1978, p. 65.

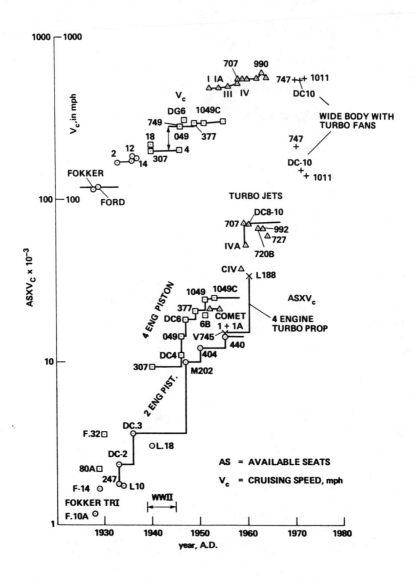

Fig. 3.3. Passenger carrying productivity as expressed by AS
          x V_c multiengine American transports.

Source: Rosenberg, Thompson, and Belsley, 1978, p. 66.

3.4). The Lockheed Electra L-188, a four-engine turboprop, exhibited an annual rate of cost decline of roughly 7 percent, while operating costs for the Boeing 707 declined at an annual rate of 8.7 percent (see Figure 3.5).   These declines in operating costs stem from modifications in aircraft design and improvements in the operations and maintenance of these complex capital goods, both of which incorporate important elements of learning in use (see Rosenberg, 1982, for further discussion).

There exists a considerable body of literature describing the improvements in productivity that have been associated with learning to manufacture a newly conceived product.(8) Indeed, in some circles the phenomenon is referred to as the "Horndal Effect," after the Swedish steelworks where, over a period of 15 years, output per man hour was observed to

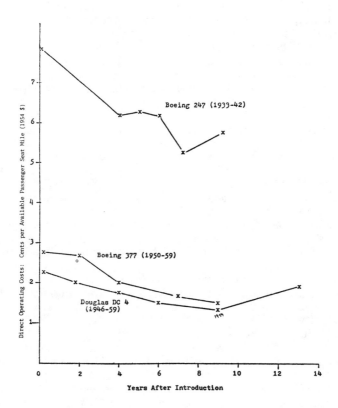

Fig. 3.4.   Beta phase direct operating cost reduction for particular piston aircraft.

Source: Phillips, 1971, pp. 38-39.

increase by about 2 percent per year even though no changes had occurred in either the plant or production techniques. The phenomenon has been further documented, not only in airframe production, but in machine tools, shipbuilding, and textiles as well.(9)

We wish to emphasize here, however, a different but related form of learning by doing. Not only does learning by doing take place in the manufacturing process as workers improve their skill in the making of the product, but, as a result of the actual use of the aircraft itself, a considerable learning process occurs that reduces the operating costs of the aircraft in use after its manufacture. Much of the learning in aircraft has been associated with the gradually growing body of experience associated with the operation of a new-model airplane.(10) The experience is, perhaps, most characteristic of complex final products with elaborately differentiated but interdependent component parts. Operating cost reductions, as we will see, depend heavily upon gradually learning more, during the actual operation of a new aircraft, about the performance characteristics of an airplane system and its components, and therefore understanding more clearly its eventual full potential. For example, it is only through extensive use that detailed knowledge is developed about engines' operation, their maintenance needs, their minimum servicing and overhaul requirements, and the like. This is due partly to an inevitable - and highly desirable - over-cautiousness on the part of the manufacturer in dealing with an untried product. As experience accumulates, it becomes possible to extend the operating life beyond original expectations.

An additional source of performance improvement following introduction of a new aircraft is the incremental improvements in various components. We may consider the history of the Douglas DC-8 as representative of this process of development; several of the events in this history are summarized in Figure 3.5.

In the DC-8 we have an aircraft that has experienced a more than 50 percent reduction in operating energy costs over its life span on a per-seat-mile basis, as well as an increase in productivity (AS x V_c) from 62,500 for the DC8-10, DC8-30, and DC8-50, to 130,000 for the DC8-61-63 series, although the basic configuration has been largely unchanged and the modifications have been relatively unsophisticated compared to differences between aircraft types. Clearly, an important set of modifications has had to do with the engines, which have progressed both to increase available thrust and decrease specific fuel consumption, thus increasing the potential payload and directly reducing operating costs. At the same time, there have been modifications to the wing profile that reduced the drag of the aircraft. With the DC-8-30, a dropped flap

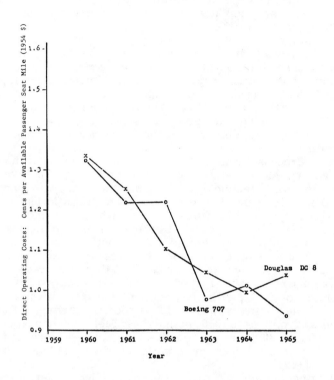

Fig. 3.5.   Beta phase direct operating cost reduction for two
             specific turbojet aircraft.

Sources: Phillips, 1971, pp. 40-41.

was added, then a leading edge extension with the DC-8-50.
Subsequent models increased the aspect ratio and repositioned
the flap.(11) Engine pylon design also underwent some modification.   These variations in the aircraft's geometry were
motivated by the drag reduction and increased fuel economy
that were achieved.   But it is clear from the figure that a
third very substantial contribution to increasing the aircraft
productivity has been the ability to stretch the aircraft,
increasing capacity from 123 seats to 251 seats.   This demonstrates the large payoff to be reaped by increasing the
internal passenger capacity, provided of course that the

aircraft can be operated sufficiently close to capacity. The interdependence of these technological improvements is perhaps obvious but requires explicit exposition. The possibilities for stretching and consequently adding payload volume and weight to the vehicle are dependent upon the incorporation of more powerful engines into the wing design.

The story of the DC-8 is quite representative of the commercial aircraft industry design philosophy. Innovations that have been incorporated within a particular vehicle and have made substantial improvements in their operating cost characteristics generally focus upon improvements in engine thrust and fuel consumption, reduction in overall drag by modification in wing design, or stretching the vehicle to increase payload capability. Although dramatic improvements in operating costs may appear to come directly from the stretching process, this process is unsustainable without the complementary development of power plant technology and, occasionally, wing technology. Engine technology in particular during the turbine era has experienced a dramatic improvement in thrust per pound of engine weight, which has increased by over 50 percent in 20 years, but even more so in fuel consumption per hour per pound of thrust. For example, in 1950, about 0.9 pounds of fuel were required for each hour-pound of thrust. By the early 1960s, this requirement, with the development of the turbofan engines, dropped to around 0.75 pounds of fuel per hour-pound of thrust. With the innovation of the high-bypass-ratio turbofan engines, fuel requirements dropped to 0.6 pounds of fuel per hour-pound of thrust. This 30 percent decline in fuel requirements has direct implications for increasing the deliverable payload of an aircraft.(12)

The phenomenon of stretching as applied to jet transports from the Comet to the 747 is a classic example of a process that is not very "interesting" technologically but is of vital economic importance.(13) As was noted above, the process reflects the basic complementarity between the performance of the engine and the airframe. Indeed, there is little incentive to improve engine design unless airframe designers know how to exploit the improvement.(14) The carrying capacity of the airplane depends, first of all, on the capacity of the engines. As engine performance is improved, exploitation of the potential requires redesign or modification of the airframe. The simplest response, as improved engines become available, is merely to stretch the fuselage and add more seats. Indeed, as this phenomenon came to be better understood, most airplanes were deliberately designed in order to facilitate subsequent stretching. Although airplane designers generally design to conform to the capabilities of existing engines, it is understood that engine performance improvements will be available within the lifetime of the model, and it is important to be in a position to exploit them. Since designers expect these future engine improvements (as well as other complementary technological improvements), they consciously attempt to design flexibility into the airplane. This applies especially to the practice of designing the fuselage in such a way as to facilitate later stretching. Such stretching has constituted an important part of the productivity improvement that has been characteristic of the beta phase. Stretching may be viewed as a process by which, as a result of accumulated knowledge and improved engine capabilities, the payload possibilities of a new airplane design are expanded to their limits. Clearly, this is an economic as well as a technological phenomenon. Modification of an original design through the stretching process, is usually motivated by the growth of passenger demand or new route opportunities.

The sources of the impressive record of inter- and intragenerational technical progress documented above are numerous. We argue below that aircraft have benefited from technical developments outside of the industry itself to an extent greater than in virtually any other major manufacturing industry. Advances in metallurgy underpinned the development of the monocoque airframe in the 1930s, while improvements in fuels, the results of research sponsored by automotive and petroleum firms, aided in the propulsion of these new designs. In the postwar period, metallic and nonmetallic composite substances have played a central role in improvements in both the airframe and the engine; again, these new materials have been developed largely outside of the commercial aircraft industry. Advances in electronics also have been of great importance. An additional important aspect of technical change in commercial aircraft is the role of government in procurement of military aircraft and in the support of research for both military and civilian applications.

IV. THE SOURCES OF TECHNICAL CHANGE

The impressive record of innovation exhibited by the commercial aircraft industry reflects the industry's good fortune as a beneficiary of at least three important external sources of innovation and/or research support; innovations in other industries, such as metallurgy or electronics, government-supported research in civil aviation, and military procurement and research support. The number and complexity of the systems that are combined in a modern aircraft design are partly responsible for the fact that, to an unusual extent, the aircraft industry has benefited from innovations and research support from sources outside the industry. The characteristics and consequences of government policies toward the aircraft industry, including procurement and research support, are discussed in greater detail in the next section. Here we simply document and discuss the extent of the inflow of technical change from external sources. The contributions of other industries to aircraft industry innovation are discussed first, followed by a consideration of the federal role, and a discussion of the sources and categories of aeronautical research and development expenditures.

Interindustry transfers of technology are widespread in advanced industrial societies that are characterized by highly sophisticated patterns of specialization and interindustry flows of components. Any purchaser of goods from a given supplier is a potential beneficiary of innovation in the supplier firm's industry. This pattern of transmission is especially common in the relationship between manufacturing firms and the firms that supply them with capital goods. As was noted above, the large number of component systems used in an aircraft has placed the industry in a position to benefit from developments in other industries. Innovations in these "supplier" industries occasionally have been motivated by an awareness of their potential applications in aircraft; in some cases, federal funds supported research in these supplier industries, based upon the potential usefulness of the innovations from these sectors for military aircraft. The important point, however, is that the aircraft industry has benefited from innovations produced by research carried out by other innovative industries.

Electronics

Over the last 25 years, the commercial aircraft industry has greatly increased its reliance upon electronics technology, particularly solid-state semiconductor circuitry. The increasing use of semiconductors was spurred by the requirements of strategic missile guidance systems in the 1950s. Compared to vacuum tubes, solid-state circuits were far lighter, more reliable, and generated less heat. The increased importance of military space projects, many of which were carried out by aircraft firms, blurred the boundaries between the electronics and aircraft industries. The resulting development of semiconductor guidance systems produced substantial benefits for commercial aircraft. The origins of this far-reaching innovation, however, were remote from the commercial aircraft industry, stemming from Bell Telephone Laboratories' efforts to improve long-distance telephony.

Exploitation of electronics technologies for commercial aircraft was rapid during the 1960s and 1970s. Air traffic control equipment had to be improved substantially, to meet increasing traffic flows of larger commercial aircraft. Communications, navigation, instrumentation test equipment, radar, and other systems were developed by the electronics industry for application in commercial aircraft. The increasing use of integrated circuits has facilitated miniaturization of a wide range of instruments.

Applications of new electronics technologies in other industries also have benefited commercial aircraft. The development of computers, greatly advanced by semiconductors, has yielded major spillovers into the commercial aircraft sector. Air traffic control and reservations computers have supported the expansion of commercial air transport. Onboard minicomputers have improved the navigation and maneuvering performance of commercial aircraft. In the development and production processes, computers also play an increasingly important role. Computer-assisted design techniques have reduced, if not eliminated, the great uncertainties about airframe performance, enabling more extensive testing to be carried on outside of the wind tunnel. Computers also are being applied to the control of machine tools in the fabrication process, substantially improving productivity.

Metallurgy and Materials Science

At least since the introduction of monocoque airframe in the early 1930s, progress in commercial aircraft design and innovation in metallurgy have been tightly linked. With the advent of the jet engine, metallurgy assumed substantial importance for developments in the power plant, as well as the airframe. Since the 1940s, metallurgical research on the behavior of metals at high temperatures has been of great importance to the development of turbine blades, inlets, outlets, and compressors for turboprop and jet engines.(15) As the disastrous attempt of Rolls-Royce to use a new composite material, Hyfil, in the development of engines for the L-1011 demonstrates, the uncertainties surrounding aspects of metallurgical and materials development have had a great impact upon commercial aircraft. Metallurgy remains a discipline in which a strong theoretical basis for predictions about performance is lacking: experimentation and uncertainty remain central. In addition, materials fatigue remains poorly understood, and very difficult to test for effectively.(16)

Important sources of metallurgical research for commercial aircraft include firms such as Alcoa, which developed duralumin in the 1920s under military contract for use in Navy dirigibles, Duralumin subsequently was employed extensively in monocoque airframes. General Electric, a major producer of steam turbines and other generation equipment requiring advanced alloys for high-speed operation, also became involved in metallurgical researches involving the development of supercharged aircraft engines, and later, jet engines. As was the case with Alcoa, military support of General Electric's supercharger materials research was of considerable importance.(17)   Additional indirect federal support for materials research was channelled through the Subcommittee on Heat-Resisting Alloys of the National Advisory Committee on Aeronautics (NACA), formed in 1941.

Government Support of Commercial Aircraft Research: the Role of NACA

The commercial aircraft industry is unique among manufacturing industries in that a governmental research organization, the National Advisory Committee on Aeronautics (NACA, subsequently the National Aeronautics and Space Administration, NASA), has long existed to serve the needs of aircraft design. Similar research facilities, support by both government and industry to carry out research on "generic" technological innovation, have been advocated recently for other industries. The argument most frequently made in support of such "cooperative" research establishments states that individual firms within a given industry face insufficient incentives and real disincentives (the free-rider problem) to carrying out the basic research necessary to support innovation.   NACA is widely viewed by industry and government observers as a success in this regard; the industry to which it was directed has exhibited impressive innovative performance.   With these concerns as background, the role of NACA in the development of commercial aircraft is worth investigating in some detail.

World War I sparked the establishment of a number of bodies intended to bring together leading academic, business, and government figures in an effort to analyze important problems of national security, frequently in the areas of industrial mobilization, research, and technology.   The National Research Council was one such body; the National Advisory Committee on Aeronautics was another, more firmly under government control than the NRC.   Established in 1915, NACA was intended to "investigate the scientific problems involved in flight and to give advice to the military air services and other aviation services of the government."(18)   Despite this early military-oriented mandate, NACA during the 1920-1935 period did not deal solely with military aircraft problems, working instead on problems of aerodynamics and aeronautics common to both military and commercial aircraft.

Using experimental facilities at Langley Field, Virginia, and Moffett Field, California, NACA functioned as an important source of performance and other test data in aerodynamics. The committee pioneered in the construction and use of large wind tunnels, completing one in 1927 large enough to accomodate full-scale airframes.   This and other facilities provided a

steady stream of test results that led to significant improvements in airframe design. The famous "NACA cowl," intended
to reduce the wind resistance of radial air-cooled engines, cut
engine drag by nearly 75 percent. NACA research also demonstrated the superior performance of airframes with retractable
landing gear, and yielded improved knowledge regarding engine positioning in the aircraft wing:

> By a comprehensive survey of the net efficiencies of
> various engine nacelle locations, the optimum position
> in the wing was found. This N.A.C.A. engine
> location principle, together with other refinements,
> had a revolutionary effect on military and commercial
> aviation the world over. It changed military aviation
> tactics, made long-range bombers possible, and
> forced the development of higher speed pursuit
> planes. In the commercial field it permitted the
> speeding up of cruising schedules on the air lines
> from 120 miles per hour of the Fords to the 180 miles
> per hour of the new Douglas planes. The overnight
> transcontinental run became possible and the air
> lines vastly increased their appeal to the public.
> Even in the midst of the depression, air line traffic
> boomed [Hunsaker, 1941, p. 139].

After 1935, NACA research was increasingly designed to
serve military needs, and such development projects largely
crowded out the activities of greater benefit to commercial
aircraft. Phillips (1971, p. 121) noted that after 1935, NACA:

> tended to shift. . . . from research that lacked a
> specific military or commercial purpose to that
> relating to specific military missions and even to
> specific military aircraft. This changed the nature
> of the aircraft industry's reliance on exogenous
> science and technology. Prior to this time, develop-
> ments in both military and commercial aircraft
> occurred from technical developments achieved with
> neither a specific military nor commercial purpose.
> After this, technical developments more often had a
> defined military purpose and new types of commercial
> planes more often had visible antecedents in military
> aircraft.

The prewar performance of NACA was achieved at a remarkably low cost, even by the standards of the time. Total
appropriations for NACA between 1915 and 1940 approximated
$25 million. It is crucial to note, however, that NACA carried
out very little research during this period that could be described as "basic" in nature. Prior to 1940, the committee

functioned primarily to provide research infrastructure for the aircraft industry, making available as it did extensive experimental design data and testing facilities. This was a very important contribution, given the modest research resources of the industry prior to 1940, but it does not resemble the type of support frequently envisioned by advocates of government-industry research cooperatives (this point is discussed further below). Indeed, one recent account of the development of the jet engine has characterized the United States prior to 1940 as a backwater of theoretical aerodynamic research, attributing the failure of American engineers to originate the concept of the jet to American ignorance of the theoretical knowledge underpinning aeronautical design (Constant, 1980).

After World War II, during which NACA work was exclusively military in character, the division of labor in aeronautical research appears to have changed somewhat. The major aircraft producers had acquired substantial in-house facilities of their own(19); NACA's infrastructure was less critical. Military support of research and development occupied a vastly more important role than had been true in the pre-1940 period. NACA declined in importance, functioning as a sponsor of more fundamental aeronautical research in the academic sphere, and continuing to conduct empirical research on a scale that was now dwarfed by military-supported activities. The committee had fulfilled an important function, however, having supported a level of research that would have taxed the resources of the commercial aircraft industry prior to 1940.

Military-sponsored Research

The final area of major external support for commercial aircraft innovation is military-supported research and military procurement. Research supported by the armed services has yielded indirect, but very important, innovational spillovers to the commercial aircraft industry, most notably in aircraft engines. From the Pratt and Whitney Wasp of 1925 to the high-bypass-ratio engines of the 1970s, commercial aircraft engine development has benefited from, and usually followed, the demands of military procurement and military support of research. In the immediate aftermath of World War I, during the 1919-1926 period, "Virtually every cent going into the development of engines" was derived from "direct payment by the government from special funds allocated to research and development (Schlaifer, 1950, p. 160). More recently, of course, the development of the first jet engine in the United States was financed entirely by the military, reflecting both the perceived military urgency of the project, and the lack of interest in development of such an engine expressed by commercial firms prior to 1940:

> In the United States neither Lockheed, where the
> first American designs of a turbojet were made, nor
> the Northrop airplane company, which proposed in
> 1940 to develop a turbo-prop, was willing to do any
> actual development at its own expense, only the
> preliminary studies being financed in this way. A
> year or two before this, some engineers in the
> Turbosupercharger group of the General Electric
> Company had proposed the development of a turbojet
> to the management of the company, but the proposal
> had been rejected [Schlaifer, 1950, p. 88].

More recently, military-supported research on power plants for
the giant C-5A transport led to the development of the high-
bypass-ratio engines that now power the wide-body commercial
transports.(20)   Fifty-five percent of the R&D costs for these
turbofan engines was contributed by the Defense Department,
while the FAA and NASA accounted for roughly 13 percent;
industry expenditures were 32 percent of the total.(21)

Direct military research support has been most important
in the propulsion area. However, the development of commercial aircraft has also benefited substantially from military
support of airframe development and production for purely
military purposes. Such spillovers became important only after
World War II, in contrast to the situation for aircraft engines.
With the advent of jet aircraft, however, airframe makers often
were able to apply knowledge gained in military projects to
commercial aircraft design, tooling, or production. In some
cases, similarities in airframe design were sufficiently
pronounced that development and tooling costs for commercial
airframes were reduced substantially. An example of this is
the Boeing 707. Boeing had developed a jet tanker to provide
in-flight refueling for the strategic bombers, the B-47 and
B-52, that the firm previously had sold to the Air Force.
Over 1,000 of the tankers, the KC-135, eventually were sold
to the Air Force. The 707 airframe design followed that of the
KC-135 quite closely, so closely, in fact, that the first
prototype 707 to be "rolled out" of the Seattle factory did not
have windows in the fuselage. A comparison of the costs
incurred by Douglas in the development of the DC-8 with those
of Boeing gives a rough idea of the financial benefits that
accrued to Boeing:

> Douglas lost $109 million in the two years 1959 and
> 1960, having written off $298 million for development
> costs and production losses up to the end of 1960.
> Boeing did not suffer so badly. They wrote off $165
> million on the 707 by then; some of the development
> cost may have been carried by the tanker program,
> which also provided a few of the tools on which the
> airliner was built [Miller and Sawers, 1968, pp.
> 193-194].

The closer is the similarity between military and commercial designs, the greater will be such external benefits reaped by the contractor. Dynamic spillover effects also are of importance; development or procurement contracts may serve to support the acquisition by a producer of new design or production skills. As was mentioned above, military contractors have occasionally chosen to produce a specific component in house, rather than subcontracting its manufacture, in order to acquire expertise in the area (this was especially true of the airframe producers and electronics components in the 1950s and 1960s) at government expense.(22) In certain cases, the costs of tooling for production of a commercial airframe may be partially borne by a government procurement contract, as in the case of the 707 and the KC-135. In addition, some of the "learning by doing" that takes place in production of a military airframe may be transferrable to commercial production.

In order to convey some sense of the importance of external sources of innovation in the commercial aircraft industry, one need only consider the epochal DC-3. As was noted above, the DC-3 represented a productivity improvement not equalled until the introduction of wide-body transports 35 years later. The aircraft's low operating costs were due in large part to its radial air-cooled engines, rated at nearly 1,000 horsepower. Miller and Sawers (1968, p. 94) noted that "The most striking feature of the progress of the decade of the 1930s was that more power was obtained from engines of the same size." In the case of the DC-3, the low weight-to-power ratio of its engines enabled transport of a larger number of passengers for a given horsepower rating than had been possible previously. The improvements in engine design referred to by Miller and Sawers were the result of government-sponsored research, and were dependent as well upon improvements in fuel, notably the addition of tetraethyl lead to aviation gasoline, that resulted from research sponsored by DuPont, General Motors, NACA, and the National Bureau of Standards.

The DC-3 airframe design incorporated numerous results of NACA research, including the cowling on the engines and the placement of the engines in the leading edge of the wing, as well as the retractable landing gear. The wing design itself incorporated several important NACA developments, as Phillips (1971, p. 117) points out:

> the wings of the DC-3, as well as those of the other
> planes of its generation, owe their origin to NACA
> and other non-commercial or non-United States re-
> search. In particular, the DC-1 had a NACA 2215
> wing section at the root - with fillets into the
> fuselage which were the results of NACA research -
> and a NACA 2209 section at the tip.

The monocoque airframe of the DC-3 was made possible only by
the development by Alcoa of the duralumin alloy. Thus, while
the design and development work that combined these components successfully into the design of the DC-3 was brilliant,
the original research underlying the perfection of many of the
crucial components had been performed or funded by institutions outside of the aircraft industry.

Industry R&D expenditures, 1945-1969

An examination of the sources and expenditure categories of
research within the overall aircraft industry (including both
military and commercial aircraft producers) illustrates more
precisely the character of research support and activities
within the industry. A useful summary of R&D data for the
postwar industry is contained in the study conducted by Booz,
Allen, and Hamilton for the Department of Transportation-
NASA study of R&D policy for civilian aviation. Table 3.7
contains comprehensive data on the sources of research funds
for fiscal years 1945-1969. Total R&D expenditures rose by
nearly 700 percent during this period, from $365 million in
1945 to roughly $2.8 billion in fiscal 1969. The most rapid
period of growth was in the 1950-1954 period, reflecting the
substantial infusion of military funds during the Korean War;
from nearly $600 million in fiscal 1950, R&D expenditures rose
to more than $2 billion in 1954. Seventy-eight percent of this
increase was accounted for by increases in military-supported
R&D. Throughout this period, even in the late 1960s, the
defense portion of total R&D expenditures never fell below 65
percent.
     Among the major sources of nonmilitary research funding,
the declining role of NACA support through the 1950s stands
out clearly; from parity with industry expenditures (which
appear as the "industry nonreimbursed" category) in the late
1940s, the NACA portion of nonmilitary research support had
dropped to less than 10 percent in fiscal 1958, immediately
prior to Sputnik and the reorganization of NACA into NASA.
Expenditures by the Atomic Energy Commission supported
research on nuclear propulsion of aircraft and space vehicles,
while the Federal Aviation Administration supported work on
avionics and (during the 1960s) engine development.
     The industry contribution to R&D remains strikingly small
in the late 1960s, despite a rapid rate of growth. "Nonreimbursed expenditures" never accounted for more than 25 percent of total R&D spending, and were below 20 percent of the
total for most of the 1945-1969 period. Industry expenditures,
however, accounted for an increasing share of nonmilitary
research expenditures during this period, reflecting the
growth of large in-house research establishments and soaring

Table 3.7.   Sources of Aeronautical R&D Funds
Annual Expenditures in Millions of Dollars.

| | Federal | | | | | | | | Private | |
| | Defense | | | | | Nondefense | | | | |
Fiscal Year	Air Force	Navy	Army	ARPA	Industry Reimbursed	NASA	FAA	SST (FAA)	AEC	Industry Non Reimbursed	Total
1945	170	124			17	30	1			23	365
1946	188	209			21	37	1			28	484
1947	182	139			28	30	1			37	417
1948	141	186			35	42	2			48	454
1949	198	160	2		54	53	2			70	539
1950	245	112	4		80	52	8			91	592
1951	308	179	14		176	62	4		7	164	914
1952	558	217	21		295	113	3		11	277	1495
1953	878	241	29		366	76	3		21	339	1953
1954	996	265	36		365	55	1		24	343	2085
1955	941	249	40		343	47	1		27	320	1968
1956	958	243	50		358	51	1		49	353	2063
1957	1037	248	57		381	50	1		79	392	2245
1958	1135	266	74		360	45	15		73	356	2325
1959	1082	268	68		319	48	28		76	339	2228
1960	896	274	49		290	32	48		69	329	1987
1961	979	247	72		293	39	45		69	306	2050
1962	1011	214	88		301	44	53	11		304	2026
1963	1333	261	104		293	66	59	19		234	2419
1964	1290	250	101		298	84	38	18		304	2383
1965	1231	244	76	9	304	102	30	21		353	2370
1966	1268	257	98	23	367	110	31	112		445	2711
1967	1058	303	104	14	452	134	35	190		565	2855
1968	1138	243	131	11	481	171	35	63		673	2946
1969	779	461	134	2	457	216	36	94		609	2806

Source: DOT-NASA, Booz, Allen and Hamilton, Table C-13.

development costs for commercial aircraft. From 42 percent of nondefense R&D spending in fiscal 1946, the industry share rose to nearly 64 percent by fiscal 1969. Military-civilian research spillovers, rather than direct federal support of nondefense research projects, were the primary means whereby federal research support affected commercial aircraft research.

Table 3.8 contains information from the DOT-NASA study on the types of research conducted by producers, breaking research activities into "basic research," "applied research," and "development" categories.(23) Perhaps the most striking finding is the small portion of total industry research (both privately and publicly funded) that goes to basic research; the basic research share of total R&D expenditures is below 10 percent throughout the 1945-1969 period, and the industry nonreimbursed share of this small fraction is below 10 percent. Public sources, primarily the Air Force, Navy, and NASA (in the 1960s) supported most of what basic research was carried on in the aircraft industry. Applied research expenditures account for a much greater share of the total; the nonreimbursed industry share of this in 1969 was 34 percent substantially above the industry share of basic research. Once again, direct military support and industry-reimbursed spending account for the majority of this class of expenditures. Development expenditures account for the largest share by far of total R&D expenditures throughout the period, never falling below 60 percent of the total. The military share of this category is once again the largest, with the Air Force share of development costs above 50 percent through the 1953-1966 period. Development expenditures comprise over 70 percent of total Air Force research support during the entire 1945-1969 period. The share of development costs accounted for by industry nonreimbursed expenditures during this period never exceeds 15 percent.

The relative shares of three major aircraft components in total research spending, avionics, airframes, and engines are given in Table 3.9. While airframes comprise the largest share of total aeronautical R&D, 40 to 45 percent, the avionics share exceeds that for engines throughout the postwar period. The categories of private industry research expenditures are given in Table 3.10; unfortunately the data do not distinguish between reimbursed and nonreimbursed R&D spending by category. Nonetheless, the relative magnitudes of the various categories are of considerable interest. These relative shares have remained remarkably stable through the postwar period, with prototype development in first place, followed in descending order by avionics, propulsion, and aerodynamics. These data reveal that the majority of R&D is expended upon the integration of these complex components, rather than their separate development, underlining the point made earlier about the high degree of systemic complexity embodied in an aircraft design.

Table 3.8. Distribution of Source of Funds by Type of Aeronautical R&D Annual Expenditures in Millions of Dollars.

| | BASIC | | | | | | | | APPLIED | | | | Federal | |
Fiscal Year	Air Force	Navy	Army	NASA	SST	Atomic Energy Commission	Industry*	Total	Air Force	Navy	Army	Aviation NASA	Aviation Administration	SST
1945	9	6	–	5	–	–	2	22	24	17	–	6	–	–
1946	9	10	–	6	–	–	2	27	26	29	–	7	–	–
1947	9	7	–	5	–	–	3	24	25	19	–	6	–	–
1948	7	9	–	7	–	–	4	27	20	26	–	8	–	–
1949	10	8	–	9	–	–	6	33	28	22	–	10	–	–
1950	12	6	–	9	–	–	9	36	34	16	1	10	–	–
1951	15	9	1	11	–	2	17	55	43	25	2	12	–	–
1952	28	11	1	19	–	2	29	90	78	30	3	21	–	–
1953	44	12	1	13	–	5	35	110	123	34	4	14	–	–
1954	50	13	2	9	–	5	35	114	139	37	5	10	–	–
1955	47	12	2	8	–	6	33	108	132	35	6	9	–	–
1956	48	12	3	9	–	11	36	119	134	34	7	10	–	–
1957	52	12	3	9	–	17	39	132	145	35	8	10	–	–
1958	57	13	4	8	–	16	36	134	159	37	10	9	–	–
1959	54	13	3	8	–	17	32	127	151	37	10	9	–	–
1960	45	14	2	5	–	15	31	112	125	38	7	6	–	–
1961	49	12	4	7	–	15	30	117	137	35	10	7	–	–
1962	51	11	4	6	3	–	30	105	141	30	12	6	1	8
1963	67	13	5	11	5	–	29	130	187	36	15	13	1	14
1964	64	13	5	14	5	–	30	131	180	35	14	16	–	13
1965	62	12	4	17	5	–	33	133	172	34	11	19	–	16
1966	63	13	5	19	–	–	41	141	177	36	14	21	–	2
1967	53	15	5	23	–	–	51	147	148	42	14	25	–	2
1968	57	12	7	29	–	–	58	163	159	34	18	32	–	1
1969	40	23	7	37	–	–	53	160	112	64	19	41	–	2

(continued)

Table 3.8. Distribution of Source of Funds by Type of Aeronautical R&D Annual Expenditures in Millions of Dollars. (continued)

DEVELOPMENT

Atomic Energy Commission	Industry	Total	Air Force	Navy	Army	ARPA	NASA	Federal Aviation Administration	SST	Atomic Energy Commission	Industry	Total	TOTAL
–	20	67	137	101	–	–	19	1	–	–	18	276	365
–	25	87	153	170	–	–	24	1	–	–	22	370	484
–	33	83	148	113	–	–	19	1	–	–	29	310	417
–	42	95	114	151	–	–	27	2	–	–	37	331	454
–	62	122	160	130	2	–	34	2	–	–	56	384	539
–	86	147	199	90	3	–	33	8	–	–	76	406	592
–	170	252	250	145	11	–	39	4	–	5	153	610	914
–	286	419	452	176	17	–	73	3	–	8	257	986	1495
–	353	529	711	195	24	–	49	3	–	15	317	1314	1953
–	354	547	807	215	29	–	36	1	–	17	319	1425	2085
–	332	516	762	202	32	–	30	1	–	19	298	1344	1968
–	356	544	776	197	40	–	32	1	–	35	319	1400	2063
–	387	591	840	201	46	–	31	1	–	56	347	1522	2245
–	358	578	920	216	60	–	28	15	–	52	322	1599	2325
–	329	541	877	218	55	–	31	28	–	54	297	1563	2228
–	310	491	726	222	40	–	21	48	–	49	278	1377	1987
–	300	494	793	200	58	–	25	45	–	49	269	1436	2050
–	303	501	819	173	72	–	32	52	–	–	272	1420	2026
–	290	556	1079	212	84	–	42	58	–	–	258	1733	2419
–	301	559	1046	202	82	–	54	38	–	–	271	1693	2383
–	329	581	997	198	61	9	66	30	–	–	295	1658	2370
–	406	656	1028	208	79	23	70	31	110	–	365	1914	2711
–	509	740	857	246	85	14	86	35	188	–	457	1968	2855
–	577	821	922	197	106	11	110	35	62	–	519	1958	2946
–	533	771	645	374	108	2	138	36	92	–	480	1871	2806

*Approximately one-half of the aggregate of industry funds are reimbursed by the government.

Source: DOT-NASA, Booz, Allen and Hamilton, Table C-15.

Table 3.9. Aeronautical R&D Funds Used by Industry,
Classified by Aircraft Component
Annual Expenditures in Millions of Dollars.

Fiscal Year	Airframe	Engine	Avionics	Total
1945	118	66	79	263
1946	153	85	102	340
1947	138	76	91	305
1948	148	82	99	329
1949	184	102	123	409
1950	212	117	141	470
1951	332	184	221	737
1952	550	306	366	1222
1953	716	397	477	1590
1954	759	422	505	1686
1955	715	397	476	1588
1956	749	416	499	1664
1957	815	453	543	1811
1958	834	463	556	1853
1959	795	441	530	1766
1960	711	395	473	1579
1961	730	406	486	1622
1962	729	405	485	1619
1963	852	473	568	1893
1964	843	468	562	1873
1965	845	469	563	1877
1966	982	546	655	2183
1967	1056	587	703	2346
1968	1098	610	733	2441
1969	1026	570	685	2281

Source: DOT-NASA, Booz, Allen and Hamilton, Table C-21.

Table 3.10. Industrial Aeronautical R&D Funds,[*] Annual Expenditures in Millions of Dollars.

Fiscal Year	Safety	Human Factors	Flight Mechanics	Structures	Aerodynamics	Propulsion	Avionics	Other	Prototype Aircraft Development	Total
1945	1	2	2	4	5	6	7	3	10	40
1946	1	2	3	4	6	7	9	5	12	49
1947	2	2	4	6	8	9	12	6	16	65
1948	3	3	5	8	11	12	15	6	20	83
1949	4	5	8	11	16	18	23	9	30	124
1950	5	7	10	16	22	25	31	14	41	171
1951	11	13	21	32	45	50	63	29	76	340
1952	18	23	35	53	76	85	107	48	127	572
1953	21	27	44	66	93	105	132	61	156	705
1954	22	27	44	67	94	105	132	60	157	708
1955	21	24	41	62	87	98	124	59	147	663
1956	23	27	44	67	94	105	133	61	157	711
1957	24	30	48	72	103	115	145	65	171	773
1958	22	29	45	67	95	106	133	61	158	716
1959	21	26	41	62	87	97	123	55	146	658
1960	19	24	39	57	82	91	116	54	137	619
1961	19	24	37	56	79	89	112	50	133	599
1962	19	24	37	56	80	90	113	52	134	605
1963	18	23	36	54	76	85	108	49	128	577
1964	19	24	38	56	80	88	112	51	134	602
1965	21	26	41	62	87	97	122	55	146	657
1966	25	32	51	76	108	120	152	69	179	812
1967	32	39	64	96	134	150	190	88	224	1017
1968	36	45	72	107	152	170	215	101	256	1154
1969	33	42	67	100	141	158	199	90	236	1066

[*]Approximately one-half of these R&D funds are provided to industry as allowable overhead expenses on government-ment procurement contracts.

Source: DOT-NASA, Booz, Allen and Hamilton, Table C-9.

139

V. THE DEMAND FOR INNOVATION IN COMMERCIAL
AIRCRAFT: THE INFLUENCE OF GOVERNMENT

The preceding section documented the substantial research support that the aircraft industry has received from the federal government during the 1925-1975 period. Since most of this research was directed to the development of military aircraft, especially after 1940, we argued that the history of technical development in commercial aircraft consists largely of the utilization for commercial purposes of technical knowledge developed for military purposes at government expense. Government intervention and support to enhance the "supply" of potential innovations thus has been substantial. This "supply-side" government influence has been joined by a substantial number of important innovations emerging from other industries for exploitation by commercial aircraft producers.

Government policies, however, have also played an important role in affecting the demand for innovation by the commercial aircraft industry. Consciously or not, the policies of the Post Office in the 1929-1934 period, and those of the Civil Aeronautics Board during 1938-1978, influenced the structure and conduct of the air transportation industry so as to provide substantial incentives for rapid adoption of innovations in commercial aircraft. Government policy toward the commercial aircraft industry is unique, we believe, in its impact upon both the supply of technical knowledge and the demand for application of this knowledge in innovation within the civilian sector. In this section, government policies toward air transportation are discussed briefly to substantiate this assertion.

The transfer of responsibility for air mail transport from the Post Office to private contractors took place in 1925 following passage of the Kelly Air Mail Act. Bids were opened to private contractors on various mail routes; successful bidders were to be paid on a weight basis. During the ensuing five years, airmail postal rates were reduced by Congress, creating a substantial increase in air mail volume, while payments to operators remained at their previous levels. The result was an increase in contractor profits. Smith (1944) states that "compensation to carriers rose from 22.6 cents an airplane mile prior to July 1, 1926, to 73.6 cents a mile for the second half of 1927. . . . by the end of 1928, however, payments were up to 92 cents a mile, and by the end of 1929 the government was paying the operators $1.09 a mile for carrying the mail" (p. 125). This period of initial prosperity for the mail contractors, many of whom were subsidiaries of commercial aircraft producers, was based largely upon mail transport. Reflecting the growth of the mail-carriage aircraft market, such aircraft as the Boeing 40 were designed primarily for mail, rather than passenger, transport.

The McNary-Watres Act of 1930, and its administration by Postmaster General Brown during the Hoover administration, constituted a policy of developing a smaller number of large trunk carriers who would derive a far greater proportion of their revenues from passenger transport. The act changed the method of computation of payments for mail carriage from a pound-mile basis to a space-mile basis; that is, payment was made whether or not mail was carried in an aircraft. In addition, extra payments were made to carriers that used multiengine aircraft, radio, and other navigational aids. The final major section of the McNary-Watres Act was to be its undoing, as it conferred substantial discretionary powers upon the Postmaster General to alter route structures or merge carriers, when "in his judgement the public interest will be promoted thereby." Brown exploited his power to restructure air transport, orchestrating the merger of Transcontinental Air Transport and Western Air Express into TWA, and working to develop a small number of financially strong, transcontinental carriers who would provide a strong market for larger, more comfortable passenger transports.(24)   While Brown's goals were partially achieved,(25) his tactics produced a furor that resulted in the Air Mail Act of 1934, mandating divestiture by aircraft producers of subsidiary transport firms, and placing the award of mail contracts on a per-ounce basis, to be awarded strictly to the lowest bider. While it represented an inefficient mechanism for the support of air transport, and Brown's administration of the act led to its demise, this set of policies toward air carriers coincided with rapid growth in passenger traffic and the introduction of the monocoque fuselage air transports, the B-247 and the DC-2, which were of great importance in the development of the commercial aircraft and air transportation industries.

Continued congressional dissatisfaction with passenger safety and regulatory policy in general within air transportation led to the establishment of the Civil Aeronautics Board in 1938. Through its issuance of operating certificates and its overseeing of airline fares, the board effectively controlled pricing policies of airlines, as well as entry into or exit from air transportation, during the 1938-1978 period. These powers were used throughout the postwar period to prevent entry into scheduled trunkline air transportation and to prevent price competition. The CAB also controlled the award of routes to airlines; in general, multiple carriers were allowed to operate in "major" city-pair markets (such as New York to Los Angeles or New York to Chicago), while less important routes often were allowed to be monopolized by a single carrier.

This regulatory environment, in which entry and price competition were forbidden and multiple carriers operated in the more profitable market segments, gave rise to a high level of service-quality competition. One result of this was a very rapid rate of adoption of new aircraft designs by the major carriers, based upon their belief that rapid introduction of state-of-the-art aircraft was an effective marketing strategy where price competition was not possible. Jordan's (1970) study compare California's intrastate air carriers (not regulated by the CAB, and subject to price competition as well as easier entry) with the interstate carriers in the rapidity of adoption of cabin pressurization and jet aircraft:

> The trunk carriers were consistently the first to introduce each innovation. In fact, they introduced all but two of the over 40 aircraft types operated by all three carrier groups between 1946 and 1965. In addition, they adopted these innovations rapidly and extensively. The local carriers, on the other hand, were slow to introduce the two innovations and their rates of adoption were low [p. 53].(26)

The drive to be first with a new design was one of the central motives for the willingness of major airlines to make early purchase commitments to airframe manufacturers, as a means of obtaining the earliest possible delivery. Service-quality competition thus fostered rapid diffusion and adoption of innovations drawing upon government-supported research, as well as supporting fierce competition among airframe manufacturers. Fruhan (1972) argued that the lack of price competition under CAB regulations was partially responsible for the wide fluctuations over time in airline purchases of aircraft, as airlines attempted to provide sufficient carrying capacity to maintain higher load factors in a given city-pair market.(27) Purchases by one carrier were matched by a competitor, resulting in recurrent binges of new equipment purchases, such as that in the early 1970s, that left airlines burdened with heavy debts, and excess carrying capacity.

CAB regulation encouraged a rapid pace of innovation and adoption within the commercial aircraft and air transportation industries. This rapid rate of innovation, however, and the associated impressive productivity growth exhibited by air transportation have come at some cost. Consumer welfare was impaired by the lack of variety in service quality and price. The result of government regulation was to restrict the range within which consumers have been free to trade off price against quality. The result was a pattern of producer competition and competitive airline investment practices that could be described as inefficient. In addition, the direction of innovation may have been affected by this regimen of regulation and service-quality competition. As was noted above, the innovation process within the commercial aircraft industry historically has involved substantial financial and design participation by major airlines in new aircraft development. The preservation by CAB regulation of the dominance by a small number of firms of transcontinental routes, generally the most profitable industry segment, made this the major market for new aircraft during the postwar period of regulation. Given the sensitivity of the design and development processes to the desires of the financially strong trunk airlines, the result has been a bias in the direction of innovation, noted by Caves (1962, p. 10):

> A very important problem not eliminated by the increasing number of competing aircraft manufacturers is that of optimal variety in types of aircraft offered on the market. In the decade when piston-engine aircraft reached their peak of development, ending in the early 1950s, the duopolistic rivalry between Douglas and Lockheed led them to concentrate on development of an aircraft that would capture the largest single market - that of airlines flying United States transcontinental or trans-Atlantic routes. Relatively forgotten were the airlines in need of large planes efficient on relatively short hops, as well as the airlines needing low-cost equipment to serve low-density routes. . . . Airlines and aircraft manufacturers are both relatively few in number; airlines seek to minimize the number of different airplanes in their fleets for efficient maintenance purposes. These facts guarantee a standing pressure for aircraft manufacturers (operating under considerable uncertainty) to bias their research and development efforts toward the largest single market, whatever the structure and conduct of the airline industry may cause that to be. As already indicated, over the years the resulting bias has normally been toward long-haul, luxury aircraft.

An example of such a "missed opportunity" is the turbo-prop engine which, as Caves and others have argued, might have been developed further during the 1950s and early 1960s so as to compensate for its deficiencies in speed (relative to the jet engine) with greater fuel economy and lower operating costs than obtained for jet aircraft. The regulatory environment of the period, however, precluded the option of offering passengers lower fares for slower transportation, thus reducing the incentives faced by the airlines for adoption of the turboprop in preference to the jet for short-range uses. While the implicit counterfactual case that is proposed here is a somewhat speculative one, it raises important issues about the nature and the distribution of the benefits of the rapid rate of technical change in commercial aircraft. One may also speculate that, had the turboprop been given the encouragement to develop that might have existed in an unregulated world, the industry would have been better equipped to absorb the impact of the dramatic rise in fuel prices in the 1970s.(28)

The impact of deregulation upon innovation in commercial aircraft will probably take some time to be felt. Airline operating costs now are dominated largely by the soaring price of fuel. It is interesting to note, however, that price competition has come to play a major role in airline business behavior, and that service quality has become increasingly differentiated, with various "no frill", advance purchase, business class, and other discounts or premiums available to the consumer. Simultaneously with these developments, one notes less competition among domestic aircraft producers in the introduction of the next generation of aircraft. No other American producer has stepped forth to offer an aircraft that will compete directly with the new Boeing designs, the 767 and the 757. This probably reflects a less intense demand by the airlines for rapid deliveries of the new aircraft, as service quality and novelty have lost their former central roles in air transportation competition.(29) The order for 60 Boeing 757s placed by Delta Airlines came only after lengthy negotiations between McDonnell Douglas and Delta. Despite Delta's strong desire to purchase a Douglas DC-11, McDonnell Douglas was reluctant to proceed in the absence of additional orders from other carriers for the proposed new design:

> Delta wanted the DC-11 in order to make certain that McDonnell Douglas would keep building airliners, thus preventing a Boeing monopoly in commercial aircraft. But when it came to making a commitment, McDonnell Douglas was timid. Soured by a profitless decade for its commercial planes, the DC-9 and DC-10, McDonnell Douglas insisted on finding a second carrier before launching the DC-11 ["The Big Deal McDonnell Douglas Turned Down," Business Week, December 1, 1980].

It is likely that McDonnell-Douglas's search for a second airline willing to make an advance purchase commitment would have been more fruitful during the era of CAB regulation than was the case in a deregulated world.

A final policy episode of considerable relevance to this discussion of federal policies affecting the demand for commercial passenger transports is the SST development program.(30) The federal SST program, aimed at developing a commercial supersonic transport, represented a sharp departure from past policies toward civilian aeronautical R&D in several important ways. The program was administrated by the Federal Aviation Administration, rather than by NASA, the agency that was historically in charge of civilian aeronautical research. Of greater (and not unrelated)(31) significance was the fact that the SST program was intended to result in the production of two prototype aircraft, as well as possible government financing or loan guarantees for start-up costs for commercial production. The program was motivated by the perceived threat to American dominance of commercial aircraft sales embodied in the British-French Concorde, as well as the desire to use technological "spillovers" from the development of the B-70 strategic bomber prototype. The prototype development contract was awarded to Boeing on the strength of a design that the firm claimed would meet the performance requirements stipulated by the FAA, in sharp contrast to the usual design and development process for commercial aircraft, in which the airlines are major participants.(32) Mounting technical and cost problems (the original Boeing prototype design proved infeasible, and major changes had to be made), as well as growing opposition from environmental groups, ultimately led the Congress to kill the program in early 1971.

The SST episode in many ways constituted an application of the military procurement model to the development of commercial aircraft; the federal government conducted a design competition and proposed to support the development efforts of the winning prime contractor. Such policies had proven more or less successful in military aircraft procurement, simply because of the largely nonmarket character of this process; the federal government was the sole domestic customer for military aircraft. It was therefore eminently sensible for the ultimate purchaser to specify in detail the operating and design characteristics of the aircraft that were to be purchased in the military market. The attempt, however, to develop a successful commercial SST through the same government procurement mechanism was almost certain to lead to designs and decisions that would suppress ordinary commercial considerations. This was, indeed, precisely what happened with the Anglo-French joint development of the Concorde.

The increasing attempt to involve governments in the aircraft development process was a result, of course, of the enormous rise in such development costs with the advent of the jet engine and its eventual introduction into commercial use. Nevertheless, it is important to note that these development costs had been successfully borne by the private aircraft sector, even including the huge development costs of the Boeing 747.(33) The Concorde experience, which proved to be a financial disaster, forcefully underlined the importance of the exercise of commercial judgments, and of bolstering the incentive to exercise such judgments in high-technology industries

such as aircraft. Establishing the purely technical feasibility
of an aircraft is very different from establishing its commercial
feasibility. It is clear that the early phases of the SST design
paid no attention to prospective operating costs.(34)   That
was, of course, also true of the Concorde, which had to be
developed at a joint cost to the French and British governments of several billion dollars.(35)   It is arguable that the
poor British performance in the aircraft industry in the post-
World War II years was aggravated by the availability of large
government subventions that dulled the commercial judgment of
development decision makers.(36)

The SST experience suggests the desirability of confining
the federal role in the aircraft industry to an indirect one,
affecting only the adoption incentives of the purchasers of
commercial aircraft and the development decisions of aircraft
manufacturers. Detailed government participation in the
specifications of design and performance characteristics creates
a structure of incentives in which hard-nosed economic considerations are likely to receive insufficient weight.

What is particularly disquieting about the SST experience
is that it appeared to be clear from the outset that the SST
was not an economic proposition. In spite of the preponderant
burden of expert testimony, there was a strong determination
inside Congress to proceed with it anyway. Aircraft manufacturers were, not surprisingly, prepared to participate in
such a program so long as it was heavily financed by the
government, but neither they nor the airlines were the major
initiating force.(37)

VI.   CONCLUSION

In concluding this discussion of federal policy and innovation
in commercial aircraft, we summarize our assessment of the role
of the federal government in affecting innovation within the
industry, and consider the degree to which other industries
could benefit from a similar fabric of government policies.
While the innovative performance of the industry suggests that
this policy framework has been successful, it is likely to be
limited in its applicability to other industries. In view of some
of the other failings of both this policy framework and the
commercial aircraft producers, a transfer to other industries of
this policy framework may not be desirable.

The crucially important aspect of federal policy throughout this 50-year period is its impact upon both the supply of
and demand for innovation. Military support of new aircraft
development provided important technical skills, knowledge,
and innovations that could be used by manufacturers in commercial aircraft. Government demand for new designs, push-

ing at the outer limits of available technologies, was no less crucial in bringing about the rapid embodiment of new technical knowledge or isolated breakthroughs in some subsystem in a new aircraft design.   The assurance of a market for a successful military aircraft gave manufacturers considerable incentives to pursue and use rapidly the technical and scientific knowledge acquired at federal expense.   This assurance of the existence and characteristics (in varying detail) of the nature of the demand for innovative technologies is of great importance in understanding the speed with which technical breakthroughs came to be embodied in new military aircraft. The modest success of such programs as the NASA Technology Utilization program, or federally funded demonstration projects aimed at increasing the supply and availability of commercially useful knowledge reflect in part the uncertainties about demand faced by the potential users of this knowledge (the NASA program also has been hampered by the often limited potential of NASA technologies for civilian applications).   In the military aircraft market, which generated considerable spillovers into commercial applications, such demand uncertainty was minimal.

The commercial aircraft market also was affected on the demand side by government policies. We argued above that the McNary-Watres Air Mail Act, and the subsequent regulatory policies of the Civil Aeronautics Board, engendered a strong demand on the part of airlines for new aircraft embodying military-spawned innovations.   While the number of commercially unsuccessful aircraft indicates that the market was not an assured one, the effect of regulatory policies was to provide a strong impetus for aircraft manufacturers to embody new technological developments quickly in innovative aircraft designs, as well as for the airlines to adopt new aircraft designs as rapidly as possible.   To a lesser extent than was true of the military market, the existence of a strong demand for "state of the art" commercial aircraft aided the rapid embodiment of new technological knowledge.

The usual justification for publicly supported research appeals to the public good characteristics of knowledge and information, arguing that the social payoffs to fundamental or basic research greatly exceed the private returns to investment in such research. Government support therefore is considered best applied to the most basic forms of research. In the case of NACA, however, established as a source of research for the aircraft industry, basic research was notably absent.   Constant (1980) argues convincingly that a major reason for the failure of any American firm to develop the jet engine prior to World War II was due to the lack of theoretical work in aerodynamics and aeronautics pursued in the United States, as opposed to Germany or Great Britain.   NACA's role prior to 1940, according to Constant, was primarily that of a provider of testing facilities and empirical data, rather than a supporter of advanced theoretical work in aerodynamics. Nonetheless, the American firms were well-placed to utilize the theoretical work in aerodynamics and the jet engine, most of which had been developed abroad in the aftermath of World War II, the result being the 707 and the DC-8, the first commercially successful jet transports. Constant attributes the postwar dominance of American firms in jet aircraft to the extremely large and highly developed domestic airline system that had evolved since the 1930s. Government agencies and policies, such as McNary-Watres or the CAB, affecting the nature of the demand for commercial aircraft, thus may have been as important as federal support of research in the development of the postwar aircraft industry.

The experience of the commercial aircraft industry underlines the importance, in designing policy towards innovation, of affecting both the supply of and demand for innovation and technical knowledge.(38) While this conclusion clearly is one of considerable generality, with obvious relevance to technology policies in other industries, it is not clear that the specific policy instruments that have been utilized in the commercial aircraft industry are appropriate or applicable in other industries. Certainly, the resource costs of these policies in the aircraft industry have been substantial. High profits and federal research support in the development and sale of military aircraft have comprised an important government subsidy to the development and manufacture of new commercial designs. Carroll (1972) argues that government contracts have been much more stable in volume, and yielded substantially higher profits, than commercial sales in the 1950s and 1960s. According to Carroll, the profitability of military sales contributed to intense competition in commercial aircraft production and sales, including possibly excessive duplication of development costs, tooling, and product lines, in some segments (e.g., long-haul transport) of the market. Resources thus may have been inefficiently allocated as a result of this implicit subsidy. Further, we have argued above that the competition between McDonnell Douglas and Lockheed may have had deleterious consequences for product safety. Finally, of course, there are the welfare costs to consumers of CAB regulation of air transportation, another element of the policy framework that has supported this high rate of innovation in commercial aircraft.

One area in which an aircraft industry policy paradigm may be of relevance is that of technologies for reducing emissions of pollutants and carcinogens from automobiles and industrial production processes. This is an area in which the performance characteristics of the technologies that are mandated by federal regulations could be clarified in such a way as to make the demand for innovation less ambiguous.

Coupled with a more substantial level of government funding of research in this area, a set of policies could be developed that would affect the supply of technical knowledge and innovations, as well as the demand for new emissions-control processes, so as to improve the state of the art in this important technology. Another area where such an approach might be useful is that of energy technologies. Here, government currently funds research extensively, in contrast to the situation of emissions-control technologies, but has done little by way of providing a clear and stable demand for energy technologies with certain specific cost and performance characteristics (indeed, until the recent moves to remove price controls on oil and natural gas, government demand policies discouraged the application of new energy technologies). By making commitments to purchase certain forms of energy at a guaranteed price, e.g., synthetic fuels for a strategic petroleum reserve, or the output of certain technologies with specific cost or performance characteristics, e.g., solar-energy sources meeting announced criteria, federal policies could provide a more effective set of "market pulls" in addition to the currently available "pushes" from extensive research funding. The essential requirement is to design policies that affect both supply and demand.

NOTES

1. Kendrick (1961) reports that output per person in the air transport industry grew at an average annual rate of 8.8 percent during the 1929-1948 period, higher than almost any other industry in his sample. Output per person grew at an average annual rate of 8.2 percent during the 1948-66 period, far higher than any other of Kendrick's industries (1973), while total factor productivity during the 1948-1966 period grew at an annual rate of 8.0 percent. Employing a different methodology in an examination of 45 manufacturing and service industries, Fraumeni and Jorgenson (1980), and Gollop and Jorgenson (1980) also conclude that air transportation has displayed one of the highest rates of growth in total factor productivity during the postwar period.

2. The Air Mail Act was the response of the newly elected Democratic Congress and the Roosevelt administration to the airmail policies of Walter Brown, Postmaster General under Hoover. Controversy erupted over the letting of air mail transport contracts, stemming from Brown's attempts to utilize these contracts as a means of influencing the development of the structure of the air transportation firms, and with this, the development of the entire aviation industry in the United States. The precise nature and impact of the Brown policies,

which were intended to move transportation companies away from exclusive reliance upon mail contracts and into passenger transport, are discussed in greater detail below.

3. P. 10. It is also important to note that very few of these civilian aircraft were multiengine transports.

4. Holley, p. 22. Phillips (1971) notes that for Boeing and Lockheed during this period, "Military orders sustained them and each attempted new commercial planes prior to World War II. Lacking the military orders - that is, in a market environment more typical of most industries - Lockheed and Boeing would presumably have failed" (p. 113).

5. According to Aviation Week and Space Technology, "The 767 subcontracting also will be devoted to sharing the risk. It will resemble the 747 situation in many respects, although in this case the major subcontractors will be required to assume a larger share of the risk, for potentially greater profits" (July 24, 1978).

6. See The Wall Street Journal, July 12, 1979.

7. These measures are employed by Rosenberg, Thompson, and Belsley (1978). The (As x V_c) measure, combining aircraft capacity and speed, may be viewed as a measure of the average "throughput capacity" of a given aircraft design.

8. The discussion in the next several pages is taken from Rosenberg, Thompson, and Belsley (1978).

9. A. Alchian, "Reliability of Progress Curves in Airframe Production," Econometrica (October, 1963) pp. 679-692; Werner Hirsch, "Firm Progress Rations," Econometrica (April 1956) pp. 136-143; Leonard Rapping, "Learning and World War II Production Functions, "Review of Economics and Statistics" (1965) pp. 81-86; Paul David, "Learning by Doing and Tariff Protection: A Reconsideration of the Case of the Ante-Bellum U.S. Cotton Textile Industry," Journal of Economic History (September, 1970), pp. 521-601; Kenneth Arrow, "The Economic Implications of Learning by Doing," Review of Economic Studies (June 1962), pp. 155-173. According to Hirsch, the U.S. Air Force "for quite some time had recognized that the direct labor input per airframe declined substantially as cumulative airframe output went up. The Stanford Research Institute and the RAND Corporation initiated extensive studies in the late forties, and the early conclusions were that, insofar as World War II airframe data were concerned, doubling cumulative airframe output was accompanied by an average reduction in direct labor requirements of about 20%. This meant that the average labor requirement after doubling quantities of output was about 80% of what it had been before. Soon the aircraft industry began talking about the 'eighty percent curve'" (Hirsch, p. 136). It is possible, of course,

that cost reductions that have been attributed to learning by doing have actually been due to other factors which have not been defined as a residual. For earlier discussions by the learning curve in the aircraft industry, see Adolph Rohrbach, "Economical Production of All-Metal Airplanes and Seaplanes," Journal of the Society of Automotive Engineers (January 1927), pp. 57-66, and T.P. Wright, "Factors Affecting the Cost of Airplanes," Journal of the Aeronautical Sciences (February 1936), pp. 122-128.

10. A parallel process, with which we do not deal, is the extensive learning that was involved in the operation and management of an entire aircraft fleet. There were many operational problems for which optimal procedures had to be developed - scheduling problems, turn-around time, dovetailing the requirements of equipment with those of personnel, and so on. Such "software" responsibilities belong to the realm of management and not technology, although the two realms are obviously interrelated.

11. While the modifications alter the aerodynamic parameters of the wing, sometimes substantially, the wing itself does not generally experience internal structural alterations. This is because of the prohibitively high cost of wing design that makes it much more economical to modify the flaps, leading edge, and wing tips. At the same time, the possibility of eventually utilizing even these add-on devices must be anticipated to some degree during the initial wind-development stage.

12. Boeing Commercial Airplane Co. (1976), p. 4.

13. The technique of stretching has a lengthy history and was applied with great success to the DC6-DC7C series as well as the Lockheed 649 to 1049H series of propeller-driven aircraft. A well documented recent example of this technique is shown in the case of the DC9 series aircraft in Business Week (pp. 95, 100, November 14, 1977) where the DC9 series has been increased in size by lengthening the fuselage from 104.4 feet (80 passengers) in 1965 to 147.8 feet (155 passengers) in 1980 in five distinct steps. In addition, modifications to the wing and power plant have enabled it to increase performance and keep abreast of the latest noise regulations.

14. The role of highly specialized producers, and the question of the optimum degree of specialization from the point of view of technological innovation, are highly important questions that are still not very well understood. Specialist producers tend to be very good at improving, refining, and modifying their specialized product. They tend not to be very good at devising the new innovation that may constitute the eventual successor to their product. They tend, in other words, to work within an established regime, but they do not usually

make the innovations that establish a new regime. Thus, the horse-and-buggy makers did not contribute significantly to the development of the automobile; the steam-locomotive makers played no role in the introduction of the diesel, and indeed expressed a total disinterest until it was finally developed by General Motors; and the makers of piston engines did not play a prominent role, in England, Germany, or the United States, in the development and introduction of the jet engine. The severely circumscribed technological horizons of the specialized producers - to some extent an inevitable "occupational hazard" - may help to account for what one recent book on the aircraft industry describes as "an apparent proclivity on the part of once successful manufacturers to remain too long with the basic technology of their original success" (Phillips, 1971, p. 91). The point is that intimate familiarity with an existing technology creates a strong disposition to work within that technology, and to make further modifications leading to its improvement and not to its displacement. Scribes may be expected to invent forms of shorthand, but not typewriters, but if improved ones show up, they will be adopted.

15. Taylor (1970) notes as central to the improved performance of high-bypass-ratio jet engines," the fan, cooled turbine blades allowing higher turbine-inlet temperature, and higher-pressure-ratio compressor" (p. 56). Central to all of these improvements were improved alloys.

16. A recent witness testified to the importance of such uncertainties: "Steiner pointed out that 'accelerated aging' tests have not proved accurate in the past. He cited the case of certain alloys that 'aged in a most peculiar manner' a few years ago. In five to 10 years, these alloys - utilized on the Boeing 707 and other transports - developed inter-granular corrosion, requiring expensive inspection procedures and replacements. With that kind of history, Steiner said, 'any sound manufacturer or financial institution would have reason to be a little timid about locking advanced composites into a primary structure which is non-removable'" (Aviation Week and Space Technology, September 12, 1977, p. 35).

17. "In this country, the early work of Sanford Moss on the gas turbine, starting in 1901 at Cornell University, eventually led to the development of the General Electric turbosupercharger. This device, first applied to aircraft engines by Rateau in France, was flown before the end of World War I. The U.S. Army's interest resulted in Moss's concentration of his efforts on the aircraft supercharger in the period between the wars. The expense was borne by the Army from 1919 to 1937. It was the proving ground in this country for improved high-temperature metallurgical development" (Badger, 1958, p. 512).

18. Statement of Dr. Joseph Ames, Hearings of the President's Aircraft Board, 1925. Ames served as NACA's first chairman.

19. Some idea of the growth in the in-house research establishments of major aircraft firms during World War II is conveyed by a comparison of date on these firms contained in the 1940 and 1946 editions of the National Research Council Survey of Industrial Research. The in-house professional staff at Douglas Aircraft grew from 22 persons in 1940 to 111 in 1946; the Glen Martin Company grew from 42 to 76 in the research department; Lockheed grew from 10 to 314; Consolidated Vultee went from 12 to 195; United Aircraft (including Pratt and Whitney, Hamilton Standard, and Sikorsky) grew from 80 in 1940 to 732 by 1946; and Curtiss-Wright went from 14 in 1940 to 159 in 1946. In view of the fact that 1940 was a boom year for the industry, owing to rapidly increasing foreign and domestic military orders, these figures are all the more impressive.

20. "[A]s often happens the airbus is the result of a technological advance that was brought about by unrelated events - in this case, the U.S. Air Force's request to the industry in 1964 for engines with double or triple the thrust of existing power plants. The Air Force required engines for a huge new military transport that eventually became known as the Lockheed C-5A" ("Why Boeing Is Missing the Bus," John Mecklin, Fortune, June 1, 1968, p. 82).

21. DOT-NASA study, p. A(9).

22. "[T]he decision taken in a small but significant proportion of such cases has been to make, rather than to buy. This tendency has been especially prevalent in the aircraft industry and other sectors of the weapons industry severly affected by technological change. Faced with serious declines in their regular business of fabricating and assembling airframes, most of the major U.S. aircraft companies decided to build up capabilities in new fields of weapons technology, especially in electronics. They assembled nuclei of engineers and scientists in the fields to be entered. At the outset, however, these groups had neither the breadth nor depth of experience available in firms already working in the particular technology. Only with actual experience in research, development, and production could the companies establish capabilities equal to those already in existence. One way to acquire such experience was for a prime contractor to 'make' the components and subsystems which otherwise would be 'bought' from established firms.

"The work done by these inexperienced in-house groups was often more expensive than it would have been if subcontracted to experienced companies. With cost reimbursement

contracts, these extra costs were paid by the government"
(Peck and Scherer, 1962, p. 388).

23. The DOT-NASA study (Vol. 2, Appendix C, p. 49) offers
the following definitions of research categories:

> Basic research is concerned with exploration of the
> unknown. It is undertaken to increase the under-
> standing of natural laws and is free from the need to
> meet immediate objectives

> Applied research is directed to the solution of a
> recognized problem. It differs from basic research
> in that it is pointed toward practical applications
> rather than toward investigation for its own sake.

> Development is the systematic use of knowledge and
> understanding gained from research and directed to
> the production of useful materials, devices, systems,
> and methods. This work includes the design, test-
> ing, and improvement of prototypes and processes.

24. Testifying in 1934 before Senator Hugo Black's Special
Committee on Investigation of Air Mail and Ocean Mail Contracts, Brown interpreted his activities in the following
favorable light (Hearings, p. 2,351):

> With the passage of the McNary-Watres Act giving
> the Post Office Department the requisite authority, it
> exerted pressure on the air mail carriers, who with
> minor exceptions had theretofore been confining their
> operations exclusively to carrying the mail, to
> transport passengers and express in order to build
> up revenues from the public and thus lighten the
> burden on the Post Office Department; and it ex-
> erted every proper influence to consolidated [sic]
> the short, detached and failing lines into well fi-
> nanced and well-managed systems, providing three
> independent transcontinental operations with ap-
> propriate north and south intersecting services,
> believing that the pressure of competition would in
> time attract public patronage, reduce operating costs
> and develop, if possible, a transport airplane
> capable, under the competitive conditions in the
> passenger and express transportation industry, of
> earning enough to pay its way without any subsidy.

25. The report of the U.S. Senate Judiciary Committee's
Subcommittee on Administrative Practice and Procedure on Civil
Aeronautics Board Practices and Procedures (1975, pp. 202-
203), noted that:

In May 1929 when Brown called together the airline industry to organize and rationalize it, 24 mail contracts were distributed among approximately 19 independent companies. Three companies received the dominant shares of the $11.2 million mail payments due at least to some extent to the excess payments required by the 1928 poundage method of contract payments: Boeing Air Transport (predecessor of United), National Air Transport (predecessor of United), and Western Air Express (predecessor of TWA).

During Brown's tenure mail payments increased to $19.4 million and the miles of air mail routes increased from 14,405 to 27,678. By the end of 1933 all but 2 of the 20 air mail contracts were held by three large holding groups: United Aircraft & Transport Co. (United), Aviation Corporation (American), North American Aviation/General Motors (TWA and Eastern). These three flew 26,675 of the 29,212 air route miles and collected $18.2 million of the $19.4 million paid to air mail contractors.

26.   Jordan (1970, p. 55) concludes that:

The California intrastate carriers' service quality actually appears to have been affected less by carrier rivalry than by the desire or need of these carriers to achieve low operating costs. The intrastate carriers contended themselves with obsolescent DC-3's and DC-4's, or the nonpressurized Martin 202, until the prices of used, pressurized piston-powered aircraft fell drastically in the early 1960's. In constrast, the turboprop Electra was adopted by PSA soon after it became available, but this was a case in which low operating costs per seat-mile offset a high purchase price. On the other hand, turbojet-fan aircraft were not adopted until a medium-range turbo-fan aircraft was developed that had relatively low operating costs for short stage lengths.

27.   This apparently counterintuitive strategy derives from the fact that, within a middle range of capacity share on a given route (roughly 20 to 70 percent), load factors and capacity increases are positively correlated for a given carrier. Airlines competing in a given city-pair market thus face strong incentives to match one another's purchases of new equipment. See Fruhan (1972), Chapter 5.

28.   Another case in support of this argument concerns the attempts of the FAA in the early 1960s to develop a short-haul passenger transport capable of replacing the DC-3, then heav-

ily utilized by local-service airlines despite its advanced years, lack of cabin pressurization, and low speed. A study of Policy Planning for Aeronautical Research and Development, prepared by the Library of Congress's Legislative Reference Service for the Senate Committee on Aeronautical and Space Sciences, noted that the FAA deemed action necessary because "While U.S. manufacturers had made a variety of studies, no design had been forthcoming. . . . The key to starting the program appeared to be the need for a single order of at least 100 aircraft with the probability of at least 100 more. The local service airlines could not produce this order and only the DOD in Government could think in such quantities" (1966, p. 238).

29. Clearly, the greater fuel efficiency of the new Boeing designs provides a powerful impetus for airlines to replace their older aircraft, such as the 727 and 707. Our point is that, whereas in the previous days of CAB regulation, airlines would have been motivated to purchase these planes both because of their fuel efficiency and because of their perceived novelty and superior passenger comfort and/or safety, in the current context the "service quality" argument is less compelling, leading to a lower level of competition among airlines for positions in the delivery queue.

30. See Eads and Nelson (1971), for a useful and critical analysis.

31. James M. Beggs, Undersecretary of Transportation, testified in 1970 before Congress that:

> As a general rule the NASA in their aeronautical program has pursued programs only up to the point of proof of concept, the idea here being that NASA and its aeronautical centers are best kept employed by advancing the technology and not by going into development programs to make specific pieces of equipment.
>   Mr. James Webb, who, of course, ran NASA for many years, was fully committed to that proposition, that NASA should pursue technology up to the point of proof of concept and not get into the development of specific air transports of any other kind of commercial program. . . . So when the decision to go forward with the supersonic transport was made, NASA was, of course, asked whether they would be interested in managing the program. And they said in line with their philosophical position they would not."[Hearings, May 11, 1970, pp. 969-970].

32. While the FAA performance specifications for the SST prototype included a stipulation that the ultimate production

version of the aircraft must be "economically profitable," primary concern during the prototype design process was directed to performance characteristics, which did not include seat-mile operating costs.

33. Boeing was able to persuade its major subcontractors to share in the development costs of the 747. It also had firm purchase commitments from several major buyers – Pan Am, TWA, Lufthansa, and BOAC. But these commitments, it is important to notice, were not valid unless Boeing was able to fulfill specifications laid down by these buyers.

34. As Congressman Yates stated during congressional hearings (Hearings, p. 955) on the SST in 1970:

> In approving the SST budget request for last year the House Appropriations Committee stated in its report that, "The age of the supersonic transport is upon us." That same theme is repeated over and over again in administration presentations before congressional committees and in debate in both Houses. It is true only in the limited sense that supersonic transport flight is now feasible from a technical standpoint. It neglects to mention the very likely possibility that the supersonic derby between the U.S. SST, and the Soviet Tu-144, and the Anglo-French Concorde may be a competition to determine which nation can lose the most money – assuming they can even develop the production model. . . . A cursory reading of the airlines letters submitted last year to the FAA shows that for every ounce of enthusiasm about speed, there is a pound of misgiving about range, payload, comfort, and economic utility. If the age of the supersonic is in fact upon us, one will have to look someplace else besides the airlines for proof of the proposition.

35. Various estimates of the Concorde's development costs have appeared. Although the numbers most commonly cited are in the range of $2.2 to 2.4 billion, they may well be much higher. According to Gillman (1977, p. 73),

> In May 1976, Professor David Henderson, newly appointed professor of political economy at University College, London, argued that the government's figure of 1.46 billion shared between Britain and France was a drastic underestimate. It had been reached by adding the yearly expenditure on the project at the current prices. If these were adjusted to 1975 prices, and interest charges of 10 percent added, then the cost of Concorde was not 1.46 billion but 4.26 billion ($6.82 billion at the present exchange rate of $1.60).

36. Eads and Nelson (1971, pp. 497-498) state that:

> In contrast to the U.S. experience, the record of the British aircraft industry has been relatively dismal. The British government has been prepared to cover up to 50 percent of the costs of launching civil aircraft designs and to assure a base market for these designs by requiring British flag carriers to purchase the resulting product, regardless of operating costs. As a result the British have rung up a string of technological success that, by and large, have been commercial failures. Even the massive infusion of government aid has not served to maintain the health of the British aircraft industry, and, with few exceptions, the aircraft have generated nowhere the hoped-for volume of export earnings.

37. During the 1970 hearings (Hearings, pp. 1,002-1,003) Senator Proxmire lamented:

> I am not very comforted by thinking about how power works in the Congress and how power works in the administrative branch. We have had tremendous arguments against the SST all along. But we have gotten sizeable votes against it in the floor of the Senate, and we were not able to stop it in the House.
> Last year we had the President's own ad hoc advisory committee on the SST which came down as hard as any group of experts I have ever seen against the SST, unanimously against it on every score. Every argument SST proponents made was knocked down by the ad hoc committee.
> In spite of that, we were not able to muster a great deal of opposition to it in the Congress. And even more shocking is the fact that the President went ahead over the decision of his own committee.

38. Nelson and Winter (1977) and Mowery and Rosenberg (1979) provide analyses of the innovation process that emphasize the importance of linking both "market-pull" and "technology-push" forces.

REFERENCES

Badger, F.S., "High-Temperature Alloys: 1900-1958." Journal of Metals (August 1958).

Boeing Commercial Airline Company, Document B-7210-2-418, October 1976.

Booz, Allen, and Hamilton Applied Research, Inc., A Historical Study of the Benefits Derived from Application of Technical Advances to Commercial Aircraft. Prepared for the joint Department of Transportation-NASA Civil Aviation R&D Policy Study (Washington, D.C.: U.S. Government Printing Office, 1971).

Caves, R.E., "Market Structure and Embodied Technological Change." In Regulating the Product, ed. R.E. Caves and M.J. Roberts (Cambridge, Mass.: Ballinger, 1975).

_____, Air Transport and Its Regulators (Cambridge, Mass.: Harvard University Press, 1962).

Carroll, S.L., "Profits in the Airframe Industry," Quarterly Journal of Economics (November 1972).

_____, "The Market for Commercial Aircraft," Regulating the Product, ed., R.E. Caves and M.J. Roberts (Cambridge, Mass.: Ballinger, 1975).

Constant, E.W., The Origins of the Turbojet Revolution (Baltimore: Johns Hopkins University Press, 1980).

Eads, G., and R.R. Nelson, "Government Support of Advanced Civilian Technology: Power Reactors and the Supersonic Transport." Public Policy (Summer 1971).

Eddy, P., E. Potter, and B. Page, Destination Disaster (New York: Quadrangle, 1976).

Enos, J.L., Petroleum Progress and Profits (Cambridge, Mass.: MIT Press, 1962).

Fraumeni, B.M., and D.W. Jorgenson, "The Role of Capital in U.S. Economic Growth, 1948-76." In Capital, Efficiency, and Growth, ed. G.M. von Furstenberg (Cambridge, Mass.: Ballinger, 1980).

Fruhan, W.E., The Fight for Competitive Advantage: A Study of the United States Domestic Trunk Carriers (Boston: Harvard Business School Division of Research, 1972).

Gellman, A.J., and J.P. Price, Technology Transfer and Other Public Policy Implications of Multi-National Arrangements for the Production of Commercial Airframes (Washington, D.C.: NASA, 1978).

Gillman, P., "Supersonic Bust: The Story of the Concorde," Atlantic (January 1977).

Goldberg, V.P., "Regulation and Administered Contracts," Bell Journal of Economics (1976).

_____, "Competitive Bidding and the Production of Precontract Information." Bell Journal of Economics (1977).

Gollop, F.M., and D.W. Jorgenson, "U.S. Productivity Growth by Industry, 1947-73." In New Directions in Productivity Research and Analysis, eds. J.W. Kendrick and B.N. Vaccara (Chicago: University of Chicago Press, for the National Bureau of Economic Research, 1980).

Hochmuth, M.S., "Aerospace." In Big Business and the State, ed. R. Vernon (Cambridge: Harvard University Press, 1974).

Holley, I.B., Buying Aircraft: Material Procurement for the Army Air Forces, Vol. 7 of the Special Studies of the U.S. Army in World War II (Washington, D.C.: U.S. Government Printing Office, 1964).

Hunsaker, J.C., "Research in Aeronautics." In Research - A National Resource, National Resources Planning Board (Washington, D.C.: U.S. Government Printing Office, 1941).

_____, "Forty Years of Aeronautical Research." In the 1955 Annual Report of the Smithsonian Institution (Washington, D.C.: U.S. Government Printing Office, 1956).

Jordan, W.A., Airline Regulation in America (Baltimore: Johns Hopkins University Press, 1970).

Kendrick, J.W., Productivity Trends in the United States, (Princeton: Princeton University Press, for the National Bureau of Economic Research, 1961).

_____, Postwar Productivity Trends in the United States, 1948-1969 (New York: Columbia University Press, 1973).

Miller, R., and D. Sawers, The Technical Development of Modern Aviation (London: Routledge and Kegan Paul, 1968).

Mowery, D.C., and N. Rosenberg, "The Influence of Market Demand Upon Innovation: A Critical Review of Some Recent Empirical Studies," Research Policy (1979).

Nelson, R.R., and S.G. Winter, "In Search of Useful Theory of Innovation," Research Policy (1977).

Peck, M.J., and F.M. Scherer, The Weapons Acquisition Process (Boston: Harvard Business School Division of Research, 1962).

Phillips, A., Technology and Market Structure (Lexington, Mass.: D.C. Heath, 1971).

Rae, J.B., Climb to Greatness (Cambridge, Mass.: MIT Press, 1968).

Rosenberg, N., A.M. Thompson, and S.E. Belsley, Technological Change and Productivity Growth in the Air Transport Industry. NASA Technical Memorandum 78505 (1978).

Rosenberg, N., "Learning by Using." In N. Rosenberg, Inside the Black Box: Technology in Economics (New York: Cambridge University Press, 1982)

Schlaifer, R., Development of Aircraft Engines (Boston: Harvard Business School Division of Research, 1950).

Smith, H.L., Airways (New York: Knopf, 1944).

Stekler, H.O., The Structure and Performance of the Aerospace Industry (Berkeley: University of California Press, 1965).

Taylor, E., "Evolution of the Jet Engine," Astronautics and Aeronautics (November 1970).

U.S. Joint Economic Committee, Subcommittee on Economy in Government, Hearings on Economic Analysis and the Efficiency of Government, Part 4, Supersonic Transport Development (Washington, D.C.: U.S. Government Printing Office, 1970).

U.S. Senate Committee on Aeronautical and Space Sciences, Policy Planning for Aeronautical Research and Development (Washington, D.C.: U.S. Government Printing Office, 1966).

U.S. Senate Judiciary Committee, Subcommittee on Administrative Practice and Procedure, Civil Aeronautics Board Practices and Procedures (Washington, D.C.: U.S. Government Printing Office, 1975).

U.S. Senate Special Committee on Investigation of Air Mail and Ocean Mail Contracts, Hearings, Pt. 4-6 (Washington, D.C.: U.S. Government Printing Office, 1934).

Williamson, O.W., Markets and Hierarchies (New York: Free Press, 1975).

4
The Computer Industry
Barbara Goody Katz
Almarin Phillips

I. HISTORICAL BACKGROUND

The emergence of the commercial computer industry is one of
the most remarkable phenomenon of the twentieth century.
The diffusion of computer technology has affected in a significant way literally all facets of modern life.   No business, no
profession, no area of economic endeavor has been untouched.
Everywhere one turns, whether in work or play, some application or influence of the computer is apparent.

   Relationships between governments and computing technology are as old as written history itself.   Herodotus (485-425
B.C.) wrote of the then ancient computing device, the abacus;
his work was subsidized by a grant of ten talents ($2,410,
1947 values) from the citizens of Athens.(1)   Much later, in
the 16th century, Jobst Burgi, a court clockmaker to Langrof
Wilhelm IV of Hess and Kaiser Rudolph II, invented a pedometer, literally a digital computing mechanism, unlike the analog
clocks of the time.(2)   John Napier, inventor of logarithms,
the slide rule, and the computing "Naperian Bones," was a
Scotsman who occupied years of his life inventing "secret
instruments of war" in defense of his country against a mis-
takenly foreseen Spanish invasion.(3)

   At the age of 19, the eccentric and precocious Blaise
Pascal invented what many regard as the first real calculating

*Barbara Goody Katz wishes to acknowledge the good intuition
of Robert B. Goody, who suspected that there was an untold
story concerning magnetic amplifiers in the history of computers.

machine in 1642. Many of these were made, including some that could add livres (0-9 digits), sous (0-19) and deniers (0-11).(4) Numerous improvements followed, but the next substantial improvements are attributed to Leibnitz. Burgi's pedometer as well as Pascal's inventions led Leibnitz to work on the logic of computing devices.(5) Leibnitz, well known for inventing the calculus, was a public servant in Mainz with considerable freedom in his employment to do research in sundry areas of thought. Among his many mechanical inventions was a calculating machine (circa 1672) that extracted roots as well as performing the functions of addition, multiplication, subtraction and division.(6) By 1694, Leibnitz' machine performed multiplication by repeated addition, with a register that could slide from the units to the tens, hundreds, and so on, positions.

The Leibnitz device is more the predecessor of the adding machine, comptometer, cash register, and desk calculator than of the modern computer. Problems in manufacturing accurate mechanical parts inhibited the manufacture of usable products, however. More than a century after Leibnitz, in 1820, Charles Xavier Thomas, of Alsace, produced a calculating machine based on many improvements to Leibnitz' concepts. This calculator was put into production for commercial use about 40 years later and perhaps 1,500 were in use by 1880.(7) By the turn of the century, a number of European - primarily German - firms were selling continually improving models of this type.

A key-driven adding machine was patented in 1850 in the United States. Another basic version of the calculator was developed by Frank Stephen Baldwin and. W.T. Odhner in 1875. In 1887, Dorr Eugene Felt patented the "Comptometer." W.S. Burroughs and Felt designed and, after 1890, sold adding machines with paper "listing" tapes. John H. Patterson introduced a cash-register variant of an adding machine in the United States in 1884. A holding company, National Cash Register, was formed in 1906. As the 20th century progressed, names such as Monroe, Sunstrand, Corona, Remington, Swift, Underwood, and Victor became associated with commercial adding machines, mechanical calculators, and billing and accounting devices. There is no evidence that, aside from being among the purchasers and involvement in the issuing of patents, governments were of particular importance in the commercial developments of this industry during the 19th century.

After 1800, and paralleling the development of adding machines, calculators, and cash registers, another related technology was developing.(8) Charles Babbage (1792-1871), a Cambridge mathematician, was impressed by errors he discovered in astronomical tables and by the potential of instructing (programming) machines by the use of punched cards such as those invented for use with the Jacquard textile looms. At the age of 20, Babbage conceived the idea of a "difference engine," which by progressively taking first, second, third, etc., differences in successive numbers until a constant difference was found, could then reverse the process (integrate) and produce the original numbers by successive addition. The potential that this "engine" had for producing accurate tables led to interest by Sir Humphrey Davies and the Royal Society. A development project supported by the British government began in 1823. Nine years later Babbage suspended the work and, in 1842, the government discontinued its support of Babbage's research. In the last nine years of the grant period, Babbage worked on an alternative concept, the "analytical engine." This machine continued the use of difference procedures, but the operations were to be controlled by punched cards of the Jacquard type. Babbage failed to produce an operating device.

The idea of using cards for data input and control was developed further, but no commercial application of devices is apparent until a few decades later. Herman Hollerith (1860-1929) wrote a doctoral dissertation on tabulating machines, worked at the Census Office from 1879 and 1883, and designed and built a tabulator that employed an electrical reading of punched cards in an electrically driven machine. The Hollerith machine was built for and used by the Census Office for the 1890 Population Census.(9) The possibility of using them in business was apparent. Hollerith formed the Tabulating Machine Company in 1896. James Powers, an engineer with the Census Office prior to 1910, invented an automatic card-punching machine and formed the Powers Tabulating Machine Company in 1911. In the same year, the Hollerith firm, by merger, became the Computing-Tabulating-Recording Company. Thomas J. Watson, Sr., having been fired by National Cash Register, took over C-T-R in 1914 and became president in 1915. The name was changed to International Business Machines Corporation in 1924. Powers' company merged with Remington Rand in 1927. The Remington tabulator read cards by mechanical means whereas the IBM machine used an electrical method of reading.(10)

A large number of computing devices in addition to large-scale tabulating equipment, adding machines, cash registers, and desk calculators appeared after 1900. In particular, further improvements in "differential analyzers" were made.(11) Burroughs and National Cash Register provided key-operated analyzers for the British Nautical Almanac Office. Burroughs created another for the U.S. Naval Observatory. In 1929, a Difference Tabulator, made and contributed by IBM, was in use at the Columbia University Statistical Bureau. An improved group of IBM calculators capable of automatic numerical integration of the differential equations of planetary motion was donated to and installed in the Thomas J. Watson Astronomical Laboratory at Columbia in 1933.(12) Vannevar Bush made further improvements in differential analyzers through his research at MIT in the 1920s and 1930s.(13)   The Moore School of the University of Pennsylvania, under contract with the Aberdeen Ballistics Research Laboratory, and loosely in cooperation with MIT, developed and built a differential analyzer for the production of firing tables as early as 1934-1935.(14)

The MIT and Moore School contracts with Aberdeen grew from the improvements in differential analyzers made by Bush. A commercially manufactured MIT machine was delivered to Aberdeen in 1935.   Bush was the leading intellectual as well as the leading entrepreneurial figure in this area of development. Bush was instrumental in the establishment, funding and staffing of the Aberdeen Ballistics Research Laboratory.   John G. Brainerd was the pivotal figure at the Moore School.(15)

In 1937, Howard H. Aiken, then a graduate student at Harvard, prepared a memorandum on digital computation devices.   This work came to the attention of James W. Bryce, an IBM scientist, and through Bryce, to Thomas J. Watson, Sr. Watson funded Aiken's project in 1939, assigning responsibility for design and construction to Clare D. Lake.   The work was at IBM's Endicott, New York, facility.   Aiken was made a Commander, USNR, after the United States entered World War II, and continued his work on the project.   The Automatic Sequence Controlled Calculator (Mark I) was completed in 1944 and donated to Harvard by IBM.(16)   Subsequent models, Mark II, III, and IV (Advanced Digital Electronic Computer) were built for the Navy and Air Force at Aiken's Computation Laboratory after the war.

Also in 1937, Dr. George Stibitz of the Bell Telephone Laboratories began work on the use of telephone relay devices in calculating equipment.   By 1940, Bush had succeeded in having the National Defense Research Committee (and in 1941, the Office of Scientific Research and Development) formed, and this group supported Stibitz' development of the Bell Relay Calculator.   A number of relay-operated computers were built for the various sectors of the armed forces by the Bell Labs.(17)

In the period 1937-1940, the budget for Army Ordnance R&D was roughly doubled, from about $1,000,000 per year to nearly $2,000,000.(18)   The National Research Defense Committee was inaugurated, the Office of Scientific Research and Development was planned, and other branches of the government (e.g., National Bureau of Standards, Office of Naval Research, Bureau of Census, National Advisory Committee for Aeronautics) were interested in improvements in computational capabilities.(19)   These interests, coupled with both the scientific improvements and the exigencies of war, broke the "stalemate . . . between the cost of the advance [in the state of the art] and the value of that advance."(20)   There had been "no fundamental advance made in commercial calculating machines" since the early 1900s.(21)

II. PLAN OF STUDY

It is with this background that the present analysis begins. The role of governmental projects in the subsequent development and diffusion of commercial computing machines is traced in considerable detail.   The narrative,   while   necessarily historical and, indeed, somewhat anecdotal and impressionistic, is fashioned rather informally around several recent theoretical turns in economics.   These are:

1. Theories explaining relationships among exogenous technological changes and market structure, as represented in work by Phillips,(22) Nelson and Winter(23) and, more on the edge of pure theory, by Dasgupta and Stiglitz(24);
2. Theories of internal firm organization and contractual versus internally controlled transactions, as represented in work by Williamson,(25) Klein, Crawford and Alchian,(26) and Phillips.(27)
3. Theories treating product quality, imperfect information, and markets, as represented by work of Akerlof,(28) Salop and Stiglitz,(29) and Cooper and Ross.(30)

It is well known that during the initial ten or fifteen years of commercial sales of computer and computer-related products and services, the technology - however measured - advanced in quantum leaps.   Governmentally supported R&D programs were both antecedent and directly related to first- and second-"generation" commercial machines as well as to significant intragenerational improvements.   The government, that is, created opportunities for new product developments, but always in a context of imperfect information, technological uncertainty, and markets characterized by consumers with varying amounts of sophistication.   The firms, of course, contributed to computer technology as commercial products were developed.   And the firms had highly varying technological and marketing strategies as well as varying attitudes toward risk.   The firms also had remarkably different records of success.   Analysis of the differences in the internal organizations of the firms is helpful in explaining the differences in technological and marketing strategies, differences in attitudes toward risk and differences in the commitments by the firms to achieving success in the computer industry.   The emergence of the industry is a complex story of many interactions and yet one in which, one hopes, a number of important conclusions are suggested.

III. ENIAC, UNIVAC, AND ERA

The big news at the Moore School of the University of Penn-
sylvania in late 1945 was that ENIAC (Electronic Numerical
Integrator and Calculator) actually ran. John W. Mauchly and
J. Presper Eckert, in a project directed by John G.
Brainerd and aided by Herman and Adele Goldstine, Arthur Burks, Carl
Chambers, and many others at the Moore School of the University of Pennsylvania, had succeeded in fulfilling the
requirements of a contract with Aberdeen Ballistics Laboratory.
ENIAC was different from previous differential analyzers, the
electromechanical Mark I and the Bell Relay computers. It
was, like Mark I and the Bell Relay computers, digital.
ENIAC, however, was fully electronic and capable of computing
at speeds several hundred times faster than that of any electromechanical or relay-type machine.(31)

ENIAC provided a great technological impetus to the
computer industry even though the original version had no
feasible commercial applications. The government, in sponsoring the many areas of research leading to ENIAC, did much
more than it had intended. A cadre of engineers and scientists with mutual interests had been developed. Among them
was John von Neumann, who became associated with the ENIAC
project in August 1944. With Mauchly, Eckert, and Herman
Goldstine, von Neumann developed the concept of the "stored-
program" computer, with logic instructions stored in memory so
that they could be modified arithmetically without a manual
resetting of thousands of switches. The Moore School group
had in fact begun design work on a stored program computer,
the EDVAC (Electronic Discrete Variable Automatic Calculator)
during the early stages of the construction of ENIAC.(32)

The cadre of people working on computer development
came from universities, various government departments, and
industry. They had frequent formal and informal contacts
with one another. MIT sponsored a lecture series on computers in October 1945, well prior to ENIAC becoming active.
The MIT group was centered in the Servo-Mechanisms Laboratory, funded by the Office of Naval Research and the Bureau
of Aeronautics. The latter had commissioned MIT to build an
advanced analog computer in 1944.

Perhaps the most significant influence on the yet to be
born industry in the 1945-1946 period came from the six-week
course, "Theory and Techniques for the Design of Electronic
Digital Computers," given at the Moore School in the summer
of 1946. This course was organized by a faculty member, Carl
Chambers, and was sponsored by the Office of Naval Research
and the Army Ordnance Department. Attendees included representatives of the Army, Navy, National Bureau of Standards,
MIT, Columbia, Pennsylvania, Harvard, the Institute for

Advanced Study, Cambridge University, Bell Labs, National Cash Register, and General Electric, among others. The Moore School conducted similar seminars and lectures in subsequent years.

Many aspects of planned, stored-program machines were discussed at the 1946 sessions. Six months later, a four-day conference was organized by Howard Aiken at Harvard, and sponsored by the Navy. There were 350 conferees, and the proceedings were published by the Harvard University Press. In addition to government and academic participants, there were representatives of RCA, Eastman Kodak, Electronic Control Company, Brush Development Company, Northrop Aircraft, Reeves Instrument Corporation, Bell Labs, Raytheon, Prudential Life Insurance Company, John Hancock Mutual Life Insurance Company, General Electric Company, Engineering Research Associates, Bendix Aviation Corporation, Marchant Calculating Machinery Company, Massachusetts Mutual Life Insurance Company, Bausch and Lomb, Western Union, Monsanto Chemical Company, Sylvania Electric, Technicord Records, Hughes Aircraft, Sperry Gyroscope Company, Clinton Laboratories, New England Power and Service Company, Arthur D. Little, Inc., Hydrocarbon Research, Inc., United Aircraft, and others. The press was represented.

IBM sponsored five conferences on computing between 1948 and 1951. Harvard repeated its conference in 1949. The Association of Computing Machinery was formed in 1948 around the "close fraternity" of people from universities, industry, and government. The first issue of the Digital Computer Newsletter, published by the Office of Naval Research, appeared in April 1949. In short, and growing directly from government support for ENIAC and related projects, there was free and open access to not just the technology of the day, but also to the many computer-related R&D projects then underway.

Well-known stored-program computers such as EDVAC (Eckert and Mauchly, Army Ordnance Contract), EDSAC (Wilkes, Cambridge, based on the Moore School course), SEAC (National Bureau of Standards, for the Census Bureau and the Navy) and IAS (von Neumann, Institute for Advanced Studies, RCA Labs, and Army Ordnance) were subsequently developed under at least partial government sponsorship. In addition, other nonprofit organizations were similarly engaged in designing and developing stored-program machines, including the University of Amsterdam, University of California at Berkeley (CALDIC), University of California at Los Angeles (SWAC), University of Frankfurt, Harvard University (Mark III), University of Illinois (ORDVAC, ILLIAC), University of Manchester, University of MIchigan (MIDAC), MIT (Whirlwind), University of Rome, University of Vienna, a Swedish university (Stockholm?), Federal High School (Zurich), RAND

Corporation (JOHNNIAC, after von Neumann), and the Naval Research Laboratory. Funding, it appears, came almost entirely from government budgets.(33)

Commercial firms participated in these activities. From 1944 to 1947, IBM developed and produced a one-of-a-kind Selective Sequence Electronic Calculator (SSEC). IBM also made a few small, special-purpose relay computers for Aberdeen and the Dahlgren Naval Proving Grounds. Standard IBM card punches, card readers, printers, and other products were used as components in many other computers. AT&T, through Bell Labs, produced both large and small relay computers for the Army Ground Forces Board, Fort Bliss, and the Naval Research Laboratory.(34) RCA was the co-developer of the IAS computer at Princeton. Dr. Jan Rajchman, an RCA scientist, is credited with development of a "selectron tube," an advanced electrostatic storage device.(35) Based on the records available, the firms that had at least the substantive technology base include, in addition to IBM, AT&T, and RCA, such firms as Bendix, Boeing, Douglas, Hughes, North American Aviation, Northrop, Raytheon, Sperry, General Electric, Westinghouse, Philco, ITT, GTE, Burroughs, Friden, Monroe, National Cash Register, Remington Rand, Royal, and Underwood.

None of these firms elected to be the first venturer. Eckert and Mauchly, who were dismissed from the University of Pennsylvania in 1946 because of their interests in commercialization of the ENIAC and EDVAC concepts, formed the Electronic Control Company in 1946 and the Eckert-Mauchly Computer Corporation in 1947.(36) Personnel from the Naval Communications Supplementary Activity formed Engineering Research Associates in St. Paul, Minnesota.(37) The Computer Research Corporation was formed by former Northrop personnel. Thomas J. Watson, Sr. offered both Eckert and Mauchly positions and a laboratory under their own management at IBM. They declined, probably because IBM would not assure them that their computers would be marketed. Eckert and Mauchly approached the Bureau of Census, which was known to be interested in a computer. Through the National Bureau of Standards (NBS), Census requested bids and, in addition to Eckert-Mauchly, found interest at Hughes Tool and Raytheon. Hughes did not submit a bid; Raytheon's bid was in excess of that of Eckert-Mauchly. The latter were awarded the Census contract in June 1946, only three months after they had left the University of Pennsylvania.

In 1947, Eckert-Mauchly received funding from A.C. Nielson and Prudential Life Insurance Company, both of which agreements included possibilities for purchases of an EDVAC-type computer called UNIVAC. Henry Straus, a Delaware racetrack owner, and vice president of American Totalizator, supplied half a million dollars of cash and notes in return for

40 percent of Eckert-Mauchly's common stock. Straus was killed in an airplane crash, local financial organizations refused Eckert-Mauchly's request for funds, and the new corporation was clearly destined for bankruptcy by 1949.

Eckert and Mauchly knew virtually everyone at any corporation that had hitherto shown interest in computers through participation in conferences, joint research, and government contracts. They contacted NCR, Remington Rand, IBM, Philco, Burroughs, Hughes Aircraft, and probably others. Remington Rand made an offer for the firm that was accepted in February 1950.

The acquisition of Eckert-Mauchly hardly reflects a confident decision on the part of Remington Rand that UNIVAC was the wave of the future. The first move by Remington Rand was, in fact, an attempt to cancel all UNIVAC contracts. The Census Bureau refused to cancel, but Prudential and Nielson did cancel after a year of unfruitful efforts at renegotiation.(38) The Census UNIVAC I was delivered in 1951, followed by sales of five more of the same machine to the Atomic Energy Commission, Air Force, Army, and the Navy Bureau of Ships.(39) Commercial deliveries of UNIVAC I commenced in 1954 with a sale to General Electric.

Government projects led to another early effort at commercial sales. The Engineering Research Associates (ERA) group, which included William Norris, started with a Navy contract for "special purpose," "highly classified" computing machinery and related work. This was almost immediately augmented by a Navy contract for what was called ATLAS I, with the understanding that variants of ATLAS I might be sold commercially. The ATLAS I was renamed ERA 1101, and the 1101 was followed by ERA 1102 and 1103. ERA was a financial failure and it too merged with Remington Rand in 1952. The ERA 1103 became the UNIVAC 1103, but deliveries did not begin until 1953.

Mauchly felt that the refusals by banks and other institutional providers of capital to lend financial support to Eckert-Mauchly was the source of much of the Eckert-Mauchly difficulties. Others have attributed the refusals to conservatism and bias on the part of Philadelphia banks toward traditional industrial and commercial borrowers.(40) After Remington Rand had acquired Eckert-Mauchly, it sought to withdraw from the commercial aspects of the venture rather than to commit heavy resources. William Norris, along with Henry Forrest, a co-venturer in ERA, felt that Remington Rand "faltered," and failed to make the "financial commitment that was necessary". Mauchly concurred in this opinion.(41)

Such views, while understandable, must be seen in a more complete context. None of the other companies, it should be recalled, would agree to a commercial venture, with or without Eckert and Mauchly, based on the then existing technologies. Many had the necessary financial resources and the requisite technology base. They, along with the banks and Remington Rand, uniformly opted not to devote resources to immediate commercialization. There are a number of reasons for private funds not being committed to commercialization at this time. First, except for Mauchly, Eckert, and some in the ERA organization, the general view prior to 1950 was that there was no commercial demand for computers. Thomas J. Watson, Sr., with experience dating from at least 1928, was perhaps as acquainted with both business needs and the capabilities of advanced computation devices as any business leader. He felt that the one SSEC machine - which was on display and in operation at IBM's corporate headquarters in New York City - "could solve all the important scientific problems in the world involving scientific calculations."(42) Watson saw only limited commercial possibilities.(43) This view, moreover, persisted even though some private firms that were potential users of computers - the major life insurance companies, telecommunications providers, aircraft manufacturers, and others - were reasonably informed about the emerging technology. A broad-based business need was simply not apparent.

Second, except for experimental and special-purpose relay and electromechanical computers such as the Bell Labs Model III, IV and V, the Harvard Mark II, the BINAC, and SEAC, none of the more advanced machines on which work had begun in the post-ENIAC period was in operation at the time. Even the ERA 1101 and UNIVAC I were not yet deliverable, although work had begun on them three or four years earlier. Computers were highly experimental and built individually on a "job shop" basis. Thus, what development there was was largely exploratory, with the main emphasis on technological advance, not on economical quantity production. No one foresaw profitable large-scale production.

Third, the experimental technologies were diverse and changing very rapidly. Exploratory development work was going on in many locations with great variation in the componentry, circuitry, performance, and foreseen uses. The unusually full interchange of information among the various researchers at the time reflects not a mystifying unawareness of appropriability problems, but rather the lack of any material aspect of the technology that had an economic value that might be appropriated. There was not a recognized, "main-stream" technology on which to focus privately financed R&D investment.

The perceived demand for computers up to 1950 was about what it had been in 1945. As private firms saw it, the only demand was from government agencies such as the Bureau of the Census, NACA, the Ballistics Research Laboratory, the Naval Proving Grounds, the Coast and Geodetic Survey, the

Weather Bureau, the White Sands Missile Range, and the like. These buyers participated in development activities and paid for most of the development costs. A number of firms invested modestly in information (research) relating to computer technology but, again with the Eckert-Mauchly and ERA exceptions, none invested heavily in physical and human capital with a defined commercialization objective in mind. The technology and the market had yet to merge in a significant way.

In a real sense, the technologist users (in government) and the technologist suppliers (in private firms) had coincident interests and were members of a cognizable "fraternity." They attempted to prevail on their respective host organizations – whether government agencies or private firms – to supply funds to meet their technological and scientific objectives. The demand, that is, was more in the form of budget requests for funds for investment in R&D - without regard to immediate economic returns on investment - than it was a demand for marketable computer hardware.(44)

IV. OPPORTUNITY AND ENTRY: THE FIRST GENERATION

The Remington Rand UNIVAC I, produced by the Eckert-Mauchly division, and the ERA 1101 were the first announced commercial computers. The UNIVAC was based on the Census order; the ERA, on the ATLAS I and related Navy work. Only three 1101s were sold and, over its entire product cycle, only 40 UNIVAC Is were installed. Nonetheless, Remington Rand was generally regarded as "the leading company" in the EDP industry in the early 1950s. The company was thought to have an "initial year to two years lead . . . by having a machine that was available and operational before other machines began to appear." The UNIVAC name became prominent enough so that it was for a time the generic term for a computer.(45)

The early marketing of the UNIVAC hardly meant that other companies were not exploring computer developments and possible entry. The attendance at computer conferences itself belies that conclusion. Government-sponsored computer research and procurement had put a number of other firms into a technological position similar to Eckert-Mauchly and ERA in the late 1940s and early 1950s. AT&T, which had completed a large-scale, electromechanical computer under Stibitz's guidance at Bell Labs, was doing large amounts of research in electronics and supplied several relay computers to the government between 1943 and 1947. AT&T, in connection with settlement of a government antitrust case, elected not to develop and market electronic computers, but instead focused on use of the same technology in telecommunications applications.

Raytheon, another of the companies following ENIAC and EDVAC developments closely, was awarded a contract to produce a computer by the Bureau of Standards on behalf of the Office of Naval Research in 1947. The computer became the RAYDAC (Raytheon Digital Automatic Computer) and was delivered to ONR in 1951. In the same period, Raytheon also produced other computers for various classified government uses. The company was regarded as "one of the prime centers of technological development [in the early 1950s] and probably [a] leader roughly parallel with the UNIVAC operation in terms of scope of competence."(46)

Despite the leading technological edge that government computer contracts and outside associations provided Raytheon, the company did not market a commercial computer. A RAYCOM computer, developed from RAYDAC, was announced for commercial sales in 1949 or early 1950, but these plans did not materialize. Raytheon saw itself as "primarily a government funded corporation" that "did not attack commercial activities in other fields very effectively." A commercial venture would require "funding from the [corporate] exchequer" in contrast to funding by government contracts.(47)

In 1955 - by which time several other companies were in the market and technology had advanced - Raytheon formed a joint venture, the Datamatic Corporation, with Minneapolis-Honeywell Regulator Company. The idea was to use the RAYCOM technology to produce and market large-scale systems. The Datamatic-1000 was first installed in late 1957 with a capacity for 481 scientific or 1,455 commercial operations per second. Only 8 or 10 D-1000s were sold, largely for straight accounting work. Raytheon withdrew from Datamatic in 1957. It continued, however, as an extremely competent developer and manufacturer of computers for the government.(48)

RCA was another company that could have made a "first move" into the commercial field. Studies of electronic computing devices had begun at RCA "as early as 1935." Government support was very important. RCA developed and delivered electronic systems for antiaircraft fire control in the early 1940s. It produced a computer, the Typhoon, for the Navy in 1947. By 1950, exploratory research was done in relation to a commercial application. Like Raytheon, RCA devoted most of its activities to classified government computer projects in these early years. RCA worked on tube development of ENIAC and other computers and began research on core memory and transistors for computer use in 1952.(49)

The BIZMAC, RCA's first commercial machine, was developed under contract with the Army. Its purpose was "stock control of replacement parts for military combat and transport vehicles." Only six BIZMAC's were shipped, beginning in late

1955. These had speeds of only 286 scientific or 968 commercial computations per second. The company acknowledged that its "major obstacle" was its own "doubts as to RCA's seriousness in the EDP business." Resources were allocated to color television, not to EDP and computers.(50)

General Electric was also a beneficiary of the technology spawned by ENIAC, EDVAC, the 1947-1950 computer conferences, and government and business contractual work. Like many other companies, it restricted its first development and manufacturing efforts to specialized systems for ordnance and military applications. The 1953 OARAC (Office of Air Research Automatic Computer) was one of these. The ERMA (Electronic Recording Method of Accounting), announced in 1956, was the first commercially available GE machine. Consonant with the risk-minimizing policy inherent with government contracting, the ERMA was developed under a $60 million contract with the Bank of America for use in check handling. Under this contract, 30 ERMAs were delivered, but GE "failed to capitalize" even on its lead in EDP applications in the banking industry.(51)

In contrast to GE, the small Consolidated Engineering Corporation set up the Electrodata Corporation in the early 1950s to develop and market the CEC 202/203. Electrodata introduced its Datatron 203/204 in June 1954, with marketing headed by a former IBM executive. This effort was based, however, on a contract with the Jet Propulsion Laboratory, which in turn had government contract support. JPL received the first machine. Six additional Datatrons went to U.S Naval Ordnance, Allstate, Socony-Vacuum, American Bosh Arma Corporation, Land-Air, Inc., and Purdue University. With an advanced Datatron 205, Electrodata had 24 installed computers and 19 unfilled orders by March 1956. It had unanticipated success in its sales. A few months later, Electrodata was acquired by Burroughs.(52)

For its part, Burroughs had begun electronic computer research in 1947. Representatives of the company were attendees at the computer conferences and Burroughs, under contract, upgraded the ENIAC by supplying a new "static magnetic memory" from its Philadelphia Research Center. Yet Burroughs was cautious in its own attempts to sell computers commercially. As late as 1953, Burroughs opined that, "[I]n business the arithmetic is usually not difficult. . . . It would be of no advantage to speed up the rate of figuring, if input, output and other peripheral operations did not keep pace. . . . [There is] the major obstacle of cost. The outlook for electronics in business, then, must be summed up in the words 'not yet'."(53)

Burroughs did nonetheless produce one UDEC (Unitized Digital Electronic Computer) for Wayne State University and upgraded a UDEC to a UDEC II in 1955. The speeds of UDEC

II were roughly those of ENIAC. Contemporaneously, and consistent with its view of the limited commercial usefulness of computers, the Burroughs E-101 was announced in 1954 for scientific and business applications. Through this period Burroughs was developing computers under defense contracts and indeed, "began to seek out defense contracts for which its facilities and capabilities were best suited and which had the greatest potential for commercial systems development." The major stimulus for commercial interest at Burroughs was their "receipt of government contracts involving precision computational and data processing equipment in the area of fire control, navigation, anti-aircraft battery evaluation, and ultimately, the guidance computer for the Atlas ballistic missile and the data processing for the SAGE intercontinental air defense network."(54)

The acquisition of Electrodata in 1956 signified the beginning of Burrough's serious effort in the commercial market. Production of the Datatron 220 began in 1957, with delivery scheduled for December 1958. Unhappily, the 220 was a vacuum-tube computer, slow in comparison to the transistorized models then nearing delivery. The unsuccessful introduction of the 220 caused Burrough's effective, if temporary, withdrawal from the market nearly simultaneously with its first serious entry. That the 220 was a vacuum-tube computer is especially significant in view of the fact that Burroughs used transistors in the Atlas guidance computer delivered to the Air Force in April 1955. The Atlas is said by some to be the first operational transistorized computer.

National Cash Register began experiments in electronics in the late 1930s and was also among those attending the Mauchly-Eckert-Von Neumann-Chambers-Aiken computer conferences. NCR performed classified electronics work for the government during World War II and, between 1945 and 1952, produced a "giant" electromechanical brain for bombing navigational purposes under government contract. As discussed below, NCR held patents on an electronic calculator that used thyratrons rather than vacuum tubes as electronic counters.

NCR entered the general-purpose computer field in 1953 through its acquisition of Computer Research Corporation, an offshoot of defense work done by Northrop Aircraft. Northrop had contracted with Eckert-Mauchly in October 1947, for the small-scale BINAC (Binary Automatic Computer). This development was in conjunction with the Air Force Snark missile project at Northrop. Northrop also made in-house efforts to develop DIDA (Digital Differential Analyzer) and MADDIDA (Magnetic Drum Digital Differential Analyzer). The latter was completed in 1949, but Northrop management responded negatively to the requests of the designers that a commercial version be developed. Northrop had just lost the "Flying Wing" contract and had experienced recent losses in other high-risk ventures.

The MADDIDA designers left Northrop, founded CRC, and obtained a contract to build a computer for North American Aviation's Navajo missile project. The commercial version of this computer was designated the CRC 101. CRC also had computer contracts with Lincoln Labs and the Air Force prior to acquisition by NCR. Ironically, Northrop funded a project for commercial computer development after the CRC founders had left, but this work never resulted in a commercial product.(55)

The CRC 102D was introduced by NCR in late 1953. This machine and the improved NCR 107 were available for delivery shortly thereafter. The 107 resembled the ENIAC in terms of computational speeds. NCR then developed a 303 that again failed to be sold because of its inferior performance. The NCR 304 was announced in 1957 for delivery in late 1959. It was called the "first all-solid state system" and was built by GE using transistorized circuits developed and produced by GE. NCR subsequently marketed a 310 computer, but it was basically the CDC 160 and was, in fact, produced by Control Data Corporation. The NCR 390 and NCR 315 of 1960 were really the first of NCR's own products in the market.(56)

Philco did not attempt entry into the commercial computer area until the mid-1950s, and did so on the basis of government contracts to develop and produce a "surface barrier transistor." From these, a contract was given for a transistorized airborne computer, the C-1000, for the Air Force. Philco then contracted to produce a follow-on, large transistorized computer for the National Security Agency. Philco modified and introduced this computer commercially in 1958 as the TRANSAC S-2000-210. This computer was also called the "first large-scale transistorized EDP system."(57) Follow-on 2000-211 and 2000-212 machines were announced, and early customers included the AEC, GE,(58) State of California, United Aircraft, Chrysler, Systems Development Corporation, Ampex, State of Israel, University of Wyoming, and the Defense Communications Agency. Core memory for the 2000s came from Ampex and some software was supplied by Applied Data Research. Philco lacked the sophisticated peripheral hardware - disc drives, tape drives, printers - as well as the sales and technical maintenance support necessary for large market penetration by the 2000 series. Ford Motor Company acquired Philco in December 1961, with the objective of getting into the space and defense sectors of computer application.(59)

As noted, Thomas J. Watson, Sr., of IBM, had contributed a differential tabulator to Columbia University in 1928. In 1933, he funded Columbia's Astronomical Computation Laboratory, including a gift of specially adapted punch-card equipment. It was Watson who arranged for IBM's design and construction work on Aiken's Mark I. The IBM SSEC was developed in the 1944-1947 period, and IBM provided punched-card input/output devices for the ENIAC. IBM personnel were active in the various conferences and IBM was the sponsor for some of these.

In 1949, IBM established an Applied Science group to perform exploratory research in possible business applications of the new technology. The head of the group, Cuthbert Hurd, was part of "upper management," but was not given to believe that an early mission was to produce a computer. Still, the Applied Science group was separate from the Pure Science division, and Hurd was a computer specialist. There was much opposition within IBM to the "long hair," "double dome" electronics scientists. Outsiders, especially Hurd's scientist colleagues, doubted that IBM would ever produce a computer.(60)

The decision by IBM to enter the industry was largely a result of the Korean War. Watson, Sr. had sent a telegram to President Truman pledging IBM's services for "whatever was needed."(61) Hurd and Thomas J. Watson, Jr., who had been following computer development since the 1946 Moore School meetings, argued that government agencies clearly needed improved computational and data-processing abilities in the war effort and, less persuasively, that business had similar requirements. In turn, they convinced a number of government agencies and contractors that they did, indeed, need better computers. Based on 30 letters of intent, all from government agencies and the defense-related work of private companies, the "Defense Calculator" project was approved in early 1951. The name itself was chosen to avoid an "IBM" label in order to minimize risk to the company's image in the business-machine field. When the prospective rental price of the computer was changed in March 1951, from $8,000 to $15,000 per month, all but six of the letters of intent were withdrawn.(62)

IBM elected to continue with the Defense Calculator project at this juncture, assuming the risk that the computers could be leased. It was decided to produce 19 machines, all alike, and their production was on an assembly-line basis rather than on a job-shop, custom-made basis. The computer was renamed the IBM 701 and was announced in May 1952, with first customer installation about a year later. The 701 was capable of 993 scientific or 616 commercial operations per second, and was produced and delivered at a rate of one per month. The 701, unlike others then in development, was produced in modules that lowered production costs and made delivery and installation quite easy. In its initial design, the 701 could not handle alphabetic characters. To the market, the 701 was an "IBM UNIVAC."

Once committed to the computer market, IBM continued immediately to improve its products and its related marketing efforts. In late Fall, 1952, and prior to first delivery of the

701, the Applied Science group proposed the IBM 650. The 650 was one of a number of designs IBM had under consideration at the time. There were only six 701 orders in hand, and the Sales and Product Planning Department forecast net sales of zero for the 650. Every 650 sold would just replace possible 701 sales in their view. Applied Science, on the other hand, forecast sales of 200 650s, mostly for scientific and engineering use.(63) After a heated internal debate resolved by a decision of Thomas J. Watson, Jr., the 650 was announced in July 1953, and first delivery was in December 1954. In the end about 1,800 IBM 650s were produced, mostly for business applications.

The IBM 650 was not the fastest of machines. It could originally output only 111 scientific or 291 commercial operations per second. The 650, however, was very flexible in its uses, carried a relatively low price, was reliable, was easy to install and maintain and, over time, was upgraded by alphabetic capability, an excellent printer, tape drives, the RAMAC disc drive, and the SOAP (Symbolic Optimal Assembly Program) assemblers for programming. The 650 was the "Model T" of computers. It was the 650 that changed IBM's image from a producer of "IBM UNIVACs" to the leader in the industry.

In September 1953, IBM announced the 702 for delivery in 1955. IBM also introduced the more specialized 610 in 1954. The IBM 705 I and II were introduced in March 1956. The 704 and a 701 with magnetic core memory came on the market the same year. The last of the nontransistorized IBM models, the 705 III was announced in September 1957. The 7070, IBM's first fully transistorized computer, was announced in September 1958.

V. SOME PRELIMINARY FINDINGS

Table 4.1 indicates the principal offerings of computers during the first generation, running from the ERA 1101 of December 1950, to the Burroughs 220 of December 1958. During this period, the attitude of a number of private firms toward investment in the technologies changed radically. Up to 1950, there were what can be termed "homogenous expectations" among the firms that a profitable commercial market did not then or foreseeably exist. While there are no data available to substantiate the fact, it is nonetheless obvious that privately financed R&D aimed at developing products for commercial sales was very low. By the mid-1950s, this was less obviously so. A number of firms – almost all of which were building directly or indirectly on governmentally financed R&D – were committing private funds and seeking to develop successful commercial products.

Table 4.2 indicates this change.   Both Tables 4.1 and 4.2 indicate, however, the high risks inherent in entering a commercial market with rapidly changing technological opportunities.   The earliest firms to attempt entry, Eckert-Mauchly and ERA (1950-1951) failed.   Among the last of the new, large, first-generation computers were the IBM 709, the Sperry Rand 1105, and the Burroughs 220.   ENIAC had been capable of 44.7 commercial computations per second, estimated by Knight to be the equivalent of 1.4 thousand operations per dollar.(64)   UNIVAC I yielded 271.4 operations per second and 6.8 thousand operations per dollar.   In contrast, the 709, 1105 and 220 computed at rates of 10,230, 5,527 and 1,616 per second, respectively, and yielded 90.8 thousand, 80.1 thousand and 129.2 thousand operations per dollar.   These late-model computers, despite their being 10 to 20 times as powerful in terms of operations per dollar as those introduced a few years earlier, were also commercial failures.

The number of companies involved in commercial sales of the principal first-generation computers was not large.   Only seven firms appear, although there are undoubtedly many others that carried on R&D for the purpose of offering a computer only to withdraw prior to attempting sales.   There were, moreover, well over 100 other one-of-a-kind and other machines completed on an experimental or contract basis for special uses.

Table 4.3 indicates that the computers that gained the greatest market success had excellent performance in terms of both speed and operating costs.   After 1954, IBM appears most frequently as the company with systems having the fastest computing time and the lowest computing costs.   Thus, as was found in the case of commercial aircraft, performance superiority appeared to be something of a necessary - though clearly not a sufficient - condition for market success in the first generation.(65)

VI.   GOVERNMENT INFLUENCES:
CORE MEMORY IN THE LATE FIRST-GENERATION AND THE SHIFT TO TRANSISTORIZED COMPUTERS

SAGE

The Servomechanism Laboratory at MIT produced one of the first computers, Whirlwind, with work beginning prior to the operation of ENIAC.   MIT also sponsored early computer conferences.

Project Whirlwind was initially commissioned to design a real-time flight simulator to teach pilots to interact with their craft and to reduce the expenses involved in building "fly and try" prototypes of alternatively designed planes.   The Aircraft

Table 4.1. Principal First-Generation Computers
Offered by Major Manufacturers, with Operating
Characteristics and Selected Production Information

Company and Year First Installed	Model	Commercial Operations Per Second	Thousands of Commercial Operations Per Dollar	Number Installed
Burroughs				
1950	–			
1951	–			
1952	–			
1953	–			
1954	204–205[a]	187.3	14.5	$\geq$ 43
1955	UDEC II	10.7	0.9	n.a.
	E-101	2.3	1.3	n.a.
1956	–			.
1957	UDEC III	20.9	1.5	n.a.
1958	220	1616.0	129.2	n.a.
1959	E-103	2.3	1.3	n.a.
General Electric				
1950	–			
1951	–			
1952	–			
1953	–			
1954	–			
1955	–			
1956	ERMA	n.a.	n.a.	$\geq$ 30
1958	–			
1959	–			
Honeywell				
1950	–			
1951	–			
1952	–			
1953	–			
1954	–			
1955	–			
1956	–			
1957	DATAMATIC 1000	1455.0	19.6	n.a.
IBM				
1950	–			
1951	–			
1952	–			
1953	701	615.7	11.3	$\geq$ 19
1954	650	291.1	45.4	1800

continued

Table 4.1. Principal First-Generation Computers
Offered by Major Manufacturers, with Operating
Characteristics and Selected Production Information
(continued)

Company and Year First Installed	Model	Commercial Operations Per Second	Thousands of Commercial Operations Per Dollar	Number Installed
IBM (continued)				
1955	702	1063.0	22.1	≥ 13
1956	705 I-II	2087.0	27.7	≥ 31
"	704	3785.0	49.9	≥ 90
"	701 (CORE)	1807.0	32.2	n.a.
"	305	96.5	15.7	≥ 1,500
1958	709	10230.0	90.8	≥ 32
1959	705 III	7473.0	99.2	≥ 111
NCR				
1950	-			
1951	-			
1952	-			
1953	102A[b]	8.4	1.0	n.a.
"	107[b]	34.4	8.8	n.a.
1954	303	8.3	1.0	0
1955	-			
1956	-			
1957	-			
1958	-			
1959				
RCA				
1950	-			
1951	-			
1952	-			
1953	-			
1954	-			
1955	BIZMAC I & II	967.9	5.5	6
Remington Rand				
1950	1101[c]	301.8	15.4	3
1951	UNIVAC I	271.4	6.8	≥ 40
1952	-			
1953	1103[d]	666.2	18.9	20
"	1102[d]	240.0	12.2	3
1954	UNIVAC 60/120	1.5	0.5	n.a.
1956	1103A	1460.0	28.5	n.a.
1957	File 0	73.2	3.0	100
"	UNIVAC II	2363.0	52.6	n.a.
1958	File I	92.0	3.8	n.a.
"	1105	5527.0	80.1	n.a.

[a]Developed and produced by Consolidated Engineering Corporation. Burroughs had produced a Lab Calculator in 1951.

[b]Developed by Computer Research Corporation. 1953 acquisition by NCR.

[c]Developed and sold by ERA prior to 1952 acquisition by Remington Rand.

[d]Developed by ERA prior to 1952 acquisition by Remington Rand.

Source: Kenneth E. Knight, "Changes in Computer Performance," Datamation (September 1966), pp. 45-46.

Table 4.2. Number of New Commercially Available,
First-Generation Computers and Numbers of
Companies Offering Such Computers, 1950–1959

Year	Number of New Computers	Number of Companies
1950	1	1
1951	1	0
1952	0	0
1953	5	3
1954	4	4
1955	4	3
1956	6	3
1957	4	3
1958	4	3
1959	2	2
Total	31	7*

*Number of companies included, 1950–1959.

Source: Table 4.1.

Table 4.3. Computers with Fastest Computation Times
and Lowest Cost Per Computation, 1950–1959

Year	Fastest Computation Time	Lowest cost per Computation
1950	ERA 1101	ERA 1101
1951	ERA 1101	ERA 1101
1952	ERA 1101	ERA 1101
1953	Remington Rand 1103	Remington Rand 1103
1954	Remington Rand 1103	IBM 650
1955	IBM 702	IBM 702
1956	IBM 704	IBM 704
1957	IBM 705 II	IBM 704
1958	IBM 709	Burroughs 220
1959	IBM 709	Burroughs 220

Source: Table 4.1.

Stability and Control Analyzer (ASCA), for which the Bureau of Aeronautics contracted with MIT in November 1944, was to be an advanced analog computer.(66) Jay W. Forrester, an electrical engineer, was chosen to head the project. The Servomechanism Laboratory had been set up in 1940 by Forrester and Gordon S. Brown and had conducted work mainly in the areas of fire control and radar systems.

A 1945 agreement between MIT and the Special Devices division of the Bureau of Aeronautics specified a budget of $875,000 for the 18 months needed for project completion.(67) The ASCA was, however, never completed. Forrester became rather discontented with the analog orientation of the machine and sought information from the Moore School group about digital circuits. The ASCA was changed to digital early in the program.(68)

The project name was changed in 1946 from ASCA to Whirlwind. Electrostatic tubes developed at MIT and produced in Sylvania's Digital Computer Laboratory were chosen for storage, although there was at least some exploration of neon (i.e., cold cathode trionode) technology. The joint MIT-Sylvania effort resulted in the 7AK7 tube, the first tube designed expressly for computers. When completed in March 1951, the Whirlwind I was far superior to ENIAC in computation times, but not nearly equivalent to the ERA 1101.(69) Without an improvement, and even with its real-time capability, the Whirlwind would not meet its targeted performance.

It was during this period that Forrester arrived at the idea of using magnetic cores for storage. His invention and subsequent use of magnetic core memory in a "Memory Test Computer" and, in August 1953, in Whirlwind II changed the character of first-generation machines. It is necessary to look backward again in time to trace this effect.

The ASCA project agreement was of the expansive type prevalent during World War II. In 1946 the research and development organization within the Navy Department was shuffled and resulted in the creation of the Office of Naval Research (ONR). The Mathematics branch of the ONR took supervisory control over the ASCA project. ONR objected to what seemed to be excessive ASCA requirements for funding. "The showdown came when MIT requested $1,831,583 for the 15 month period between July 1, 1948, and September 30, 1949. This request was doubled ONR's proposed allocation."(70) Two million dollars had already been spent.

At this junction, support for Whirlwind was found in the Air Force. The Air Force wanted to create a continental air defense system and tapped an MIT faculty member to aid in structuring the problem. A colleague told him of Project Whirlwind which by then had an advanced-design computer capable of the real-time application that would be needed. In 1950, Whirlwind became a prototype and test facility for SAGE at the newly established Lincoln Laboratory.

The SAGE system, a cooperative effort between the United States and Canada, was designed to interpret radar information. If alien aircraft were detected, the system was to select the appropriate interceptor aircraft and determine antiaircraft missile trajectories. The system had to store and process large amounts of information. It also required the coordination of several computers in real-time calculations. MIT recognized the enormous complexity of SAGE. The Whirlwind of 1950-1951 would have to be modified to become "a reliable, repeatable, practical design" with an objective to "manufacture, install and maintain several dozens of the systems - systems of unprecedented complexity which employed heretofore unproven technologies."(71) In 1952, the Air Force authorized MIT to solicit proposals from a number of companies to aid in design and in implementation. Serious discussions were undertaken with RCA, Raytheon, Remington Rand, Sylvania, and IBM.

IBM had never been involved in so large and complex a project.(72) The same must have been true of the other companies. Senior IBM personnel became embroiled in internal debate, including "a day long meeting chaired by Mr. Watson, Sr., in the Board Room which resulted in no progress whatsoever toward a decision."(73) The IBM Defense Calculator (the 701) was not yet completed and SAGE was enormously more complicated and risky. IBM nonetheless submitted a proposal.

In October 1952, MIT selected IBM to work with Lincoln Labs in the design of SAGE. In April 1953, IBM received a prime contract from the Air Force for more detailed design. A contract was awarded to IBM in September 1953, "to design, fabricate, support and maintain two prototype computers for the SAGE system."(74) The RAND Corporation (and a spinoff, Systems Development Corporation) was responsible for applications programming. Burroughs was responsible for modifying the radar waves into digital signals and designing the grid patterns, the operators' consoles, and the display units. Site construction was provided by Western Electric.

There was obvious risk to IBM in becoming involved in SAGE despite the government funding:

> Many of the concepts had been tried only in a laboratory. There was no guarantee IBM could hire the numbers of people that would be needed to carry out its responsibilities. Failure to deliver the computers successfully, because the project was so massive, could have led to adverse financial repercussions and damage to IBM's reputation. . . . a mistake in computation might result in accidental destruction of one of our country's own airplanes, with the resultant financial exposure and publicity such an accident might entail. All of us were

concerned in 1953 about diversion of key engineering
and systems persons and Applied Science persons
who were barely completing the design of the 650,
701 and 702. Moreover, IBM would need to construct a completely new factory to build the SAGE
computers and all of us in the highest management
group wondered what would happen if the contract
were cancelled in midstream.(75)

SAGE was successful. It was one of the first computers
to have random-access magnetic core memory. IBM's participation in SAGE "led to reduced manufacturing costs, improved
reliability and serviceability and reduced size and power
requirements" for their computers. IBM innovations made in
connection with SAGE include:

1. Techniques to manufacture core memory rapidly, inexpen-
   sively, and reliably.
2. Computer-to-computer telecommunications.
3. Real-time simultaneous use by many operators.
4. Keyboard terminals for man-machine interaction.
5. Simultaneous use of two processors to improve reliability
   and serviceability.
6. Ability to devolve certain functions to remote locations
   without interfering with the dual processors.
7. Use of display options independently of dual processors.
8. Construction consisting exclusively of printed circuit
   boards.
9. Inclusion of an interrupt system, diagnostic programming,
   and maintenance warning techniques.
10. Associative memory development.(76)

It is uncontested that "the experience which IBM gained
from its work on the SAGE system was significant to the future success of the company.(77) Why MIT choose IBM over
the other companies is not easily discernible from available
records. An important reason may be that IBM had by that
time elected to produce the 701 on an assembly-line basis. In
any case, IBM received the contract and built on it. The 704
and 705 computers were announced in 1954. While some 701s
were then installed, delivery of the 702s had not begun. Both
the 704 and 705 used SAGE-related developments. In particular, the 704 and 705 used core memory in place of tube memory. The 701 was later redesigned to provide core memory.
Deliveries of the 705, 704, and 701 (core) began in 1956. The
704 was regarded as a "major technological improvement" and a
"creative masterpiece."(78) It was accompanied by the FOR-
TRAN programming language and, as Tables 4.1 and 4.3 show,
the 704 and 705 had a major market impact.

The 305 RAMAC (Random Access Method of Accounting
and Control) of 1956 introduced the concept of the disc drive.
The access time of the disc drive in the 305 was 200 times

faster than that of the tape drives then available.(79)   The
709 was announced in early 1957 and delivered to customers in
1958.   While generally compatible with the 704, the 709 was
again four times faster than the 704 and provided many other
technological and user improvements.   As was true of the
other post-1958 vacuum-tube machines, it quickly became apparent that the 709 would not be commercially successful
because of the advent of large-scale transistorized computers.

LARC

In 1954, the University of California Radiation Laboratory
(UCRL) (now the Lawrence Livermore Laboratory) requested
proposals for a computer to be some 100 times faster than the
UNIVAC I they were using.   The call for proposals went to
IBM, Remington Rand, and others.   Within Remington Rand,
the Philadelphia (Eckert-Mauchly) group was unaware of the
existence of the proposal until six months after it had been
sent.   The St. Paul (ERA) group received the request and
began to prepare a response on its own.   Eckert was infuri-
ated.   He wanted his group to respond to the proposal.   Beyond the usual rivalries, he was particularly interested in this
contract because "he thought that the company had to develop
solid-state technologies for the next commercial large scale
systems following the UNIVAC I computer."(80)   The UNIVAC
II had not been announced.   Improvements in the solid-state
magnetic   amplifier   technology   (officially   referred   to   by
Remington Rand as FERRACTOR™) that had been developed in
the early 1950s in an effort to improve computer speed and
reliability seemed to offer a promising route.

The Eckert group won the right to make the sole Remington Rand proposal for the LARC (Livermore Automatic Research Computer).   The proposal was presented in April 1955,
following the IBM presentation, and Sperry Rand was given
the contract in September.(81)   Eckert did not conceive of
LARC as a wholly new computer.   The Air Force Cambridge
Research Center Computer (AFCRC), for which a contract had
been signed between the Air Force and Remington Rand in
1954, was the first completed Remington Rand Computer to use
solid-state magnetic amplifier (FERRACTOR) technology.   It
had been tested at Remington Rand's Norwalk Lab earlier that
year.(82)   When the LARC contract was signed, it was anticipated that the AFCRC computer would be a small processor for
the new computer and that a "coil-gating" technique to improve
the speed of the FERRACTORS would suffice.   The idea was to
employ some transistors of an early type but to rely mainly on
magnetic amplifiers.   The available transistors were regarded
as expensive and not very reliable.(83)

The final LARC specifications were established in March
1956, and completion was planned for February 1958.   After

the starting date but before the specifications were frozen, it became clear that no improvements in the FERRACTORS would be sufficient to obtain the required speed. Attempts were made to use the medium-speed transistors that were commercially available but even those were too slow. Toward the end of 1955 Herman Lukoff, the chief engineer for LARC, recalled that

> We started hearing rumors about Philco's development of a new high speed transistor, something called a surface barrier transistor (SBT). . . . A visit to MIT was promising. I was introduced to a young engineer by the name of Ken Olsen (now president of Digital Equipment Corporation) who had obtained some of the new surface barrier transistors from Philco and was using them in lab experiments. He verified that the transistors were fast, ten to thirty times faster than contemporary transistors. Philco called them 30 MHz units.(84)

The final specifications replaced the magnetic amplifier logic with surface barrier transistors. Also, use of the AFCRC computer as an input-output processor was eliminated.

When the contract was signed, it was recognized that the $2.85 million contract price might not cover the entire cost. Eckert, however, was convinced that improved solid-state components were mandatory to all future Univac computers and that any cost overruns would be justified by subsequent sales improvements. In his opinion, the LARC contract would help fund much of the necessary work on new technology. By the time the specifications were frozen, the cost of the LARC was projected to be at least double the agreed-upon price. To spread the development costs, Univac contracted for the construction of a second LARC for the David Taylor Model Basin Project.

LARC was designed to be a modular system that could be expanded by the addition of peripheral equipment. In terms of its own component construction, however, it was not modular. Noise and dense circuit-packaging difficulties caused pressing problems in the LARC construction. Computers were used to aid in the packaging design. These did not solve all of the difficulties:

> Prior to fabrication, several engineers resigned because they believed that it would be impossible to wire the backboard and they didn't want to be associated with a failure. Fortunately, the wiremen weren't aware of the fact that it couldn't be done, so they went right ahead and completed the work.(85)

Memory development was also a problem in LARC. The memory needed a switch capable of rapid handling of heavy currents, but the Yourke current switch had not yet been invented. The LARC memory had to rely on a more costly solution involving high-current, slow transistors and special diodes that were developed exclusively for LARC. Also, magnetic cores with the appropriate cycle time were unavailable, causing Univac to develop and manufacture unique cores.(86)

LARC was delivered in 1960. The following year the second LARC was delivered to the David Taylor Model Basin (now the Naval Ship Research Development Center at Carderock, Maryland). All specifications were met by the LARC, but the 1958 delivery date was obviously missed. The total cost of LARC has been estimated at close to $19 million, rather than the $2.85 million originally set.(87)

The possibility of Sperry Rand's marketing of LARCs was discussed in late 1957. By then, UNIVAC II was being delivered and plans were underway for the File Computer and the UNIVAC 80/90 Solid State Computer. It was estimated that only eight to ten other LARCs could be sold even though LARC turned out to be faster than the IBM 7090, the Philco 2000-210, and nearly as fast as the Philco 2000-211 transistorized computer.(88) The sales campaign for the LARC consisted of full-page ads in The New York Times and The Wall Street Journal. Further:

> A group of aerospace executives was flown in from the West Coast to see the LARC computer. However, by the time the computer was delivered, the Remington Rand Univac management had had such a belly full of past grief that they were in no mood to move forward. A decision was reached to carry many of the LARC concepts forward into a new system known as the UNIVAC III computer.(89)

The major achievement of LARC was its contribution to system concepts. LARC pioneered in multiprocessing, contained an input-output controller that was a forerunner of the input-output portion of modern operating systems, had independent ferrite core storage to decrease the system access time, had four levels of storage with different speeds, capacities, and costs, had a CPU instruction overlap feature that enabled the computer to operate from different instructions coincidently and included an electronic page recorder to reduce the need for paper output. Concepts developed for the LARC were later incorporated in other machines.(90)

STRETCH

After IBM lost the LARC contract to Sperry Rand, it proposed what was essentially the same machine to AEC's Los Alamos

Laboratory. The proposal was accepted in November 1956, and the computer was designed under the name STRETCH. From IBM's point of view, the objective was to "stretch" the state of the art "to build the fastest possible machine," "exploring the unknown and rethinking and redesigning almost every aspect of earlier IBM computer systems."(91) This objective was set even as the 704 was in its early stage of installation and was even concluded in terms relative to the 704. The IBM 7030, as STRETCH was eventually called, was to be "100 times more powerful" than the 704.(92)

It was recognized at the outset that transistor technology would have to be advanced to fulfill STRETCH requirements.(93) While the Los Alamos contract was for only $3.5 million, IBM initially projected engineering costs of $15 million and manufacturing costs of $4.5 million for the first machine.(94) The STRETCH was delivered to Los Alamos in April 1961. It was about 37 times as fast as the 704 in scientific computations and about 168 times as fast in commercial computations.(95) IBM may have suffered direct losses of as much as $40.7 million on the project.(96) But, STRETCH:

1. Utilized radically new parallel architecture, permitting several operations to be performed simultaneously.
2. Employed SMS (Standard Modular Systems) component technology.
3. Employed printed circuit cards and improved back-panel wiring.
4. Included an 8-bit byte.
5. Resulted in greatly improved transistors and the means for manufacturing them.
6. Had a common mode for attaching peripherals.
7. Emphasized alphabetic characters.
8. Combined fixed and variable word-length operations.
9. Used a combination of decimal and binary arithmetic.(97)

The first 7090, IBM's entry into commercial transistorized computers, was delivered to the Air Force for the DEWLINE air defense system. This was in April 1959, about a year and a half before STRETCH was completed. The 7090, however, "became the vehicle by which the componentry of the STRETCH system (including transmission, circuits, pluggable units, cards, frames, power supplies and memories) became a part of the IBM product line."(98) The components of the 7090 were STRETCH components and the engineers who worked on the 7090 came directly from the STRETCH project. The 7090 was perhaps one-third to one-eighth as fast as STRETCH, but it was more than twice as fast as the Philco 2000-210. Further, the 7090 was designed to be used in conjunction with other IBM computers, particularly the transistorized 1620 and the 1401, and quickly evolved into the "family" concept. The derivative 7070, 7080, 1410, 7040, and 7044 all appeared in the next three and a half years.

ALTERNATIVES TO THE VACUUM TUBE

The speed, reliability, power requirements, physical size, and heat-generating properties of vacuum tubes were acknowledged to be severe technical limitations to computers from the start. A variety of alternative technologies were pursued before the transistor was universally adopted in the 1958-1960 time frame. The choice made by the participants in the nascent computer industry with respect to investments in technological alternatives to vacuum tubes were crucial in shaping the industry. The participants pursued a variety of technologies; some were dead ends, some were temporarily successful, and others, primarily in the transistor category, were longer-term successes. Again, it is necessary to go back in history.

Thyratrons

Thyratrons are hot filament gas tubes that are able to handle more current with less physical size than vacuum tubes. Generally, a smaller number of thyratron tubes are necessary to perform the same action as would be required if vacuum tubes were used. Both types of tubes have the disadvantage of large size. Thyratrons were used as counters in the 1930s, first in England and then in the United States. Research on thyratrons at the Cavendish Laboratory in Cambridge, England, had an important impact on research at National Cash Register in the late 1930s. A working model of an accumulator using thyratrons was completed at NCR in December 1939. NCR held patents on the first electronic calculators using thyratrons in electronic counters.(99)

Owing to the disadvantage of large size, NCR research stressed miniaturization of thyratrons. Government involvement in NCR's thyratron technology was coordinated by Warren Weaver in the Office of Scientific Research (OSR) of the National Defense Research Council (NDRC). The overall thrust of the OSR research efforts was directed to building an electronic differential analyzer. Information on specific attributes of the thyratrons developed by NCR was shared among members of a committee formed by Weaver to exchange technological information concerning the development of an artillery computer for NDRC. Members of the committee included the Armour Corporation, MIT, NCR, Eastman Kodak, Bell Labs, and RCA. An NCR thyratron electronic calculator was demonstrated at Enrico Fermi's project at the University of Chicago. Aderdeen Proving Ground also obtained an NCR thyratron electronic calculator. J.P. Eckert was aware of the NCR thyratron counter and, indeed, evaluated it along with the RCA ring counter and Lewis ring counter for ENIAC. Eckert chose to adopt neither, instead designing his own decade ring counter.(100)

A computer based on thyratron technology was planned at NCR prior to World War II. This computer was to have been able to add, subtract, and multiply, and divide. NCR referred to it as patent Model #3754. NCR filed a patent for a binary computer capable of addition and multiplication in March 1942. A revised version, using fewer tubes, was also patented.(101) During the war, NCR worked exclusively with the Navy.

By 1950, NCR recognized that the thyratron approach could not compete with vacuum-tube, stored-program computers. Representatives of NCR negotiated with Eckert and Mauchly prior to the acquisition of Eckert-Mauchly by Remington Rand. NCR also considered purchasing Engineering Research Associates. Finally, NCR bought Computer Research Corporation (CRC), the spinoff from Northrop Aircraft. Prior to the CRC acquisition, NCR had in development a machine called NEAM (National Electronic Accounting Machine). The NEAM was:

> essentially a paper tape analog of an IBM tab system. It had a paper tape input, magnetic tapes for intermediate storage, and they were going to have a sorter and a collator and do everything on paper tape with a magnetic tape intermediate storage. It was incredible because at that time at CRC we were recording 124 bits to the inch, and they were recording 16 bits to the inch in magnetic recording, and that's how they were going to build an electronic accounting machine.(102)

The NEAM, which probably was to have used thyratrons, was never built, but one NCR machine did use thyratrons. It was called the Computronic and was not marketed until 1959. Why the Computronic was introduced at this late date is a mystery. NCR personnel indicated that by this time they had realized the dominance of transistors and, indeed, the NCR 304, introduced in November 1959, was a small, transistorized machine. The 304, was noted, was built by General Electric. The NCR-built Computronic used miniature thyratrons and consisted of a multiplier tied into a bookkeeping machine. NCR sold 4,245 of the machines at a price of $18,000. Commenting on the miniature thyratron tubes, not on the Computronic itself, the developer noted:

> And, I think we finally did get more reliability out of gas tubes than you'd ever expect to get. Of course, we were always ridiculed pretty much by these other people about the reliability of gas tubes, and they thought we were on the wrong track.(103)

Why did NCR fail to take advantage of its early opportunity in the field of electronic calculators? In the 1930s NCR had "very elaborate computing development activity in electronics and never exploited it." In fact, it never got out of the NCR laboratory.(104) The National Cash Register Company offers a particularly intriguing industrial "might-have-been." NCR management, like most others at this time, was not interested in new automatic computing devices, but only in improving its existing line of office equipment. Further, the original Navy-funding R&D at NCR focused for many years on the thyratron and largely skipped the first generation of vacuum-tube computing devices.(105)

A large-scale electronic computer, the Colossus, built by the British as a cryptoanalytic machine during World War II, used thyratron rings as counters.(106) A computer completed in 1956 at Pennsylvania State University, the PENNSTAC, which was similar to the IBM 650 and used an IBM 650 magnetic drum for storage, was also constructed using thyratrons.

Trionodes

Trionodes are miniature bi-stable neon gas tubes. A trionode consumes far less power than a vacuum tube, is only a fraction of its size, and has a longer expected lifetime. The trionode was developed at Northrop Aircraft under government-funded research in the late 1940s. With the addition of resistors, this tube functioned as a flip-flop.(107)

For a period of time in the late 1940s the neon-tube technology was being pursued as an alternative to vacuum-tube technology at Northrop in connection with the Air Force SNARK missile guidance project. While the Incremental Slope Computer project and the MADDIDA project were underway at Northrop with the conventional vacuum-tube technology, others at Northrop, particularly Glenn Hagen and Charles Williams, were paralleling these projects with their own trionode technology research. Eventually the neon-tube technology was abandoned and the Northrop patent rights were sold to the Walkirt Corporation. While some computer components (e.g., accumulators) were made of trionodes, no computer with trionode logic was made.

Magnetic Amplifiers

The existence of magnetic amplifiers and the knowledge of their properties was within the purview of the international scientific community before World War II. Germany made direct use of these solid-state devices, but they found very limited application in most other countries. The limitation of tonnage on battleships provided for in the treaties of World War I spurred the Germans to investigate technologies intensely in an effort to increase the reliability of naval firings with minimum weights:

It seems that it was somewhat more than difficult to introduce this new (magnetic amplifier) technology to the American engineering storehouse, and it took the activities of our erstwhile enemies, the Germans, to sell the United States on the idea of the use of magnetic amplifiers. When it was learned that the reliability and maintenance-free operation predicted for these circuits were being obtained in the German war machine [magnetic amplifiers were used on German battleships to control the firing the guns], considerable activity was instigated in the United States to develop these circuits for application here. First, new and improved magnetic materials and rectifiers were developed for the existing circuits, and research and development was carried on to determine the most advantageous circuit configurations for various applications.(108)

After the war, many applications were seen for magnetic amplifiers. These include their use in servoamplifiers, temperature-measuring devices, regulators of speed, voltage, and frequency, d-c amplifiers and modulators, frequency reducers and multipliers, audio- and radio-frequency amplifiers, trigger and multivibrator circuits, delay lines, and memory devices.(109) R. A. Ramey published a paper in 1952 dealing with magnetic amplifiers in computers.(110)

Magnetic amplifiers had many advantages over the vacuum tube. They were substantially more rugged, had much longer lives, and were basically "sensitive, high-gain, high-speed, versatile devices capable of delivering large quantities of power efficiently."(111) Reliability was claimed to be the foremost advantage of the magnetic core amplifiers:

> There is nothing to wear out and the actual components - coils, resistors, and perhaps, metal rectifiers - are few in number. The devices operate direct from an a.c. supply, without intermediate high-tension rectifiers, and their efficiency is high, often exceeding 90%. Physically, magnetic amplifiers are small and they can be made very robust. They are relatively insensitive to temperature changes and have no warming-up time.

> Magnetic amplifiers operate by virtue of the fact that the inductance of an iron-cored or ferrite-cored inductor can be varied by changing the magnetic state by means of a single current.(112)

The general disadvantage of magnetic amplifiers was a slow response to signal changes. Sometimes their weight is also a problem. In Ramey's view:

> Transistors, by virtue of their efficiency, small size and fast response are obviously the magnetic amplifier's biggest rival. However, until transistors are

equally reliable, drift-free, able to handle large power, and operate at higher temperatures, they will not be a serious threat to the magnetic amplifier in many applications. In any case, the two devices might profitably be combined in a single equipment. This has already been done to a limited extent.

The other device which competes with the magnetic amplifier is the thyratron. This is efficient and capable of handling high power, but suffers from the usual disadvantages of thermionic valves. The two devices are sometimes used together.

The junction transistor is a potential rival in that it is compact, efficient, mechanically strong and (probably) reliable. It has the additional advantages of light weight, a greater input resistance, and a higher speed of response. At present, the transistor's temperature limitations, and the rather restricted range of available types are sufficient to exclude it from this field (control systems), but if stable transistors capable of controlling adequate amounts of power become available, the heyday of the magnetic amplifier will be over. In less critical applications, the magnetic amplifier is at a disadvantage by reason of the lack of a standard range of transductors.(113)

Research by U.S. computer companies in the area of magnetic amplifiers was conducted most actively at Remington Rand, although Logistics Research Corporation, Burroughs, Raytheon, and IBM were all to some extent also involved. Apparently IBM was aware of Remington Rand's use of magnetic amplifiers as both amplifiers and components of data-processing circuits. Information about Remington Rand's magnetic amplifier research was publicly available in the Digital Computer Newsletter and in other public sources as well.(114)

Remington Rand and Logistics Research Corporation are the only computer companies in the United States that built and attempted to market a computer using magnetic amplifiers. In all, however, at least eight different models of magnetic amplifier computers were delivered. These machines were: (1) the AFCRC (Air Force Cambridge Research Center Computer) made by Remington Rand; (2) the UNIVAC Solid State Computer (80-column card and 90-column card versions, marketed in Europe as the UCT, Universal Card Tabulating Machine); (3) the Sperry Rand STEP Computer; (4) the Sperry Rand X308; (5) the Burroughs Lab Computer; (6) the ALWAC-800, made by Logistics Research Corporation; (7) the Elliott 802, made by Eliott Brothers, Ltd., London; and (8) the EMAL-2, made by Polish Nuclear Research Institute and Warsaw Technical University.

The Burroughs Lab Computer was the first of these. It was operating at Burroughs beginning February 21, 1951. This machine was later designated the Philadelphia Lab Computer and was used to solve both industrial and engineering problems. Wayne State University received a Lab Computer in 1953.(115) Burroughs continued its commitment to magnetics after 1954, establishing a research facility in Paoli, Pennsylvania, which had as one of its focuses the development of magnetic components.(116)

The AFCRC computer was finished in June 1955 and installed in May 1956. The government-funded research and development at Remington Rand on the AFCRC ran from 1950 to 1955. New types of magnetic amplifiers were developed and, as work began on LARC as well as continuing on the AFCRC computer, the Univac Division of newly merged Sperry Rand began consideration of commercial models employing similar technologies.(117)

The development of FERRACTORS by Univac was acclaimed in the trade press:

> Magnetic amplifiers are beginning to replace electron tubes in high speed digital computers. The FERRACTOR, a magnetic amplifier capable of operating at frequencies as high as 2.5 mc, represents an increase in power-gain band width product an order of magnitude over that previously considered practical with magnetic circuitry.(118)

> The first high-speed low price electronic computer utilizing magnetic amplifiers throughout, instead of filament tubes, has been announced by Remington Rand Division of the Sperry Rand Corporation. The computer employs an entirely new principle by using "micro-ferractor" magnetic amplifiers which are no larger than the rubber erasers at the end of ordinary lead pencils. The "ferractors" will perform accurately at temperatures from 60 degrees below zero Fahrenheit to 222 degrees above zero, and are the result of five years of laboratory research. The computer opens up an era in which filament tubes and transistors will be outmoded by devices of this kind. The proto-type was completed last June, and present production plans will make the "micro-ferractors" filled computer available early in 1957.(119)

A description of the Univac magnetic computer was contained in the IRE Convention Record in New York, 1956.(120)

Despite the optimistic announcements and widespread publicity, it was not until December 1958 or early 1959 that the commercial magnetic amplifier computers were actually sold in the United States. This was the UNIVAC Solid State 80/90 I.(121) The same magnetic amplifier computers, sold as the UCT, were installed and operating in Europe in August 1958.

At about the same time, the UNIVAC Solid State computer was being developed in Philadelphia by the Eckert-Mauchly group, the File Computer was being developed in St. Paul by the ERA group. In fact, the File Computer was announced in January 1955, a year before announcement of the magnetic amplifier computer. The first File Computer was delivered in August 1956, more than two years before the first Univac Solid State Magnetic Amplifier computer was available in the United States.

The magnetic amplifier technology was spawned by government-sponsored research. Still, the extent to which Sperry Rand and, to a lesser degree, other firms relied on that technology rather than switching to transistors is not the result of governmental insistence or persuasion. Table 4.1 shows the vacuum-tube File 0 and File 1 computers to be slow and relatively expensive to operate. Despite this, it appears that the Univac division was, in addition to being late in developing a transistorized offering, preoccupied with the suppression of in-house competition between the File Computer and the SS80/90 model.

According to Herman Lukoff, who was director of research and advanced techniques for Univac, the UNIVAC 80/90 solid-state computers:

> got caught in the rivalry that developed between the St. Paul and Philadelphia divisions. A UNIVAC File Computer, developed by St. Paul and in roughly the same price range as the UNIVAC Solid State Computer, was already being marketed. There was a great concern that the "solid-state" would interfere with the File Computer sales. Consequently, the "Solid State" was held back from the market place and not delivered in this country until the summer of 1959, although it was marketed in Europe several years earlier. The late entry of the UNIVAC Solid-State computer resulted in a shortened life span for the project, although 500 systems in various forms were sold. By 1960, it was clear that the transistor was here to stay, and magnetic amplifier technology would not survive.(122)

This concern for the magnetic amplifier technology occurred long after it had been recognized that it was the transistor, not the magnetic amplifier, that would shape the second generation:

> In 1953, Univac realized that the days of the vacuum tube were numbered. About 90% of all computer maintenance problems were due to the vacuum tube and it simply had to be replaced. But with what? Univac pioneered in the development of solid state elements and one of these was the magnetic amplifi-

er. It found its way into the Univac Solid State Computer and was characterized by moderate speed and high clock power. Transistors were commercially available in the early/mid-fifties but only in the moderate speed range. The first breakthrough in high speed transistors occurred with the development of the Surface Barrier Transistor (SBT). Univac knew that the time for use of transistors in computers was fast arriving and that it had to develop the technology.(123)

Univac utilized surface barrier transistors in the LARC - because magnetic amplifiers would not yield the required performance - but was not committed to using only transistors in its commercial products until much later.

The persistance with which Sperry Rand clung to the magnetic amplifier is remarkable. At first, "The Remington Rand designers . . . used magnetic amplifiers . . . when they thought transistors were not yet practical."(124) This is confirmed by Herman Lukoff:

> I can remember the day that Pres Eckert gathered all of the engineering personnel together at Alden Park Manor to discuss future plans for the company. He stated that the transistor was not yet a practical alternative; therefore, we would cast our lot with the magnetic amplifier. Several weeks later, a week long course in magnetics was organized and all engineers were requested to attend so they could be updated on the new magnetic amplifier technology.(125)

The Eckert-Mauchly group was not alone in pursuing magnetic logic technology at Univac. The old ERA group in St. Paul constructed four X308 magnetic amplifier computers for classified delivery.(126) And prior to that ERA investigated magnetic amplifiers as an alternative for transistors for the Athena ICBM guidance system that was delivered to the Air Force in 1957. ERA actually built parallel working models of the Athena guidance system and only after completing the two working models was the final choice made to build the computer with transistors.(127)

Four months after the Univac SS 80/90 announcement, IBM announced the 1401, a transistorized replacement for the 650. According to Brock:

> The 1401 was in the same price range as both the 650 and the SS80 but had much better performance than either of them. For example, the 1401 could read 800 cards per minute compared with 250 for the IBM 650 and 450-600 for the SS80. The add time of the 1401 was 230 micro-seconds compared with 510

for the SS80 and 700 for the IBM 650. By the Knight calculations, the IBM 1401 had over twice the number of commercial operations per dollar that the SS80 had. Consequently the SS80 was not competitive with the 1401 and its effective life as a computer to expand Sperry Rand's market share was limited to a little over a year.(128)

The STEP computer, a modification of the Solid State 80/90, was announced in August 1960. As late as 1961, J. P. Eckert discussed adding an improved tape speed-up program and core memory to the STEP computer and said, "This is, however, the 'last drop' that can be squeezed from the U.S.S.C. and we must not lose sight of this."(129)

VII. THE TRANSISTOR AND ITS INTRODUCTION INTO COMPUTERS

Hindsight makes it clear that a number of companies were detoured in the search for alternatives to the vacuum tube. NCR, in pursuing thyratrons and, after acquiring CRC, magnetic amplifiers, made strategic technology errors. With respect to Univac, it can be put another way. Had the transistor technology failed to develop successfully, Univac's leadership in magnetic amplifiers would have given Sperry Rand great advantages.

The discovery of the transistor at Bell Telephone Laboratories in 1947 may have really been a rediscovery. According to Professor W. Gosling,(130) Julius Lillienfeld applied in 1925 for a Canadian patent for what today would be called a junction field-effect transistor. In 1927, Lillienfeld filed a patent for a bipolar transistor and, in 1928, he filed for an insulated gate field-effect transistor. A. H. Wilson wrote a paper in 1931 in which he stated the quantum mechanical theory "that related motion of electrons in metals to a comprehensive theoretical explanation of insulators and semiconductors."(131) The appearance of Wilson's paper caused heightened interest in semiconductors and their potential role in electronic communications and as rectifiers, but there was no view that the theory could be immediately applied:

> The implications of the Wilson theory were not evident to research workers in the field although between 1935 and 1939 the theory of semiconductor physics was advanced by Frenkel and Davydov in the Soviet Union, Mott in England and Schottky in Germany. A major problem was that the semiconductor materials available during the 1930's were too impure to provide an opportunity to link theory with experiment.(132)

In 1923 Karl K. Darrow, a research physicist at Bell, began publishing a series of papers in the Bell System Technical Journal. These papers summarized meetings of the American Physical Society and the current state of the art. A paper published in the Journal in 1927 by two Bell physicists, C. J. Davisson and L. H. Germer, on electron diffraction, was a seminal piece. In 1937 Davisson received a Nobel prize in physics.

William Shockley, an MIT Ph.D., may have been attracted to Bell Labs in 1936 owing to Davisson's presence there. James Fisk, whom Shockley had known when they were both graduate students, joined Bell in 1939 and, before becoming president of the Labs, was responsible for research in physics. Another member of the soon-to-be famous group at Bell had been one of the first of the group to join. Walter Brattain began working at Bell in 1929. Brattain was attracted to Bell Labs because of his knowledge of the Bell System Technical Journal and, in particular, because of the paper by Darrow.(133) John Bardeen, a Harvard Ph.D., joined Bell after World War II, but he had known Shockley and Fisk in Cambridge in the 1930s. Bardeen, Shockley, and Fisk had all studied under John H. Van Vleck.

Government-sponsored research played a key role in the emergency of the transistor. In the 1930s, and especially during the war, Bell Labs and various universities and industrial companies became involved in a variety of governmentally sponsored projects related to crystal detectors. The work on radar in the late 1930s falls into this category. The link between radar and computers is important:

> Radar had some of the elements of the computer; it had timing circuits, which ultimately became fundamental to the modern computer. Some of the other related equipment like the Loran even had decade (vacuum tube) counters associated with them. It was a rudimentary form of digital computer.(134)

Wartime crystal detector work was fundamental in advancing the state of the art. Work on increasing the purity of a semiconductor, germanium, was undertaken at Cornell University. Work on producing another high-priority semiconductor, silicon, was undertaken at the University of Pennsylvania in conjunction with E. I. Dupont deNemours. Several members of the physics department at Purdue University were conducting a systematic study of the properties of germanium. General Electric was also engaged in this research. Overall coordination of the semiconductor crystal research was handled by the MIT Radiation Laboratory.(135) Cooperation and information interchanges among the researchers were common:

> The communication among these institutions helped create the mutual awareness of the potential of

semiconductor materials that contributed to the ultimate success of the Bell effort. For instance, on April 9, 1945, only three months before Bell Labs issued the "Authorization for Work" on solid-state materials, representatives of the participating academic and industrial laboratories convened at Bell Labs for a "Meeting on Germanium Crystals."(136)

A separate division of the American Physical Society devoted to solid-state physics was formed in 1947. William Shockley, although not personally involved in any wartime semiconductor research, was at the forefront of the movement to open new channels of communications for those working in metals. When the transistor effect was announced, numerous symposia and conferences relating to the transistor were held:

> Bell Labs' first important policy was not to keep transistor information secret. Not only was it not kept a secret, but we actively expounded the art as well as the science of practicing the technology. Several seminars were held in the early 1950's where we effectively told all we knew about transistor technology. The whole tone of open information exchange within the emerging semiconductor industry was set by Bell system policies of patent licensing and publication. . . . The semiconductor industry's remarkable, almost overnight, growth is due in large measure to relatively open information exchanges.(137)

Governmental groups and the Association for Computing Machinery also sponsored open conferences on transistors. J. H. Felker, of Bell Labs, gave a paper on a transistorized computer at the Eastern Joint Computer Conference in Philadelphia in December 1954. This was prior to the LARC and STRETCH projects. The paper provided a full discussion of the high-speed point transistors used in the TRADIC (one of the first large-scale transistorized computers) as well as some insights into the newer junction transistors:

> The point contact transistor has been the fastest transistor we have had to work with. It also has been the transistor we had in quantity and has been reliable. The first junction transistors were not very reliable. That situation has improved enormously in the last year. Junction transistors that are available are not as fast as point contact transistors, but I think it became clear to us about a year ago that the future is with junction transistors rather than point contact. There are two things

that convinced us of this. One was that physicists aren't interested in the point contact. They don't understand and won't work on the point contact transistor so it will never be improved.

The junction device obeys the mathematics that they understand, and this is a very real thing. . . .

The other thing which is equally significant is that the junction transistor is now becoming faster than the point contact.(138)

Those who participated in the discussion of the paper included representatives of Electro Data Corporation, Remington Rand, Westinghouse Electric Corporation, Sperry Gyroscope Company, Federal Telecommunication Laboratories, IBM, Airborne Instruments Laboratory, Armour Research Foundation, ERA Division–Remington Rand, North American Aviation, University of Rochester, and Bendix Aviation Corporation.

That a transistorized computer was a technical possibility was not new in 1954. But the existence of a technological possibility is quite different from a clear choice in design. At the same conference in which Felker discussed the TRADIC transistorized computer, Jay W. Forrester discussed component reliability. He was not convinced of the immediate advantages of transistors over vacuum tubes:

I would caution against feeling that any magic will suddenly solve the dilemma of electronic unreliability. We have heard the transistor proposed for the elimination of failure now attributed to vacuum tubes. The transistor does look promising. I would caution against considering it a panacea. Vacuum tubes in some computer applications have a failure record as low as any thus far proven for transistors. The transistors will improve with time but, on the other hand, in a factory will not be made with the loving care given to the first laboratory models.

With the proper use of marginal checking (this means raising and lowering the voltage beyond normal levels and replacing tubes which are defective in those ranges) the vacuum tube presents no serious problems except from open welds and short circuits. Again, with the proper attitude toward reliable electronics, these difficulties could be greatly reduced.

For computer use, the transistor is not so interesting for its small size and power consumption as for the unproven possibility that it can be more free of intermittent changes in performance than vacuum tubes.

For future trends it seems that the electrostatic tube, regardless of type, is but a transient on the stage and that it is scheduled to be replaced in the next few years by new developments in solid state physics. A strong contender is the 3-dimensional magnetic core storage array with a good possibility for ferroelectric storage.(139)

It is not clear which transistorized computer was the first in operation. It is clear, however, that government was heavily involved in the development process and that there was no preemption of the technology by any research lab or by any firm. The first large-scale computer using transistors is generally acclaimed to be the TRADIC, built at Bell Labs and finished in January 1954.(140) The work on the TRADIC was done under an Air Force contract. The high-speed point contract transistors were manufactured by Western Electric, which had begun to produce transistors commercially as early as 1951. The TRADIC was said to perform as well as the majority of vacuum-tube computers then available.(141)

The Burroughs Atlas Mod 1-J1 Guidance Computer used surface-barrier transistors in direct-coupled transistor logic. It is on display at the Smithsonian Institution, labeled "the first operational computer to use transistors rather than vacuum tubes." It was also built under an Air Force contract. Although the Atlas Mod 1-J1 was delivered in April 1955, it was not operational until September 1957. The Athena ICBM Guidance System, delivered to the Air Force in 1957 by Sperry Rand, was, as noted above, compared to a magnetic amplifier system prior to the decision to use transistors in the production model. The Athena has also been referred to as "an early high-reliability transistorized computer system."(142) The Athena, produced by the ERA division of the company, may have been operational before the Atlas.

There were other early, special purpose transistor computers. The North American Transistorized Differential Analyzer (NATDAN):

was the first full-transistorized or semiconductor-based computer to be built as something other than a prototype. . . . It was operating in 1953. The funding was aimed at building something that was compact, low in power requirements and reliable. Transistors fit the bill very nicely and tubes were very unreliable. There may have been other military projects that did the same. We couldn't talk about the project to the outside world, and it was frus-

trating for some of us to go out and hear people a
year or two later talking about building the first
transistorized machine.(143)

A Ramo-Wooldrige machine, the RW-300, is said to have
been the first transistorized digital computer used for process
control.   It was operational in 1958 and purchased by Tex-
aco.(144)   The Lincoln Labs CG-24 had transistor logic and
core memory and was operating in 1957.(145)   One company,
ALWAC, is said to have offered a commercial, transistorized
computer as early as 1956 or 1957.   One of these was sold in
Sweden.(146)

VIII. SECOND GENERATION COMPUTERS

Table 4.4 provides a list of the principal second-generation,
commercial transistorized computers.   As was true of the
vacuum-tube machines introduced only a few years earlier,
government-funded research was critical to the development of
the second-generation computers.   Contrary to the situation of
the early 1950s, however, many companies elected to carry
government-funded research into commercial ventures.
The Philco 2000-210 TRANSAC was announced in January
1957.(147)   Philco had developed the surface-barrier transistor
under government contract three years earlier and the first
TRANSAC was built for the Air Force.   The TRANSAC was
miniaturized under a Navy contract.   Philco, by its own estimates, was "a year or so ahead of most companies in the
development of big transistorized computers."   While plans for
a commercial version were unveiled in 1958, the "Philco computer effort was small and poorly financed. . . . The first
complete 2000 delivered [in early 1960] was a 211, which had
already changed from the surface barrier transistor of the
original model 210 to the faster MADT transistors."(148)   As
Table 4.4 indicates, the Philco 2000-211 was a fast, economical
computer.   Deliveries, however, were delayed after its 1958
announcement and, as noted above, it was unaccompanied by
comparable supporting peripherals and extended software.(149)
After Ford Motor Company acquired Philco in December 1961,
computer work continued for a time with introductions of the
smaller TRANSAC S 1000 and 2000-212 in early 1963.   Ford
then "decided to get out of . . . the computer business."(150)
RCA developed the junction field-effect transistor in 1951.
The RCA 501 computer was announced in December 1958, following the late and unsuccessful BIZMAC tube entry.   The
501, while among the first available transistorized models, was
small and had slow and unreliable peripherals.(151)   Many 501s
were unsold and went to storage.

Table 4.4. Principal Second-Generation Commercial
Computers, 1958-1963.

Computers	Introduction Date	Thousands of Commercial Computations	
		Per Second	Per Dollar
Philco 2000-210[a]	November 1958	28.7	511.8
GE 210	June 1959	5.1	226.5
NCR 304	November 1959	2.4	98.4
IBM 7090	November 1959	45.5	443.0
RCA 501	November 1959	1.9	73.1
CDC 1604	January 1960	20.4	374.0
Librascope 3000	January 1960	25.3	315.7
Univac SS80/90 I[b]	January 1960	0.5	61.1
Philco 2000-211	March 1960	55.7	827.7
Univac LARC[c]	May 1960	40.5	186.8
IBM 7070	June 1960	5.1	123.2
CDC 160	July 1960	49.6	17.6
IBM 1401 (Mag Tape)	September 1960	1.6	135.2
IBM 1401 (Card)	September 1960	1.0	208.1
PDP-1 (Mag Tape)	November 1960	2.2	90.4
Honeywell 800	December 1960	23.8	352.8
Sylvania 9400	Late 1960	49.6	461.1
Univac SS80/90 II[b]	January 1961	3.0	210.9
Bendix G20 & 21	February 1961	17.1	565.9
RCA 301	February 1961	1.1	119.6
GE 225	March 1961	7.1	555.8
IBM 7030 (STRETCH)[d]	May 1961	631.2	1,311.6
CDC 160A	July 1961	1.8	246.7
IBM 7080	August 1961	30.9	350.0
RW-530	August 1961	5.1	302.0

(continued)

Table 4.4.   Principal Second-Generation Commercial
Computers, 1958-1963 (continued)

Computers	Introduction Date	Thousands of Commercial Computations	
		Per Second	Per Dollar
IBM 7074	November 1961	31.7	616.9
IBM 1410	November 1961	4.6	289.2
Honeywell 400	December 1961	2.8	197.2
Univac 490	December 1961	15.1	375.3
Univac 1206	December 1961	17.7	750.8
Univac 1000 & 1020	December 1961	3.3	218.4
NCR 315	January 1962	11.5	752.1
Univac III	June 1962	22.8	617.8
SDS 910	August 1962	2.4	587.3
SDS 920	September 1962	5.0	325.8
Univac 1107	October 1962	76.1	948.3
IBM 7094	November 1962	95.9	842.0
IBM 7072	November 1962	8.7	301.2
Burroughs B5000	December 1962	15.9	522.2
ASI 420	December 1962	11.1	493.9
Burroughs B200	December 1962	0.5	71.7
RW-400	December 1962	11.2	140.2
RCA 601	January 1963	58.9	816.1
Philco 2000-212	February 1963	84.2	722.3
CDC 3600	June 1963	74.9	849.4
Philco 1000 (TRANSACT S1000)	June 1963	104.	685.2
IBM 7044	July 1963	23.4	561.6
Honeywell 1800	November 1963	57.8	1,028.5

[a]Military computer; not commercially available.

[b]Magnetic amplifier computer; not transistorized.

[c]One-of-a-kind, Livermore computer; see text, pp. 37-41.

[d]Los Almamos computer; see text, pp. 41-42.

Source: Kenneth E. Knight, "Changes in Computer Perform-
ances," Datamation (September 1966), pp. 45-46. Dates given
by Knight are not always in accord with those given by other
sources.

There were doubts as to RCA's "seriousness in the EDP Business" in late 1959,(152) even though the company continued to be engaged in related government contract work in connection with the BMEWS project. The RCA 301 and 601 were announced in April 1960, but the 601 was not delivered for nearly three years. The 601 was a "disaster," with extremely high production costs and many technical problems in accomplishing the announced functional capabilities.(153) Marketing of the 601 actually stopped before first delivery. Efforts turned to enhancing the 301, although RCA's reputation in computers had been badly hurt by then. Despite a professed management belief that the computer business was "what RCA should be in," there was, as noted, "a greater total effort in television . . . than there was in the computer."(154)

Table 4.4 shows the GE 210 as the first available transistorized computer. At the same time, GE was building the processor for the NCR 304. One account is that "NCR just handed [GE] the transistor technology and the entire NCR logic and concept designs."(155) A process controller, the GE 312, was developed into the GE 225 of March 1961. The 225, while intended for scientific use, found substantial business uses owing in fact to its GECOM compiler, a predecessor to COBOL. Later GE models (not shown in Table 4.4) include the 215, the 235, and the DATANET 30. The latter was highly innovative as a communications controller and the initiator of commercial distributed processing systems.

While some of the later models were innovative and popular, on the whole the GE computers did not achieve commercial success. Outside suppliers were used for much of the peripheral equipment.(156) There was "an inadequate allocation of resources, both human and physical, to the business." GE was "very strong in basic technology and background and experience . . . [and] relatively naive when it came to the discipline of manufacturing large electronic systems . . . or bringing them to market."(157)

Philco, RCA, and GE eventually left the commercial computer field. In the same period, however, others entered. Table 4.4 shows the CDC 1604 among the first of the transistorized computers. The Control Data Corporation (CDC) was formed in mid-1957 by a group from the ERA contingent of the UNIVAC division of Sperry Rand. The formation of CDC occurred shortly after the controversy between the ERA and the Eckert-Mauchly groups concerning the LARC project, and the internal decision by Sperry Rand to award the proposal to the Eckert-Mauchly group. In the view of the individuals who started CDC and others acquainted with the operations, Sperry Rand, had among other things:

1. "faltered at a crucial time when it had a chance to take over the computer market";

2. Suffered from top management that was "too old," "auto-
    cratic, ironwilled," "never really understood the busi-
    ness," and practiced nepotism in management succession;
3. Failed to make the "financial commitment that was neces-
    sary";
4. Been "unwilling to take risks";
5. Failed to "mount an adequate sales effort, and did not
    choose to create the kind of an organization [needed] to
    meet the market . . . for that class of high technology
    machine";
6. Did not allocate "the resources to do the kind of respectful
    program," "choose not to put proper monies in the Univac
    Division";
7. Experienced "heedless budget-cutting, managerial infight-
    ing and a series of wrong-headed decisions," and tolerated
    internal "turmoil . . . confusion, indecision, conflicting
    orders, organization line breaches, constant organizational
    change, fighting and unbridled competition between divi-
    sions."(158)

CDC quickly developed a 1/10 scale model of a transistorized computer, announced it as the 1604 in April 1958, and
subsequently received governmental funding for R&D and the
manufacturing of the computer.(159) Initially CDC obtained
peripherals from Ampex, Analex, IBM, and Ferranti. As Table
4.4 shows, the 1604 was installed in January 1960. It was
followed shortly thereafter by the smaller 160 and 160A, the
processing units of which were, in addition, exclusively
licensed to NCR for U.S. banking and retail trade sales and
nonexclusively for foreign sales. CDC also opened service
bureaus. Early contracts with the Defense Department led to
development of a large solid-state computer, "many times more
powerful" than the CDC 1604 or IBM 7090. The CDC 3600 was
announced in May 1962, and delivered to Livermore Labs in
mid-1963. The much more powerful CDC 6600 was also delivered first to Livermore.(160) Control Data made a number of
acquisitions in 1963, including Bendix's computer business,
and extended its capabilities in peripherals, industrial process
controls, and distributed processing.

Government R&D contracts and government purchases of
computers were important in the early successes of CDC, but
willingness to take risks was in reality probably the safest
course for a small company with limited resources competing in
the high and fast-moving technology: "To have played it safe
would have meant one of two things: 1) being too late in the
marketplace with a new product, or 2) having a good marketable new product but being unable to capitalize on the demand
before our giant competitors moved in with a similar product."(161)

Table 4.4 shows the introduction of Digital Equipment Corporation's PDP-1 before the end of 1960. DEC was, like CDC, founded in 1957. A venture capital firm provided $70,000 to Kenneth Olson and Harlan Anderson, both of whom had worked on Whirlwind and SAGE at the Lincoln Labs of MIT. While the term was not then in use, many of DEC's early products were in the nature of "minicomputers" used in conjunction with other, larger computers as input-output, control and memory devices in time-sharing systems. This changed with time, however, as the PDP-1 gave rise to the PDP-4 (1962), PDP-5 (1963), PDP-6 (1963), PDP-7, and PDP-8 (1964). While many of DEC's computers were sold to government agencies and used in research by other users who were governmentally funded, DEC did not itself rely directly on government R&D projects. From the beginning DEC produced state-of-the-art products with a wide range of flexible, adaptable, fast, and efficient computers and related peripherals. (162)

Scientific Data Systems was another entrant. SDS was formed in 1961 by Max Palevsky, who had previously been with Bendix and, after that, had organized the Packard Bell Computer Corporation. Palevsky left Packard Bell because, "The ideas had about how to proceed in the computer industry required much stronger backing from the parent company which they could not provide. . . . [It] was very difficult working under a management that really knew nothing about the industry itself."(163) The SDS 910 was delivered less than a year after the company was founded; the SDS 920, shortly thereafter. SDS was rather unique in that it was primarily a design and assembly operation, with parts purchased OEM from other manufacturers. By the end of 1963, the SDS 930 and 9300 had been announced. The SDS 92 (1964) was called the first "to use monolithic integrated circuits." The SDS 940 (1965) was an improved 930 and provided time-sharing capabilities. In 1965-67, SDS introduced its third-generational Sigma Series in direct competition with the IBM 360 series. The Xerox Corporation entered commercial computer market by acquiring SDS in 1969.(164) SDS became XDS.

Honeywell also entered the commercial computer field with the second generation. The Honeywell 800 (1960) was delivered to about 100 customers. The 1800 (1963) was designed to be a larger, compatible upgrade of the 800. At the same time, a "third-generation" Honeywell 200 was announced and this proved to be an extremely successful product. A "family" of 200s - the 120, 1200, 2200 and 4200 - emerged.(165) The Honeywell 200s replaced many of IBM's 1400 series computers in the small-to-medium-sized segment of the market as improved hardware, software, and peripherals were added. The Honeywell 8200, designed as the most powerful of the line and one that would merge the 200 and 800 lines, was not successful, however.

There were other companies venturing into computer production in the early 1960s. Sylvania and Bendix introduced the S9400 and G20 and 21 models in 1960 and 1961, respectively. General Precision offered its Librascope 3000, intended primarily for reservation systems. Computer Control had a DDP series; Thompson-Ramo-Wooldrige had an RW series. Hughes Aircraft had a number of machines on the market also. General Mills developed an AD/ECS 37 and an ASI 420.(166) Burroughs, which had produced the Atlas Guidance Computer - one of the earliest transistorized computers to become operational - did not have a transistorized commercial product in the market until the introduction of the B5000 in December 1962. Burroughs introduced the smaller B200 series at about the same time.

It was noted above that there was significant government involvement in IBM's STRETCH project. The 7090, 7070, 7080, 7074, 7094, 7072, 7044, and 7010 grew from STRETCH. The complementary 1400 and 1600 computers appeared in the same time frame. Table 4.4 shows the first Sperry Rand transistorized computers as the Univac 490, 1206 and 1000 and 1020 models, all in late 1961. The Univac III and 1107 became available only in mid-1962. While Sperry Rand appears to have had somewhat the same technological opportunities and governmental support as any other firm, the magnetic amplifier "detour" and internal management problems had created important obstacles to early success in the second generation.

IX. LOOKING BACKWARD

While governmentally sponsored research and development and government procurement continued to influence the general advance in technology after the appearance of second-generation computers, the influence of government on particular firms was less obvious after 1960-1962. A possible exception to this was the involvement of GE after 1964 in MIT's Project MAC. Here the purpose was to improve the state of the art in time sharing, with funding from the ARPA Project of the Department of Defense. Earlier work at Dartmouth funded by GE was in a similar vein.

As a consequence of MAC and the Dartmouth work, GE for a time probably had a lead in commercial applications of distributed processing. The concept of a "computer utility" became popular. For its part, however, GE failed to produce the "third-generation" mainframe necessary to take advantage of this lead. The GE 635 that was used at MIT was not itself sufficiently advanced and the GE 645 that was subsequently developed did not materialize into a successfully marketable product.(167)

From 1964-1965 on, distributed processing, time-sharing, and real-time applications of computers were important aspects of the continuing development process in the industry. With this, there were vast improvements in memory, control, and input-output peripherals. Partly as a result of various governmental interests in miniaturization, increased speed and improved process-control techniques, an increasing number of "minicomputers" and "intelligent terminals" appeared. The "old" transistor developed into monolithic "chips." Random-access memory gave rise to "virtual memory" computers that ostensibly operate as though there were no bounds to memory capacity. Microprocessors with advanced chip technology appeared. Computing technology changed so rapidly that in a very short period of time a VLSI (very large scale integrated circuit) hardly larger than a fingernail had more logic than the IBM 701 or Univac I.

In the process of change, a number of companies withdrew. As noted, SDS was acquired by Xerox in 1969. GE sold its mainframe computer business to Honeywell in 1970 but, probably based on its continuing expertise from the Dartmouth and MAC experience, retained its time sharing services and continued as a joint venturer in the Honeywell Informations Systems (HIS) operation until 1976. Honeywell acquired the mainframe aspects of the Xerox computer activities in 1976 although Xerox subsequently made a substantial success of its XCS work in business computer systems. Philco, as noted, sold to Ford. In 1971, RCA solid it computer division to Sperry Rand.

The story of mergers and failures is but a small part of the industry's history in this period. As the technology became better known to both producers and users, and as the technology turned more and more to the addition of computational capabilities through various sorts of upgrade, add-ons, and peripherals, a "plug compatible manufacturer" (PCM) segment of the industry arose. Firms such as Telex, California Computer Products, Memorex, and Storage Technology were founded. Minicomputers and microprocessors were successfully developed and sold by firms such as Data General, Prime Computer, Perkin-Elmer (Interdata), Harris, Wang, Tandem Computers, Datapoint, Magnuson (with Fairchild), Hewlett Packard, Texas Instruments, National Semiconductor, Intel, and others. Amdahl, emphasizing the plug compatibility of its central processing units, entered the mainframe business.

X. A PERSPECTIVE FROM THEORY

Much that occurred in the computer industry after 1945 is susceptible to theoretical interpretation. At the level of the "technology and market structure" theory,(168) the role of an exogenously driven technology in shaping an endogenously determined structure is clear through the 1950s. Salient aspects of this scenario are:

1. Until at least 1951-1954, all significant aspects of R&D were government funded. The pervasive and nearly homogeneous view in private enterprise that considered investment in R&D was that negative returns would result. The primary reason for such expectations was the perceived absence of a commercial market. It had been demonstrated that electronic computation was technically feasible, but not that there existed a significant customer base.

2. The first attempts at commercialization were essentially by-products of projects that directly or indirectly were motivated by governmental interests and governmental funding. The initial commercializations, that is, were "technology driven" in the sense that it was primarily the cadre of technologists within government, industry, and the universities who dominated the decisions for exploratory product development. Budgets justified by military considerations carried most of the costs. Outside of government agencies and firms engaged in government contracting, there were few discernible customers. Marketing and sales departments were not clamoring for a computer to sell, and production departments were not maintaining that they could efficiently produce computers. The "established," nontechnical elements of businesses either ignored or took negative positions with respect to commercialization. Computers represented a "near-technology" - feasible but impractical.

3. The change from "near technology" to an "in being" (and "becoming") technology is a process which, for an ex ante point of view, is very risky and, from an ex post point of view, is necessarily characterized by many errors. Some products are introduced too soon, only to be made obsolete by other, superior developments. UNIVAC I may be of this class. Other products with technologically excellent characteristics were offered with inadequate marketing or financial support, or without supporting peripheral and complementary products. The ERA 1101 and 1103 were early examples. The Philco 2000/3000 series of 1960 are later examples. Perhaps because of what is ostensibly risk aversion but in fact turns out to be an assured strategy for failure when technology is changing rapidly, some products appear that are technologically outmoded at the time of their introduction. The early NCR and RCA machines fit into this mold, as does the Burroughs 220 of 1958.

Errors arise also from pursuing what turn out to be the wrong technological alternatives.   The principal illustration here is that of the magnetic amplifier and the consequent delays in Sperry Rand's having a state-of-the-art, transistorized commercial product.   Such errors may become compounded when both the technologists and the management of a company become so wedded to what is in fact the wrong alternative that the error remains for a long time uncorrected.   Along with Sperry Rand, the early Burroughs work in alternatives to the vacuum tube is illustrative of this behavior.

Another form of error is misjudgment of the rapidity of technological change.   An initially successful product may be retained for too long, or the same thing, the introduction of superior products may be purposely delayed so as to "preserve" the market for an older, existing machine.   The delay in introducing the SS 80/90 in the United States in order to run out the "product cycles" of the Univac II and File Computers is a case in point.   A related error may arise from attempts to squeeze more performance from "stretching" an existing model rather than discarding it in favor of an entirely new variety.

4.   One aspect of the change from a "near technology" to an "in being" technology is a shift from the "technology driven" status of the market to one in which commercial expectations more and more "drive the technology."   In the case of the computer industry, by 1955 many businesses recognized the possibility of profitably "dipping" into the governmentally generated technological opportunities.   It was broadly recognized by 1955 that there was a potential commercial customer base.   Privately financed R&D - to be sure, especially as an augmentation to a governmentally financed project - was seen as a risky and yet potentially profitable venture.   As this occurred, something akin to a Schumpeterian "swarming" of entrepreneurs developed.   Thus, in the late 1950s and early 1960s, and in addition to the continued development and marketing efforts of the first firms, others such as Bendix, Philco, Thompson-Ramo-Wooldridge, CDC, Honeywell, Sylvania, RCA, GE, General Mills,SDS, and DEC appear.   As is true of the early market entrants, these firms are prone to make what appear ex post to be significant technological, marketing and financing errors.   Some succeed; others do not.

5.   Unlike most new industries, the technology of the computer industry did not "settle down" after a few years. The computer did not become relatively standardized, with substantially similar functions and capabilities across manufacturers' types and with differentiation based largely on well known and minor technological options.   Success did not become critically dependent on static production efficiency combined with excellence in marketing and service.   Partly because of a continuing R&D program financed by the govern-

ment, and partly because of derivative work in universities, other nonprofit research organizations and, to a degree, privately financed, exploratory R&D by the computer companies themselves, there was a continuing stream of "near technologies" offering potential commercial exploitation. As a consequence, while the structure of the industry can in large measure be regarded as endogenously determined by the process of technological change and the patterns of innovation and diffusion, it cannot easily be seen as a process that induces or approaches an equilibrium structure. Structure is continuously subjected to major and minor shock effects as more or less important technological changes occur.(169)

6. The continuity of unpredictable "near technologies" gave rise in the case of the computer industry to a "swarming" of firms into areas that the more established firms tended for some time to ignore. Thus, the minicomputer, microprocessor, distributed processing, intelligent terminal, and OEM and PCM peripheral segments of the industry saw successful (and unsuccessful) entry by literally dozens of new firms after the mid-1960s. The technological opportunities these firms were exploiting were again derivative of the same science base as that yielding advances in central processing units and large-scale computing systems. The large systems, however, did not encompass and fulfill nearly all of the opportunities.

A closer examination of this history suggests that much more than pure chance and serendipity were involved in changing the governmentally created "near technologies" to commercial realities. While conventional microeconomics pays no attention to the internal organization of firms - even to the point of suggesting no useful definition of what is and is not a "firm" - there is a growing and important literature pointing to the fact that the internal organizational structure of firms may be as important as the external structure among firms in determining market performance.(170) Carried further, if differences in internal organizational structures affect the performance of firms, they also become at least a part of the explanation for changes in market structure and market performance.

In the context of an organization adapting to and using technological opportunities, there is more to be considered than simply whether a firm is vertically integrated or whether it is organized as an "M-form" or "U-form" firm. Whatever its formal table of organization, the firm must be capable of:

1. Internally transferring information with changing content and varying ambiguity among and between those specializing in the technologies embodied in the products, the technologies involved in producing the products, the marketing of the products, and the financing of the development, production and marketing activities.

2. Taking coordinated actions and resolving conflict among these functional specializations while simultaneously maintaining an internal reward system such that those who "lose" in the resolution of conflicts do not create dysfunctional organizational behavior.(171)
3. Receiving, processing, and acting on externally generated information regarding technology, markets, and finance, including information concerning actual and potential rivals.
4. Responding organizationally to unforeseen changes in circumstances, whether of internal or external origin.

In Williamson's terms, the firm must be capable of dealing with and mitigating the effects of "bounded rationality," "information impactedness," and "opportunism." Hierarchical organizations that facilitate the vertical flow of information and unified (internalized) actions are superior to bilaterally arranged transactions among separate firms. The latter fail to encompass costs and benefits external to the transactors.

The information-transfer aspects of this activity cannot be overemphasized. Coordinated, heirarchical processes - particularly when they transpire in a changing technological and economic environment - cannot be effective unless there is a generation and flow of reliable (not necessarily complete) information. Further, the information must flow horizontally and vertically within the firm as well as between the firm and various elements of its external environment. The most critical management function is, in a generic sense, information processing.

Since the "native tongues" of the several specialist groups in an organization differ, the "coded language" of the firm includes parochial dialects for interspecialization communications. Efficient management requires that those making basic decision to be capable of conversing and thinking in the basic language and its several dialects. Language facility as well as language development depend on frequent usage.(172)

Paradoxically, formal organizational structures often reflect the communications and information-transfer requirements of past circumstances, with little ability to cope with new events. Information may be cast in a language inappropriate for meaningful processing as it is channeled into routinized communications networks predicated on a fixed table of organization. Information may also be "filtered" if it follows routinized channels. Klein points to this in his arguments that dynamic efficiency (i.e., efficient handling of technological opportunities) requires "open" or "random" communications within firms rather than a highly structured system.(173)

Table 4.5 summarizes the organizational characteristics of the principal mainframe firms during the first ten years of commercial involvement in the computer industry. While obvi-

Table 4.5.   Organizational Characteristics of
Principal Firms in the Computer Industry,
First Ten Years in Industry.

Firms	Involvement of Top Management in Computer Activity	Coordination of Finance, Marketing and Technology in Computer Activity	Conflict Resolution Facility in Computer Activity	Perceived Importance of Commerical Computer Involvement
Remington Rand (Sperry Rand)	Low	Low	Low	Medium/High
AT&T	Medium	High	High	Low
IBM	High	High	High	High
Burroughs	Low/Medium	Low	–	Medium
Raytheon	High	High	High	Low
NCR	Low/Medium	Low	Low	Medium/High
RCA	Low	Low	–	Low/Medium
Bendix	–	–	–	Low/Medium
Philco	Low	Low	–	Low/Medium
GE	Sporadic (Low/Medium)	Low	–	Sporadic (Low)
Thompson-Ramo-Wooldridge	High	Medium	–	Low
CDC	High	High	High	High
DEC	High	High	High	High
Honeywell	Low/Medium	Medium	–	Medium/High
SDS	High	High	High	High
Xerox	Low/Medium	Low/Medium	–	Low/Medium

ously somewhat impressionistic, the purpose is to assess the extent to which the firms' internal organizational structure and mix of activities were conducive to the efficient development and innovation of computers.(174) All firms had access to the same basic technologies, although some depended more than others on governmentally supported research. Similarly, some were in the industry from its commercial inception in 1951-1953, while others entered subsequently.

To some extent, the "Involvement of Top Management in the EDP Activity" of the firm reflects the extent of diversification. That is, a firm with many distinct product lines is less likely to have the top corporate executives devoting large fractions of their time to any one line, whether or not that is the computer activity. Moreover, this "span of control" problem may arise regardless of whether the firm is organized on an M-form or U-form basis.

There is also a relationship between product diversification, the "Perceived Importance of the commercial computer Activity" and the "Involvement of Top Management." If the computer line were the only line, its perceived importance would necessarily be high and top management would ordinarily be heavily involved. The converse need not be true. Top management could be closely involved, coordinate well, and be active in conflict resolution and yet choose not to introduce a commercial product. Raytheon may be an example.

Table 4.5 suggests inaccuracy in two generalizations that have hitherto been advanced. First, it seems not to be true in the computer industry that diversification across unrelated product lines is so effective in reducing overall R&D risks that it makes the firm more effective in innovation.(175) The diversification also tends to reduce management participation in each of the lines and to increase the degree of routinized channeling of internal communications and control. Where adaptation to rapidly changing technological opportunities requires frequent top-management overview and coordinated responses across the various functional activities of the firms, diversification may result in failure in the high-technology area even while - or because - the survival of the firm itself does not depend on success in that area. Risk reduction for the entire enterprise can be achieved by reducing the commitment to and reliance on the high-technology aspect of the business. This is a choice that GE and RCA exercised.

Second, it is not obviously correct that entry by merger is less effective than de novo development. IBM, it is true, did not develop through acquisitions. In contrast, merger was the means of entry for Remington Rand, Burroughs, NCR, and Xerox, and they did not perform particularly successfully for at least some time. Yet RCA and GE did not depend on mergers as the primary entry vehicle, and they too fared poorly. Alternatively, Honeywell and CDC improved their performance after a series of mergers. To the extent that mergers are associated with decentralization of a sort that relegates the computer portion of a business to a position of infrequent or routine management review, divorces the finance, production, marketing, and R&D functions, or creates opportunities for risk reduction through emphasizing the application of resources to other product lines, mergers may indeed reduce the firms' ability to participate effectively in the high-technology area. Mergers that create geographic separation between units requiring close coordination also tend to reduce the overall effectiveness of the firm. Remington Rand and Burroughs both experienced such a problem. On the other hand, and as in the case of the CDC and Honeywell mergers, if the purpose and effects of mergers is to strengthen and complement the computer-related portion of the business and signals increased commitment of managerial attention and resources, the mergers may have beneficial effects.

The advantages of vertical integration are not in the formality of the organization chart but rather in the unification of multidimensional decisions through time. When technological opportunities abound, "adaptive sequential decision making" requires unified horizontal as well as vertical organizational integrity. The several functional divisions of a firm see the costs and benefits of development projects and innovation in different contexts. Opportunistic behavior by these divisions takes the form of attempts to warp the projects toward the achievement of their own subgoals and constraints. Moreover, development and innovation involve a process of defining and redefining technical, financial, production, and marketing interrelationships in a time continuum. Coordinated control, not the voluntarism of contracting or consensus management, is essential. This need must be seen in context, however. Thus, CDC could initially design and assemble its computer products on the basis of purchased parts and subassemblies - so long as its products were well within the state of the art. The contracting involved nonidiosyncratic capital and frequent transactions with the several parties having near-equivalued alternatives.(176) CDC was at the time a small firm driven by a unified management. Similarly, DEC could start as a nonintegrated supplier of circuitry components. On the other hand, NCR's decision to have GE produce the 304, using an NCR logic design and some of GE's advanced transistors, was a failure. There was no unified control over development at either NCR or GE, no possible way to establish open communications among the parties, no clear designation of the responsible party, and strong inherent properties leading to opportunistic behavior.

In short, pressing the state of the art in a complex, commercial venture requires a unified vertical transactions milieu as well as coordinated decisions among the traditional functional finance, marketing, production, and R&D areas. A classic illustration of this appears in the history of IBM. In 1959-1961 period, IBM was developing a follow-on product line for the 7000 computers. It was to be known as the 8000 series.(177) This development was going on at a time when the company was marketing 15 different processors, including the successful 7090 and 1401. It was anticipated that other companies would introduce computers with superior performance. The 8000 series computers were much more advanced than the 7000s and 1401s models and, after having spent "many millions of dollars" on development, IBM was about to announce the 8000s.

Some people within IBM were of the view that the 8000s, while they would find a quick market, were not sufficiently advanced and not sufficiently interrelated as a "family." The internal debate became known to top management. A full review of all aspects of the program was ordered. This review resulted in a decision to scrap the entire 8000 project. A "New Product Line" (NPL) effort was immediately inaugurated. This development had a different technological and marketing focus and required an enormous additional financial commitment. The 360 series emerged from the NPL work. The 360s were much more advanced than the 8000s, but were not available for the market until 1965. By then, and as anticipated, a number of other manufacturers had achieved considerable success in displacing the earlier IBM computers. It is literally inconceivable that the sequence of decisions relating to the 8000 and 360 series would have been made and that the results would have been successful in the absence of a vertically integrated structure with coordinated management of the several functions. IBM literally "bet the company" on its 360 decision. By reference again to Table 4.5, the effective use of a rapidly developing quasi-exogenous technology appears to require a "High" score in all of the categories.

A question remains about the existence of uninformed buyers and the possibility of "ripoffs." If "information impactedness" implies a large number of uninformed buyers and a small number of informed sellers, the possibility may arise for the latter systematically to discriminate among buyers to their own advantages. The discrimination could conceivably take the form of selling products with preferred combinations of price and quality to an informed segment of buyers, while "ripping off" the uninformed by selling an inferior quality at a high price.(178) This could theoretically occur if the uninformed segment observes the informed segment purchasing at relatively high average prices, infers from these prices that price is a proxy for quality, and yet has no independent way to judge those circumstances when, acting on such inferences, low-quality goods are opportunistically being passed off on them as high-quality goods.(179)

If this kind of behavior is observable in the computer industry, it probably is due to different causes. The appearance of inferior computers was more likely the result of management and technological errors than of intentional deception. Static analysis is inadequate to portray the market situation of the uninformed buyer when technology affords new market opportunities. In the first place, as commercial products were introduced in the early 1950s, there were few buyers and virtually all of those that existed were, if not uninformed, highly uncertain with respect to what functions a computer might perform, how reliable it might be, and the economic costs and benefits of using one. A market had to be created, and the problem of sellers was to reduce uncertainty to the point that sales could be made. There was adequate interchange of information - correct or incorrect information - among the principal potential buyers so that the ex post perceived failure of quality to match price would quickly become known.

While specific product attributes were unknown to many buyers, there was a keen recognition on the part of some sellers and some buyers that those attributes would change rapidly over time. The wide and changing menu of technological opportunities made this inevitable. A seller failing to recognize this, or a seller recognizing it but intentionally attempting to "rip off" customers, would lose in a dynamic setting requiring demand growth for that seller's products. A greater amount of inferior products sold at one point in time would reduce the amounts that might subsequently be sold.(180) This is particularly true when there is no possibility of preempting the technology or otherwise preventing the subsequent appearance of improved products to a customer group with increasing information about the products.

The behavior of the successful sellers of early computers tended to reflect the fact that most customers were uninformed and risk-averse. The option of leasing rather than purchasing gave buyers the opportunity to shift ownership and technology risks to the sellers. Rather than being caught with a machine the capital value of which was highly uncertain because of both unknown-functional capabilities and technological obsolescence, the buyer (lessee) could shift that risk to the seller. Again, the possibility of a "ripoff" is minimized if the buyer (lessee) does not take title and can, under known contractual contingencies, cancel the lease. While there is an "insurance premium" included in the lease terms, the existence of a number of lessors provided an element of competition in the lease market.(181) Unlike conventional insurance in which the likelihood of occurrence of the event being insured against is independent of the behavior of the insuror, the only way the lessor can reduce the risk that the premiums will not cover his exposure is to lease a product with customer preferred price-quality (price performance) characteristics.

The need to coordinate marketing with product quality attributes, improving customer information, and the pace of technological change is obvious. In addition to the widespread use of leasing, the practice of "bundling" - particularly the inclusion of maintenance and software services - was commonly employed by the computer firms through at least the mid-1960s. There was more reason for this than the protection of the assets of the lessor and the ostensible prevention of entry. Leasing and bundling reduced risk from the buyers' point of view and made the risk of product failure a more acceptable one. Throughout the period covered here, recognition of the need for consumer information and follow-on support (e.g., maintenance and software programming assistance) was an important aspect of overall program management. An integrated sales and customer support staff that was highly product- (i.e., computer-) specific was necessary.

XI. CONCLUSIONS

Government activities have had a profound influence on the computer industry. In the period 1945-1960, contracts between governments and private firms often opened broad technological opportunities for particular firms. In some instances the affected firms seized these opportunities; their consequent growth and success can be attributed to their having had such opportunities. Conversely, other firms did not realize the potential that governmentally sponsored R&D and product demand made possible.

After 1960 - after the "second generation" of commercial computers were on the market - the government had less direct and less significant effects on established firms. The basic technology was by then widely diffused and a broad commercial market with large numbers of reasonably informed buyers had developed. Government interest in improved technology for special purposes continued and augmented private R&D expenditures. Government projects gave rise to the entry of new firms and to new products by existing firms. For the most part, the success or failure of a firm in a particular governmentally sponsored project was of less consequence in this period, however. Technological opportunities existed in commercial markets whether or not a firm had a major government project. The swarming of firms into the market - sometimes based on a government contract and sometimes independent of direct government support - reflects these commercial possibilities. An industry had emerged.

Viewed in terms of the firms through which government contributions to technology eventually reached the market, the diffusion appears more rapid among those with internal organizational structures such that the technological opportunities are recognized and coordinated among all aspects of the firms' operations by cognizant high-level management. Where the latter exists, and where the management elects to use the technology as the vehicle for measured success, performance (and the rate of technology diffusion) is superior to that of other firms lacking such attributes. Through such differences, industry structures are fashioned. It is not the industry structure that determines the rate of technological progress but rather the technological progress - and firms' adaptions to and use of technologies - that determine industry structure.

NOTES

1. Irven Travis, "The History of Computing Devices," Theory and Techniques for Design of Electronic Digital Computers, lectures given at the Moore School, University of Pennsylvania, July 8 - August 31 1946, p. 2-1; "Herodotus," Encyclopaedia Britannica (1947); more generally, see Herman H. Goldstine, The Computer from Pascal to Von Neumann (Princeton: Princeton University Press, 1972). See also a twelve part series of articles by Marguerite Zientara, "The History of Computing," Computer World 15 (1981), beginning June 29, 1981, p. 14.

2. Travis, The History of Computing Devices, p. 2-2.

3. Ibid.; and "Napier, John," Encyclopaedia Britannica (1947).

4. "Calculating Machines," Encyclopaedia Britannica (1947). For additional detail, see Zientara "The History of Computing," Part 1, "The Life of Blaise Pascal".

5. Travis, The History of Computing Devices.

6. Ibid.; and "Leibnitz, Gottfried Wilhelm," Encyclopaedia Britannica (1947).

7. "Calculating Machines."

8. Ibid., "Tabulating Machines" and "Babbage, Charles," Encyclopaedia Britannica (1947). See also Zientara, "The History of Computing," Part 3, "Charles Babbage: Man Before His Times."

9. "Calculating Machines"; "Tabulating Machines"; Goldstine, The Computer, p. 65. See also Zientara, Part 5, "William Burroughs: Liberation from Calculation," and Part 6, "Herman Hollerith: Punched Cards Come of Age."

10. Goldstine, The Computer, pp. 65-70; International Business Machines Corporation v. United States, 298 U.S. 131, at 133 (1936).

11. Travis, The History of Computing Devices.

12. "Calculating Machines"; Goldstine, The Computer, pp. 109-110; W. J. Eckert, "The IBM Department of Pure Science and the Watson Scientific Computing Laboratory," Educational Research Forum (International Business Machines Corporation, Endicott, N.Y.: August 25-29, 1947), p. 31; W. J. Eckert, "Punched Card Techniques and Their Application to Scientific Problems," Journal of Chemical Education 24 (February 1947): p. 54.

13. J. H. Curtiss, "A Review of Government Requirements and Activities in the Field of Automatic Digital Computing Machinery," lectures given at the Moore School, University of Pennsylvania, July 8 - August 31, 1946, p. 29-2.

14. Travis, The History of Computing Devices; Goldstine, The Computer, pp. 89-98.

15. Goldstine, The Computer, pp. 89-98, 129.

16. Ibid., pp. 111-118; Howard H. Aiken, "The Automatic Sequence Controlled Calculator," lectures given at the Moore School, University of Pennsylvania, July 8 - August 31 1946.

17. Goldstine, The Computer, p. 115.

18. Ibid., p. 129.

19. Curtiss, "A Review." See also Nancy Stern, From ENIAC to UNIVAC: A Case Study in the History of Technology (Bedford, Mass.: Digital Press, 1981) for discussion of the roles of the National Bureau of Standards, Census, and the National Research Council in this period.

20. George Stibitz, "Introduction to the Course on Electronic Digital Computers," lectures given at the Moore School, University of Pennsylvania, July 8 - August 31 1946.

21. Ibid., emphasis added.

22. Almarin Phillips, Technology and Market Structure: A Study of the Aircraft Industry (Lexington: Heath, 1975).

23. Richard R. Nelson, and Sidney G. Winter, "Forces Limiting and Generating Concentration under Schumpeterian Competition," Bell Journal of Economics (Autumn 1978).

24. P. Dasgupta, and J. E. Stiglitz, "Uncertainty, Industrial Structure and the Speed of R & D," Bell Journal of Economics (Spring 1980).

25. Oliver E. Williamson, Markets and Hierarchies: Analysis and Antitrust Implications (New York: Free Press, 1975).

26. Benjamin Klein, Robert G. Crawford, and Armen A. Alchian, "Vertical Integration, Appropriable Rents, and the Competitive Contracting Process," Journal of Law and Economics (1978).

27. Almarin Phillips, "Organizational Factors in R & D and Technological Change; Market Failure Considerations," in Research, Development and Technological Innovation, ed. D. Sahal (Lexington: Heath 1980).

28. George Akerlof, "The Market for 'Lemons': Qualitative Uncertainty and the Market Mechanism," Quarterly Journal of Economics (August 1970).

29. Steven Salop, and Joseph Stiglitz, "Bargains and Ripoffs: A Model of Monopolistically Competitive Price Dispersion," Review of Economic Studies (1977).

30. Russell Cooper, and Thomas Ross, "Prices, Product Qualities and Asymmetric Information," CARESS Working Paper No. 81-04, University of Pennsylvania (February 1981).

31. For actual speed comparisons, see Curtiss, "A Review," p. 29-22. Kenneth E. Knight, "Changes in Computer Performance," Datamation (September 1966), pp. 40-54, provides more detailed comparisons of time and cost of computations for commercial and special purpose computers ranging from the Harvard Mark I (1944) through the "Librascope L 3055" (December 1963).

32. Curtiss, "A Review." Alan Turing, the British mathematician, advanced the stored program concept at about the same time. For detail see Simon Lavington, Early British Computers (Bedford, Mass.: Digital Press, 1980), pp. 23-35, and Zientara, Part 8, "Alan M. Turing: From Theory to Reality."

33. There was common knowledge among the involved scientists and engineers about the projects, as well as a great deal of interchange of technical development details. See Curtiss, "A Review," and Cuthbert Hurd, "Direct Testimony," Defendant's Exhibit 8951, U.S. v. IBM, 69 Civ. 200 (SDNY), pp. 9-10.

34. Curtiss, "A Review," p. 29-20.

35. Ibid.; Jan Rajchman, "The Selectron," lectures given at the Moore School, University of Pennsylvania, July 8 - August 31 1946.

36. See Stern, From ENIAC to UNIVAC, especially Chapter 5. This section relies on Stern.

37. E. Tomash, and A. A. Cohen, "The Birth of an ERA: Engineering Research Associates, 1946-1955," Annals of the History of Computing 1, 2 (October 1979): 83-97.

38. Both Prudential and Nielson subsequently entered contracts with IBM for what was to be the IBM 701 computer. See below, however, for the circumstances that led IBM into the decision to produce the 701. During the period under discussion here, and up to late 1950, IBM showed no interest in producing a commercial computer.

39. W. F. Sharpe, The Economics of Computers (New York: Columbia University Press, 1959).

40. Carole C. Greenberg, The Role of Commercial Banks in Regional Economic Development: Philadelphia, 1945-1970 (Unpublished Ph.D. dissertation, University of Pennsylvania, 1975), p. 140ff.

41. U.S. vs. IBM, 69 Civ. 200, Defendant's Exhibits 280, 305, Transcript at 5721-23; Mauchly Deposition, pp. 18-28, 37-38.

42. Hurd, "Direct Testimony," p. 18.

43. Sharpe, The Economics of Computers, p. 185.

44. Phillips, "Organizational Factors," pp. 116-117, notes that "[Scientists] typically belong to professional organizations in their own scientific and technologically defined disciplines. . . . The inducement to the firm [the incentive for it] to contribute [to the disciplinary goals] comes from the potential usefulness of the knowledge it gains about exogenous science and technology. The inducement to the [scientist] comes from the reward received from the firm and those received [in recognition and honor] from the professional organization. The former tends to be low [the firm does not contribute much to personnel engaged in professional organizations] when the firm perceives that the benefits to be derived from the discipline are low. From the point of view of [the scientists], a high degree of congruence between the firm's reward system and the reward system of the professional association encourages high levels of contributions to both organizations." In the present context, the congruence was higher within the governmentally budgeted groups than within private firms in the 1950s. The latter contributed little to computer R&D in this time period.

45. "Historical Narrative," U.S. v. IBM, Defendant's Exhibit 14971, p. 22. This book-length exhibit was prepared by Franklin M. Fisher, Richard B. Manche, and James W. McKie.

46. Ibid., p. 182, from testimony of Richard Bloch, Raytheon.

47. Ibid.

48. Ibid.

49. Ibid., pp. 191-193, from testimony of Beard, Crago and Hurd and Plaintiff's Exhibit 6091.

50. Ibid, pp. 193-202, with added testimony of McCollister.

51. Ibid., pp. 203-205, from testimony of Weil.

52. Ibid., pp. 212-216, from testimony of McCollister.

53. Ibid., pp. 217-224, from testimony of Withington and McCollister and Defendant's Exhibits 10282, 10283.

54. Ibid., p. 222.

55. These episodes are recounted in interview with Harold Sarkissian by Robina Mapstone, Smithsonian Institution History of the Computer Files (September 11 and December 21, 1972) and interview with Glenn Hagen by Robina Mapstone, ibid. (November 7, 1973).

56. Ibid., pp. 227-237, from testimony of Oelman and Hangen.

57. Ibid., pp. 238-240, from Defendant's Exhibits 7512, 5421, 5642 and testimony of Goetz and Fernbach.

58. For lease to the G.E. Atomic Power Equipment Department.

59. "Historical Narrative," p. 240, from testimony of Arjay Miller.

60. Hurd, "Direct Testimony."

61. Ibid., p. 22.

62. Ibid., pp. 22-29.

63. "Historical Narrative," p. 41. See also Cuthbert Hurd, "Computer Development at IBM," in A History of Computing in the Twentieth Century, ed. N. Metropolis, J. Howlett, and Gian-Carlo Rota (New York: Academic Press, 1980), pp. 407-409.

64. Knight, "Changes in Computer Performance."

65. Cf. Phillips, Technology and Market Structure.

66. Interview with Stephen Dodd, Jr. by Richard R. Mertz, Smithsonian Institution History of the Computer Files (December 9, 1969).

67. Kent C. Redmond, and Thomas M. Smith, "Lessons from Project Whirlwind," I.E.E.E. Spectrum 14, 10 (October 1977): 52.

68. Stern, From ENIAC to UNIVAC, p. 200. For more detail and a more recent account, see Kent C. Redmond, and Thomas M. Smith, Project Whirlwind: The History of a Pioneer Computer, (Bedford, Mass.: Digital Press, 1980).

69. Knight, "Changes in Computer Performance."

70. Redmond and Smith, "Lessons from Project Whirlwind," p. 56.

71. Hurd, "Direct Testimony," p. 67.

72. Ibid., pp. 56-92.

73. Ibid., p. 57.

74. "Historical Narrative," p. 70.

75. Hurd, "Direct Testimony," p. 57.

76. Ibid., pp. 58-60.

77. Ibid., p. 61.

78. Ibid., p. 80.

79. For a history of the remarkable 305, see Mitchell E. Morris, "Professor Ramac's Tenure," Datamation 27 (April 1981): 195.

80. Herman Lukoff, From Dits to Bits . . . A Personal History of the Electronic Computer (Portland, Oregon: Robotics Press, 1979), p. 145.

81. Remington Rand and the Sperry Corporation merged on June 30, 1955. A consolidated computer group, the Univac Division, was created at this time.

82. Ibid., p. 138.

83. Ibid., p. 146.

84. Ibid., p. 150.

85. Herman Lukoff, "Were Early Giant Computers a Success?," Datamation (April 1969).

86. Ibid., p. 79.

87. Lukoff, From Dits to Bits, p. 170.

88. The LARC was delivered in May 1960; the Philco 2000-211, in March 1960. Knight estimates their respective speeds in scientific computations at 142,000/second and 105,844/-second. In commercial computations, the figures are 40,450/second and 55,740/second. No commercially available computer approached these speeds.

89. Lukoff, From Dits to Bits, p. 177.

90. Ibid., p. 178.

91. "Historical Narrative," p. 128.

92. Ibid.

93. The possibility of using magnetic core logic rather than transistor logic was investigated but rejected. See U.S. v. IBM, 69 Civ. 200, Defendant's Exhibit 4767, p. 19.

94. "Historical Narrative," p. 130.

95. Knight, "Changes in Computer Performance."

96. "Historical Narrative," p. 132.

97. Ibid., pp. 133-135.

98. Ibid., pp. 137-138.

99. Interview with J. Desch and R. Mumma by Henry Tropp, Smithsonian Institution History of the Computer Files (January 17, 1973), p. 18.

100. Ibid., p. 51; Stern, From ENIAC to UNIVAC, pp. 59-60.

101. Desch and Mumma, interview, pp. 85-129.

102. Interview with Jerry Mendelson conducted by Henry Tropp, Smithsonian Institution History of the Computer Files (January 3, 1973), p. 72.

103. Desch and Mumma, interview, pp. 85-129.

104. Mendelson, interview, pp. 71-72.

105. Henry Tropp, "The Effervescent Years: A Retrospective," I.E.E.E. Spectrum (February 1974), p. 76.

106. Brian Randell, "An Annotated Bibliography," Annals of the History of Computing, 1, 2 (October 1979): 1441; B. Johnson, The Secret War (London: British Broadcasting Corporation, 1978).

107. Interview with Glenn E. Hagen, conducted by Robina Mapstone, Smithsonian Institution History of the Computer Files (November 7, 1973), pp. 4-13; Interview with Charles Williams, Smithsonian Institution History of the Computer Files (January 5, 1973), pp. 2, 14-19; Desch and Mumma interview, pp. 160-162.

108. R. A. Ramey, "Magnetic Amplifier Circuits and Applications," Electrical Engineer (September 1953), p. 793. Ramey is an authority on magnetic amplifiers and assigned important patents relating to these devices to Remington Rand. This article is the third in a three-part series devoted to magnetic amplifiers. At the time the series was written, Ramey was the manager of the Magnetic Development Section of Westinghouse Electric Corporation, Pittsburgh, Pennsylvania.

109. Ibid., p. 794.

110. R. A. Ramey, "The Single-Core Magnetic Amplifier as a Computer Element," Transactions AIEE 71, part I (1952): 442-226.

111. R. A. Ramey, "Magnetic Amplifier Circuits and Applications," p. 794.

112. R. A. Ramey, "Magnetic Amplifiers: Principles, Advantages and Limitations," Electronic and Radio Engineer 34 (April 1957): 118-119.

113. Ibid., pp. 118, 123 (emphasis added). Note that this was written two or three years before the LARC project began, but already reflects the view that magnetic amplifiers and transistors could be used in the same equipment. The Air Force AFCRC, completed in April 1956, was based on the same concept.

114. For example, "Magnetic Computer Has High Speed," Electronics (August 2, 1957), p. 160.

115. Office of Naval Research, Digital Computer Newsletter 3, 2 (July 1951): 1. "This machine, which has a magnetic drum memory and teletype input-output facilities, was assembled entirely from general-purpose electronic building blocks. Almost all of those general-purpose units belong to the line of equipment known as Pulse-Control Units. . . . Each Pulse-Control Unit is a standard logical component, such as a flip-flop, gate or pulse delay circuit, and is equipped with input and output buffers. Wave forms on the coaxial cables which interconnect units are restricted to two standard types: 0.1-microsecond pulses and two-valved d-c control voltages having 0.2 microsecond switching time. Use of pulse-control units permitted assembly of the computer directly from logical diagrams without the usual intermediate engineering steps."

116. Digital computer Newsletter, 6, 3 (July 1954): 2.

117. Digital Computer Newsletter 8, 1 (January 1956): 10.

118. "Magnetic Computer Has High Speeds," Electronics (August 1, 1957), p. 156.

119. Digital Computer Newsletter 8, 1 (January 1956): p. 10 (emphasis added).

120. A. J. Gehring, L. W. Stowe, and L. D. Wilson, "The Univac Magnetic Computer: Parts I, II and III," IRE Convention Record (1956), pp. 109-111.

121. While the history is not clear, it appears that the SS90 was the first model delivered in the United States. It presumably used a 90-column card. The 80-column card UNIVAC SS80 was delivered in June 1959, and could simulate the IBM 650. See Gerald Brock, The U.S. Computer Industry: A Study of Market Power (Cambridge: Ballinger, 1975), pp. 14, 92. The SS80/90, like LARC and AFCRC, did contain some surface barrier transistors, but its main components were magnetic amplifiers. See Saul Rosen, "Electronic Computers: A

Historical Survey," Computing Surveys 1, 1 (March 1969): 19; Digital Computer Newsletter 8 1 (January 1956): 10; Herman Lukoff, "Were Early Giant Computers a Success," p. 78.

122. Herman Lukoff, From Dits to Bits, p. 139.

123. Lukoff, "Were Early Giant Computers a Success?," p. 78.

124. Rosen, "Electronic Computers," p. 20.

125. Lukoff, From Dits to Bits, p. 137.

126. Morton Norman, former ERA employee, private interview with Barbara Goody Katz.

127. Digital Computer Newsletter 10, 2 (April 1958): 4.

128. Brock, The U.S. Computer Industry, p. 93.

129. J. P. Eckert, document dated April 7, 1966, Defendant's Exhibit 8, p. 2, U.S. v. IBM. Discussion of other magnetic amplifier computers can be found in Digital Computer Newsletter, 9 (October 1957): 1-3; ibid. 11 (January 1959): 8; ibid. 5 (July 1953): 8-10; ibid. 11 (April 1959): 12; ibid. 3 (October 1951): 5; Hidetosi Takahasi, "Some Important Computers of Japanese Design," Annals of the History of Computing 2 (October 1950): 334; R. W. Marczynski, "Early Polish Digital Computers," ibid. 2 (January 1980): 37-48.

130. W. Gosling, "The Pre-History of the Transistor," The Radio and Electronic Engineer 43, 11 (1973): 10.

131. Charles Weiner, "How the Transistors Emerged," I.E.E.E. Spectrum 10 (1973): 26.

132. Ibid. See also G. L. Person, and W. H. Brattain, "History of Semiconductor Research," Proceedings of the IRE 43 (1955): 1794-1806.

133. Weiner, "How the Transistors Emerged," p. 26.

134. Interview with Irving Reed conducted by Robina Mapstone, Smithsonian Institution History of the Computer Files (December 19, 1972).

135. Weiner, "How the Transistor Emerged," p. 28.

136. Ibid.

137. Morgan Sparks, "Morgan Sparks Reflects on 25 Years of Transistors," Bell Laboratories Record (December 1972), pp. 343-344.

138. J. H. Felker, "Performance of TRADIC Transistor Digital Computer," Proceedings of the Eastern Joint Computer Conference (Philadelphia, December 1954), p. 48.

139. Jay W. Forrester, "Digital Computers: Present and Future Trends," AIEE-IRE Computer Conference No. 10 (February 1952), pp. 112-113. Notice Forrester's reference to core memory. Recall that core memory was being developed at MIT, along with IBM, in connection with SAGE.

140. Felker, "Performance of TRADIC," p. 42.

141. Ibid., p. 48.

142. E. Tomash, and A. A. Cohen, "The Birth of ERA: Engineering Research Associates, Inc., 1946-1955," Annals of the History of Computing, 1, 2 (October 1979): 93.

143. Interview with Richard Tamaka conducted by Robina Mapstone, Smithsonian Institution History of the Computer Files (January 4, 1973, pp. 10-11.

144. Interview with Montgomery Phister conducted by Robina Mapstone, Smithsonian Institution History of the Computer Files (February 21, 1973), pp. 43-44.

145. Irving Reed, et al., "Logical Design of CG-24 (A General Purpose Computer)," MIT-Lincoln Lab Technical Report No. 139 (April 15, 1957), p. 51.

146. Digital Computer Newsletter 9, 3 (July 1957): 1; interview with Glenn Hagen, p. 34.

147. Digital Computer Newsletter 9, 1 (January 1957): 8.

148. Rosen, "Electronic Computers," p. 22.

149. "Historical Narrative," pp. 238-240.

150. Ibid., quoting testimony of Arjay Miller (Transcript, p. 85186).

151. Ibid., p. 195-198 quoting testimony of McCollister (Transcript, pp. 9541-9543).

152. Ibid., from Plaintiff's Exhibit 114, p. 5.

153. Ibid., quoting testimony of McCollister (Transcript, pp. 9622, 9543-9544).

154. Ibid., quoting testimony of McCollister (Transcript, p. 9255) and Beard (Transcript, p. 8717).

155. Interview of Mendelson, p. 74.

156. U.S. v. IBM, Plaintiff's Exhibit 320, p. 4.

157. Ibid., and testimony of Reginald Jones (Transcript, pp. 8752, 8869, 8874, 8875-8876), Weil (Transcript p. 7010).

158. "Historical Narrative," pp. 105-123, citing, inter alia, William Norris (Defendant's Exhibits 305, 272, 284); John W. Lacey (Defendant's Exhibit 13526, at pp. 43-45, 91); Business Week (Defendant's Exhibit 105).

159. Ibid., citing Norris (Transcript, p. 5608) and Defendant's Exhibit 271.

160. Ibid., citing Defendant's Exhibit 284, 13666, 331; Plaintiff's Exhibit 355, and Norris (Transcript, pp. 5615-5616).

161. Ibid., citing Norris, Defendant's Exhibit 284, pp. 4-5.

162. See, generally, ibid., pp. 670-690.

163. Ibid., p. 692.

164. Ibid., p. 711.

165. Ibid., pp. 619-637.

166. Knight, "Changes in Computer Performance," p. 46.

167. "Historical Narrative," pp. 505-512, citing testimony of Withington, Wright and Weil.

168. Phillips, Technology and Market Structure; Nelson and Winter, "Forces Limiting and Generating Concentration."

169. For a description of an endogenously determined equilibrating structural mechanism, see Dasgupta and Stiglitz, "Uncertainty." See also Richard Levin, "Toward an Empirical Model of Schumpeterian Competition" (unpublished manuscript, 1981). If Nelson and Winter type simulations were run with randomly changing technology shocks over time, the structural tendencies would be continuously upset, leading to varieties of entry, exit and market share consequences.

170. In particular, Williamson, Markets and Hierarchies.

171. See Albert O. Hirshman, Exit, Voice and Loyalty: Responses to Decline in Firms, Organizations and States (Cambridge, Mass.: Harvard University Press, 1970).

172. Phillips, "Organizational Factors in R & D and Technological Change."

173. Burton Klein, Dynamic Economics (Cambridge, Mass.: Harvard University Press, 1977), p. 164.

174. Oliver E. Williamson kindly made available the corporation files of the Center for the Study Organizational Innovation in connection with this portion of the paper.

175. For a related discussion, see Steven Globerman, and James Diodati, "Market Structure, Internal Organization, and R & D Performance in the Telecommunications Industry," Quarterly Review of Economics and Business 20, 4 (Winter 1980).

176. On these points, see Oliver E. Williamson, "Transaction Cost Economics: The Governance of Contractual Relations," Journal of Law and Economics 22 (October 1979); and Benjamin Klein, Robert G. Crawford, and Armen A. Alchian, "Vertical Integration, Appropriable Rents, and the Competitive Contracting Process."

177.   "Historical Narrative," pp. 269-280.

178.   For a discussion, see Cooper and Ross, "Prices, Product Qualities and Asymmetric Information," and Salop and Stiglitz, "Bargains and Ripoffs."

179.   In addition to the references in note 178, see Sanford Grossman, and Joseph Stiglitz, "On the Impossibility of Informationally Efficient Markets," American Economic Review, 70 (June 1980); Richard Kihlstrom and Leonard Mirman, "Information and Market Equilibrium," Bell Journal of Economics 6, 1 (Spring 1975); and C. J. Sutton, "The Effect of Uncertainty on the Diffusion of Third Generation Computers," Journal of Industrial Economics 23, 4 (June 1975).

180.   See Dennis Smallwood, and John Conlisk, "Product Quality in Markets Where Consumers are Imperfectly Informed," Quarterly Journal of Economics (1979).

181.   In the early years of introducing a radically new product, the most important alternative (and competitive force) is for buyers to do without the product.   This was a key element in Schumpeter's treatment of technological change in The Theory of Economic Development.

5
Agriculture
R.E. Evenson

Government programs directed toward the discovery and diffusion of improved agricultural technology have been in place for more than a century in the United States. The Patent Office set up an agricultural division in 1839 to encourage the importation of improved seeds from Europe and to stimulate breeding work to improve both crops and animals. Several state governments(1) established colleges of agriculture before the Land Grant College Act of 1862 which provided funding for a college in each state. In 1887 the research functions of these colleges were institutionalized by the Hatch Act which provided federal funds to each state for agricultural research. The U.S. Department of Agriculture also established a number of research laboratories prior to 1900.

Today, 52 State Agricultural Experiment Stations and a number of USDA research centers form the nucleus of research and development directed toward plant and animal improvement. The farm sector itself undertakes little in the way of formal R&D. Farm-input supply industries (farm machinery, farm chemicals, animal health) and postharvest industries (food and fiber processing), however, do undertake significant R&D that influences technology employed in the sector. Available data indicate that the public and private investments in the production of new agricultural technology are of roughly equal magnitude with a high proportion of the public support being provided by state governments.

The overall performance of the agricultural sector in terms of productivity gains has been impressive when judged against national and international standards. Measured productivity gains in the sector have been more rapid than for any other broadly defined sector of the economy over the past 50 years. Internationally, the United States is the world's dominant exporter of agricultural commodities and its comparative advantage is growing.

This chapter reviews the evolution of the system of public and private research institutions relevant to the sector and the evidence linking the economic performance of the sector to investment in research and related activities. Section I provides a capsule description of the basic economic characteristics of the agricultural sector. It documents the productivity performance and the changing political and economic character of the U.S. farming sector to provide background for the remainder of the chapter. Section II discusses invention and research and development in the private sector. Section III describes the evolution of the public sector and its political support base. Section IV reviews studies of the contribution of agricultural research to productivity. Section V discusses several current policy issues.

I.   THE AGRICULTURAL SECTOR IN THE U.S. ECONOMY

The agricultural sector has changed greatly in character over U.S. history. From its basically agrarian origins, the U.S. economy has become decreasingly agricultural throughout its history. The share of the total employed labor force in agriculture has fallen steadily from 45 percent in 1900 to 3.5 percent in 1980. The sector contributed 18 percent of GNP in 1900 and contributes roughly 4 percent today. Massive rural-to-urban migration has taken place over the period. The sector is one of the most capital-intensive sectors in the economy today. The share of labor in total costs is roughly 21 percent, far below the comparable share for almost all manufacturing sectors (excluding land, it is 27 percent). Today the agricultural sector is one of the major export sectors of the economy, accounting for more than 20 percent of U.S. exports in recent years.

Much of the decline in the relative importance of the sector is due to a transfer of activities undertaken on the farm to the industrial sector. The retail value of food in 1980 was $262 billion while the farm value of food was only $80 billion. Roughly the same ratio holds for the $60 billion in nonfood farm products. The sector also purchases some $70 billion from the farm supply industries. The "agribusiness" sector including the farm production sector thus accounts for approximately 14 percent of GNP today.

Table 5.1 provides an historical sketch of a number of indicators of change in the sector. The data show that the number of farms peaked around 1920 and has declined rapidly since. Total land farmed has been roughly constant since the 1930s while average farm size has increased more than threefold. Farm tenancy has declined, but nonfamily corporate ownership has expanded rapidly in recent years. The sector is still, however, basically a family enterprise sector.

Table 5.1.   General Characteristics of
U.S. Agricultural Sector.

	1880	1900	1920	1940	1960	1978
Land						
Number of Farms (thousands)	4,009	5,737	6,448	6,097	3,962	2,330
Land in Farms (millions)	536	838	956	1,061	1,176	1,048
Acres per Farm	134	146	148	174	297	450
Tenants (percentage)	n/a	23.3	27.7	29.4	14.5	12.0
Nonfamily Corporate ownership (percentage)	n.a.	n.a.	n.a.	n.a.	n.a.	11.2
Production Values						
Farm Products ($1975) (billions)	–	–	19.25	32.4	41.2	56.4
Marketing ($1975) (billions)	–	–	24.75	45.9	84.0	118.0
Exports ($1967) millions	–	4,689	4,551	1,910	7,489	12,834
Imports ($1967) millions	–	1,972	4,388	5,572	6,515	6,526
Crop Production ($1967) millions	–	9,574	12,647	14,202	21,438	31,109
Livestock Production ($1967) millions	–	13,509	17,125	22,663	33,065	43,505
Income Per Farm (1975 Dollars)						
Gross Farm Sales:						
Crops	–	–	16,635	13,343	28,895	36,727
Livestock Products	–	–	14,885	26,422	35,997	49,552
Government Payments	–	–	–	4,136	1,334	2,545
Net Farm Income	–	–	19,395	24,678	22,840	23,419
Income from Nonfarm Sources (as % of total farmer income)	n.a.	n.a.	n.a.	36	37	53
Ratio: Income of Farmers divided by Income of Nonfarmers	–	–	–	–	55	91
Prices (1910-1914 = 100)						
Paid by Farmers	–	–	212	98	275	638
Received by Farmers	–	–	212	100	239	524
Parity Ratio (Prices Received/Prices Paid)	–	–	99	81	80	70
Cost Shares						
Labor	–	–	.000	.428	.265	.160
Land	–	–	.185	.184	.194	.216
Machinery	–	–	.118	.147	.250	.313
Chemicals	–	–	.021	.034	.058	.096

Production data show that farm production has grown at a modest rate while the marketing-processing component of the food-fiber sector has expanded rapidly. This reflects increasing specialization and industrialization of a number of activities formerly undertaken in the farm sector.

Farm-income data show that net farm incomes have not increased rapidly and that nonfarm sources account for more than half of the real income of people classified as farmers today. This is somewhat misleading, however, because many farmers have realized huge capital gains on assets owned (chiefly land) in recent years. In fact in every year since 1970, capital gains to farmers have exceeded net farm income and in the late 1970s exceeded it by a factor of two or more. The government has played an active interventionist role in U.S. agriculture that is only partly reflected in the government payments data reported. During the 1950s and 1960s, the "farm problem," i.e., low net farm incomes (not counting capital gains) of most family farmers relative to nonfarm household income was an important political and public problem. Price support programs and the cumulation of large publicly held stocks of commodities were the order of the day. This changed in the 1970s with increased export demand from the developing world and the socialist nations.(2)

"Parity," the ratio of prices received by farmers to prices paid, was an important political bargaining point during the farm program debates of the 1950s with many campaign promises to restore the ratio (with a favorable 1910-1914 base) to 100 percent of parity via price supports. The fact that this index is really an index of productivity or efficiency gains (if properly calculated) was never really fully understood by politicians.

The cost-share data in Table 5.1 provide some insight into the changing technology employed in the sector. The decline in the labor share and the increase in the machinery share reflect the extensive mechanization of the sector since 1920.

Table 5.2 provides a more useful characterization of changing technology in agriculture. The labor productivity measures in Table 5.2 show truly extraordinary gains for most commodities. Given the apparent ease of substitution of machines for labor and the growth in the use of other inputs, these labor productivity indexes are not really comparable to similar indexes measured in the nonagricultural sector.

For crops, however, land productivity (yield per acre) is a more reasonable index of productivity change. It is influenced by the use of farm chemicals, but the mix of labor and machines generally does not affect it greatly. The data in Table 5.2 show that yields have increased in all crops after the 1935-1939 period. The classic study by Zvi Griliches of the returns to hybrid-corn research was based on the yield changes up to 1957.   As the table shows, corn yields have increased dramatically since then.   In 1979, average corn yields were over 100 bushels per acre.   This, of course, was due to increased fertilizer application to some extent, but much of it is attributable to the several generations of new hybrid-corn varieties produced by both public- and private-sector research in recent years.

Livestock productivity indexes are a little more difficult to interpret because feed inputs have increased.   Modern dairy cattle are generally heavier than dairy cattle in the 1930s and consequently consume more feed.   Nonetheless, the more than doubling of milk per cow since the 1930s is partly attributable to breeding and management improvements.

A number of total factor productivity calculations have been made for U.S. agriculture.   The U.S.D.A. reports a Laspeyres weighted index on a regional basis.   A more appro-priate Divisia-type index has also been constructed for states and regions.(3)   Table 5.3 reports a summary of annual rates of change in the Divisia TFP index by region by 5 year peri-od.(4)

The time periods are relatively short (three-year averages are used as beginning and ending values) and some weather variation exists in these data.   Nonetheless, the degree of correlation between regions over time is not so high as to suggest rapid technology diffusion between regions.

Over the 1940-1978 period, the leading regions in produc-tivity growth were the Delta, the Southeast, and the Pacific. The Appalachian region clearly comes off worst.   The Pacific and Southern Plains regions have done best in the 1970s.   The national data show the rapid gains of the late 1930s and the 1950s.   Interestingly, the Pacific region has tended to lead other regions in terms of productivity growth.   (Aggregated data for the North Central and Southern Regions are not en-tirely comparable with the data for the ten regions).   The Delta and Southeastern regions performed particularly well in the 1950s, suggesting that they were catching up to more advanced regions.(5)

II.   INVENTION AND PRIVATE-SECTOR RESEARCH
IN AGRICULTURE

Most agricultural-producing firms (farms) are small.   Very few engage in formal R&D, although most do undertake experimen-tation associated with screening new technology for cost effectiveness.   It would be reasonable to say that perhaps one-fourth of the time of a typical family farmer is devoted to searching, screening, and experimentation with new technolo-gy.   This entails attending meetings and programs offered by

Table 5.2.   Selected Productivity Indicators:
U.S. Agriculture.

Selected Crops: Labor-hours Per Unit of Production and
Related Factors,United States, Indicated Periods, 1915-80[1]

Crop and item	1915-19	1925-29	1935-39	1945-49	1955-59	1965-69	1976-80[2]
Corn for grain:							
Hours per acre	34.2	30.3	28.1	19.2	9.9	5.8	3.5
Yield-bushels	25.9	26.3	26.1	36.1	48.7	78.5	96.0
Hours per 100 bushels	132	115	108	53	20	7	4
Sorghum grain:							
Hours per acre	....	17.5	13.1	8.8	5.9	4.2	3.8
Yield-bushels	....	16.8	12.8	17.8	29.2	52.9	53.8
Hours per 100 bushels	....	104	102	49	20	8	7
Wheat:							
Hours per acre	13.6	10.5	8.8	5.7	3.8	2.9	2.8
Yield-bushels	13.9	14.1	13.2	16.9	22.3	27.5	32.0
Hours per 100 bushels	98	74	67	34	17	11	9
Hay:							
Hours per acre	13.0	12.0	11.3	8.4	6.0	3.8	3.4
Yield-ton	1.25	1.22	1.24	1.35	1.61	1.97	2.22
Hours per ton	10.4	9.8	9.1	6.2	3.7	1.9	1.5
Potatoes:							
Hours per acre	73.8	73.1	69.7	68.5	53.1	45.1	36.8
Yield-cwt	56.9	68.4	70.3	117.8	178.1	212.8	263.6
Hours per ton	26	21	20	12	6	4	3
Sugarbeets:							
Hours per acre	125	109	98	85	51	33	25
Yield-ton	9.6	10.9	11.6	13.6	17.4	17.5	20.0
Hours per ton	13.0	10.0	8.4	6.3	2.9	1.9	1.2
Cotton:							
Hours per acre	105	96	99	83	66	30	7
Yield-pounds	168	171	226	273	428	484	472
Hours per bale	299	268	209	146	74	30	7
Tobacco:							
Hours per acre[3]	353	370	415	460	475	427	248
Yield-pounds	803	772	886	1,176	1,541	1,960	1,997
Hours per 100 pounds	44	48	47	39	31	22	12
Soybeans:							
Hours per acre	19.9	15.9	11.8	8.0	5.2	4.8	3.6
Yield-bushels	13.9	12.6	18.5	19.6	22.7	25.8	29.2
Hours per 100 bushels	143	126	64	41	23	19	12

[1]Labor-hours per acre harvested, including preharvest work
on area abandoned, grazed, and turned under.
[2]Preliminary.
[3]Per acre planted and harvested.

Economic Research Service.

Table 5.2.   Selected Productivity Indicators:
U.S. Agriculture (continued).

Livestock: Labor-hours Per Unit of Production and Related Factors,
United States, Indicated Periods, 1915-80

Kind of livestock and item	1915-19	1925-29	1935-39	1945-49	1955-59	1965-69	1976-80[1]
Milk cows:							
Hours per cow	141	145	148	129	109	78	42
Milk per cow (pounds)	3,790	4,437	4,401	4,992	6,307	8,820	11,323
Hours per cwt. of milk	3.7	3.3	3.4	2.6	1.7	.9	.4
Cattle other than milk cows:							
Hours per cwt. of beef							
produced[2][3]	4.5	4.3	4.2	4.0	3.2	2.1	1.3
Hogs:							
Hours per cwt. produced	3.6	3.3	3.2	3.0	2.4	1.4	.5
Chickens (laying flocks and eggs):							
Hours per 100 layers	....	218	221	240	175	97	53
Rate of lay	....	117	129	161	200	219	239
Hours per 100 eggs produced	....	1.9	1.7	1.5	.9	.4	.2
Chickens (farm raised):							
Hours per 100 birds	33	32	30	29	23	14	11
Hours per cwt. produced[3]	9.4	9.4	9.0	7.7	6.7	3.7	2.8
Chickens (broilers):							
Hours per 100 birds	....	....	25	16	4	2	.5
Hours per cwt. produced[3]	....	....	8.5	5.1	1.3	.5	.1
Turkeys:							
Hours per cwt. produced[3]	31.1	28.5	23.7	13.1	4.4	1.3	.4

[1] Preliminary.
[2] Production includes beef produced as a byproduct of the
milk-cow enterprise.
[3] Live-weight production.

Economic Research Service.

Source: 1981 Agricultural Statistics.

Table 5.3. Average Annual Rates (in percent) of TFP Change by Region, 1925–1970.

Region	1930–1936	1935–1941	1940–1946	1945–1951	1950–1956	1955–1961	1960–1966	1965–1971	1970–1978	1940–1978
1. Northeast			3.1	2.5	2.2	2.3	2.6	1.5	.4	13.6
2. Lake States			.3	1.3	2.6	3.0	2.8	.8	2.7	13.5
3. Corn Belt			.2	2.0	4.0	1.2	3.2	.7	1.8	13.1
4. Northern Plains			1.8	1.1	2.5	5.9	4.5	.6	1.0	16.4
North Central (2,3,4)	-.1	7.8	0.9	2.0	3.6	3.9	4.9	1.1	1.8	–
5. Appalachain			.4	.4	2.6	1.9	2.1	1.0	1.5	9.9
6. Southeast			5.9	-1.0	3.7	4.6	2.1	2.3	1.3	18.9
7. Delta			3.5	-1.7	4.0	5.1	4.1	3.2	2.0	20.2
8. Southern Plains			.4	1.5	.9	6.6	2.1	2.6	3.0	17.1
South (5,6,7,8)	1.5	4.9	.8	.3	3.4	5.7	3.3	3.0	2.0	–
9. Mountain	1.2	7.0	-.1	.6	3.6	3.6	2.3	2.3	.8	13.1
10. Pacific	2.4	.3	1.7	3.0	4.9	1.4	1.6	4.0	3.0	19.6
United States	-0.2	4.7	0.2	1.3	3.9	3.9	3.5	2.6	1.7	17.1

Source: Landau and Evenson (1973).

240

the public extension service and by the State Agricultural
Experiment Stations, assessing the literature generated by the
public extension service and private-input supply firms, and
visiting input suppliers.

In the course of experimentation associated with the
search for and screening of technology, a certain amount of
technology adaptation and modification takes place. Most of
this adaptation qualifies as "subinvention," but the farm sector
has also generated a considerable amount of genuine invention
over the years.

The bulk of the new technology employed in the sector is
thus produced outside the farming sector. A substantial part
is produced by input supply firms. The farm machinery and
farm chemical industries include a number of large R&D-
intensive firms as well as many smaller innovative firms. The
postharvest industries also include a number of large firms
with R&D activities. Although the food and beverage industries are not generally considered high-technology industries,
a considerable amount of technology change has taken place in
these industries. (They will not, however, be a major focus
of this chapter.)

Farmers do produce biological technology. In fact,
virtually all animal breed improvements are produced by farmers, although they are aided greatly by artificial insemination
firms and public-sector research. A number of important crop
varieties were produced by farmers, particularly in the early
part of the century. In recent years, changes in plant-
breeding patent protection laws have stimulated increased
private breeding activities. In 1970 a strengthened Plant
Varietal Protection Act was passed. The act provides for a
type of patent protection to breeders of sexually produced
crops. The conventional patent system provides only for
protection to asexually propagated crops. (Interestingly the
PVPA is administered by the U.S. Department of Agriculture,
not by the Patent Office, which did not expect it to be used
extensively.) Since 1972, some 1,200 applications have been
processed with roughly 800 PVPA certificates issued. In a
major crop such as soybeans, 35 or so new full-time private
soybean-breeding programs have been added to the dozen or
so public programs. The PVPA requires that new varieties be
made available to all breeders through a national gene bank.
It also allows farmers to save their own seed from year to year
without infringement penalties.

Firms supplying inputs and purchasing outputs have
played a major role in the development of farm technology.
These include machinery dealers who provide repair services
and engage in some subinvention. Blacksmith shops and small
custom engineering shops also fall in this category. On the
biological side, this category includes firms specializing in
technology production and sale: seed companies, horticultural

supply companies and artificial insemination firms. Feed and fertilizer suppliers and veterinarians also contribute. Many of these suppliers conduct and facilitate the experiments that are important to any technology-development process. Their success depends on selling new products to farmers and they are continually obtaining feedback from farmers regarding the products they sell. They pass this information on in turn to manufacturers and thus "articulate" the demand for invention. Manufacturing firms are not very important in producing biological technology except in animal-feed manufacturing. (The recent boom in private biogenetic research promises to change this.) They are, of course, quite important in mechanical and chemical technology.

Table 5.4 summarizes National Science Foundation data on R&D spending related to the sector for recent years. These data are very crude and comparison with public-sector expenditures is made difficult because little "development" research is done in the public sector. A later comparison will show that private spending accounted for roughly one-fourth of total R&D relevant to agriculture in the 1950s but that this had risen to approximately 40 percent in the 1960s and 50 percent in the 1970s.

Table 5.4. Research and Development Expenditures by Private Industrial Firms of Relevance to U.S. Agriculture (in Millions of 1959 dollars).

	1952	1958	1960	1965	1970	1975
Production Oriented:						
Agricultural Chemicals (SIC 287)	31	35	27	52	67	75
Farm Machinery (SIC 352)	31	58	72	78	60	66
Product Oriented:						
Food & Kindred Products	61	72	88	107	118	133

Source: National Science Foundation, "Research and Development in Industry 1970," NSF 72-309 for 1960, 1965, 1970; Latimer, (1962) for 1952-1958.

These data do not provide much insight into the character and type of private-sector research. Fortunately, we have data on patenting that reveal much more, particularly regarding the historical role of the private sector. Patents have limitations as measures of inventive inputs or of inventive output. Nonetheless, they provide to the knowledgeable reader a generally consistent basis for assessing the extent of invention in a sector. It was earlier noted that the U.S. Patent Office instigated public-sector involvement in agricultural research. It was, of course, instrumental in encouraging private invention as well.

In fact, it would appear that the patent system was working quite effectively in stimulating invention in mechanical and chemical technology fields relevant to agriculture. Thousands of patents had been granted to private inventors in agricultural research. Further, the inventive base was broad. Patents were granted to inventors in all states with varied backgrounds (including a number of illiterate inventors). Tables 5.5, 5.6, and 5.7 provide a summary of patent data in three major mechanical invention fields, planters and seed drills, grain threshers, and plows, and give some insight into this invention.

The data show the number of patents granted by decade by the state of origin of the inventor. They also show (in parenthesis) the number of these inventions that were assigned to a corporate entity at the time of the patent grant. This is a good proxy for corporate invention.

The reader will note two phenomena in these tables. The first is the steady growth in assignment reflecting the development of the farm machinery industry. The second is the regional pattern of invention. As settlement proceeded westward, we observe inventions emerging from a region roughly 50 years or so after settlement of the region. We also observe patenting, particularly assigned patenting, tending to be located where the farm equipment firms were located. In the period prior to 1880 or so, a large number of small firms producing tillage equipment were in business. Danhof (1967) reports that 800 distinct models of plows were advertised for sale in the northern United States in 1880. Many of these small firms or shops started their businesses around a particular invention.

During the 1880s and 1890s, the industry consolidated rapidly. The large firms (McCormick Deering, John Deere, Case, Allis Chalmers, Minneapolis Moline, and others) in the industry were located in the Midwest. These firms often purchased the assets, including patents, of small firms as they expanded.

The second phenomenon revealed in the tables is that those regions with the earliest inventions are the first to exhibit declines in patenting activity. By the late 1800s the

Table 5.5.   Patenting in Planters and Drills.   Patent Class: Subclass, 111: 1 to 89.

Time Period	New England	Middle Atlantic	Eastern Corn Belt	Western Corn Belt	Lake States	Appalachia	South	Plains States	Mountain States	Pacific States	Foreign	Canadian
Pre-1830												
1830-39		5										
1840-49	14	31	7			6	1	1				
1850-59	20	103	98	66	25	9	19	8				
1860-69	10	181 (1)	282	408	69 (3)	17	19	9		2 (1)	1	1
1870-79	21 (1)	126 (3)	247 (15)	467 (19)	81 (10)	107	70	43	1	9 (1)	3	4
1880-89	31 (1)	101 (10)	263 (42)	631 (82)	102 (19)	125 (4)	160	207 (15)	14 (2)	27	7	7
1890-99	10 (1)	99 (8)	216 (58)	339 (69)	102 (12)	110 (13)	155 (1)	211 (26)	8	13	10	13 (3)
1900-09	4	46 (9)	149 (44)	393 (94)	131 (30)	94 (9)	135 (1)	149 (9)	15 (3)	15 (1)	18 (1)	12 (2)
1910-19	3	43 (7)	99 (28)	312 (75)	90 (29)	63 (6)	82 (4)	133 (7)	22	28 (6)	14 (1)	14 (1)
1920-29	4	14 (2)	37 (11)	81 (35)	23 (5)	28 (3)	18	43 (2)	9 (1)	17	13	6
1930-39	6	29 (9)	66 (29)	126 (57)	51 (23)	32 (10)	11	59 (11)	15 (2)	26 (5)	25 (2)	13 (6)

Numbers in parentheses are patents assigned to firms.

Table 5.6. Grain Threshing, Patent Class: Subclass 130: 2, 13, 21 to 29

Time Period	New England	Middle Atlantic	Eastern Corn Belt	Western Corn Belt	Lake States	Appalachia	South	Plains States	Mountain States	Pacific States	Foreign	Canadian
Pre-1830	4	6				3						
1830–39	15	35	4	3		23	4					
1840–49	5	36	6	1	5	4	1					
1850–59	6	43 (1)	20 (1)	9	3	8	5	1				
1860–69	12	96 (8)	55 (5)	26 (1)	15 (1)	3	3	2		1		1
1870–79	7	77 (3)	80 (4)	24 (2)	45 (1)	6	4	2		7 (1)	1	2
1880–89	4	55	70 (11)	29 (5)	40 (9)	4	2	5	1	36 (2)	7 (1)	10
1890–99	2	33 (4)	44 (15)	39 (7)		1	1	19	3	8	7 (1)	4 (1)
1900–09		22 (3)	37 (9)	51 (9)	78 (11)	1	2	71 (3)	5 (1)	23 (2)	7	13 (1)
1910–19	1 (1)	6 (2)	16 (1)	26 (7)	25 (5)	7	2	51 (2)	11 (1)	20 (5)	12	15 (2)
1920–29		5 (1)	16 (6)	31 (9)	22 (6)	3	1	35 (3)	8 (1)	7	6 (1)	15
1930–39		5 (1)	5 (2)	24 (19)	13 (1)	1	1	27 (4)	3 (1)	20 (10)	17 (3)	4

Numbers in parentheses are patents assigned to firms.

245

Table 5.7. Plows, Patent Class: Subclass, 172: 133-203

Time Period	New England	Middle Atlantic	Eastern Corn Belt	Western Corn Belt	Lake States	Appalachia	South	Plains States	Mountain States	Pacific States	Foreign	Canadian
Pre-1830	7	61	7			11	5					
1830-39	9	60	15	1	1	18	3		1			
1840-49	7	45	20	7	2	11	5					
1850-59	11	65	30	32	8	30	46	2		1		
1860-69	43	177	153	294	51	68	76	10		62	8	3
1870-79	36	96 (1)	121	123	44 (3)	90	74	30		46	3	4
1880-89	20	58 (3)	80	94 (4)	39 (1)	37 (2)	58	85	4	13	2	3
1890-99	14 (1)	36 (8)	31 (2)	67 (8)	18	17	53	80 (1)	4	21 (1)	8	3
1900-09	5 (1)	26 (3)	38 (5)	74 (7)	24 (3)	22 (1)	71 (3)	84 (2)	15 (1)	33	11	4
1910-19	5	17 (3)	30 (7)	55 (16)	21 (2)	27 (2)	51 (1)	74	26 (1)	33 (3)	7	10
1920-29	2	5 (1)	21 (7)	34 (6)	20	22	29 (2)	47 (3)	22	26 (2)	8 (1)	5
1930-39	1	7 (5)	9 (4)	23 (10)	12 (3)	4	17	25	16	11 (1)	2	5 (1)

Numbers in parentheses are patents assigned to firms.

New England and Middle Atlantic states appear to have lost their initial comparative advantage in invention.

A stylized story of the invention process in these technology fields could be characterized as follows:

1. During an initial period (sometimes lasting for three or more decades) invention is sporadic. Most of this invention is produced by individual inventors who, by reason of specialized experience, are adept at solving problems. I would characterize this invention as "wildcat" invention in much the same way that specialized high-risk oil exploration is wildcat in nature.

2. A point is reached where the pieces begin to fit together around one, or (often) more, technology "cores." Further development and commercialization is now undertaken and major investments in inventive activity, pilot production, and so on, are made.

3. Each technology core now provides strong disclosure effects that enable other inventors to make inventions and improvements (increments to the core). There are also cross-core disclosure effects that can speed up the invention process.

4. With an active core process underway, the scope is opened up for "adaptive" or derivative invention. In agriculture the settlement of new regions opened up tremendous scope for modification of plows, planters, etc., to new soil, climate, and economic conditions.

5. Industrial organizations and markets now come to be critical to further development. There is a tendency for one or at most two cores to become dominant commercially. This has two effects. First, it eliminates invention incentives associated with inferior cores. Second, it causes the elimination of firms based on inferior cores and is a force leading to consolidation.

6. The cycle may then reach a new equilibrium with a slow rate of further invention and high industrial concentration. Most of the highly original, high-risk invention is left to wildcat inventors, with the industry concentrating on refinements of the going core and process inventions.

This stylized sequence of invention is generally consistent with the data in Tables 5.5, 5.6, and 5.7. Much of the early invention in technology fields defined much more narrowly than those of Tables 5.5, 5.6, and 5.7 is sporadic. The earliest invention of grain harvesters and of many other machines occurs many years before the earliest commercialization. I would consider most of this to be wildcat invention. It is of special interest to note, however, that wildcat invention is not just an historical phenomenon. By the early 1900s many technology fields in agriculture had reached stage 6. The agricultural machinery (and agricultural chemical) industries were concentrated with several large firms dominating production.

Yet every new agricultural implement to be commercialized since 1900 has been invented and commercialized by independent wildcat inventors (and in a few cases by the public sector).

The phenomenon of adaptive invention is an important one in U.S. agriculture. As settlement of new regions took place, adaptive invention followed (by roughly 40 years, it seems). Early regions had exhausted much of their adaptive invention scope and because their adaptations were now part of the core, they lost their comparative advantage to the frontier inventors. The "Yankee ingenuity" so important in the early history of inventions did not generally survive this process. Of course, over time, many technology cores were revealed to be inferior to substitute cores, resulting in the cessation of patenting. Perhaps as many as two-thirds of the current U.S. patent classes are considered "dead art" in testament to this phenomenon.

Table 5.8 reports a more complete time series for 13 rather broadly defined technology fields that provide more recent data. The table shows the high levels of patenting activities in the late 19th century and the generally steady decline in patenting activities throughout the first half of the 20th century. Only mechanized harvesting showed increased patenting activity during this period.

The past three decades have shown increased patenting activity in most of these technology fields. Invention by foreign firms has increased during this period. In agricultural machinery, patents granted to foreign inventors increased from 17 percent in the 1960s to 33 percent by the late 1970s. In agricultural chemicals, the proportion of patents granted to foreigners rose from 28 percent to 50 percent. The proportion of agricultural machinery patents granted to U.S. individuals has stabilized at roughly 30 percent since the late 1960s. For agricultural chemicals it has stabilized at only 5 percent. The public sector produces only 1 to 2 percent of total patents in these fields. It is somewhat more active (3 to 4 percent) in postharvest technology fields that are not directly considered here.

III. THE EVOLUTION OF THE PUBLIC SECTOR RESEARCH SYSTEM

The Patent Office had little reason to be concerned that mechanical and chemical invention was not taking place when it established the agricultural division in 1837. This division was responding to a need for stimulus to biogenetic invention that was not effectively covered by patent laws. It was also responding to political pressure. The farm press of the time

Table 5.8. Patenting Activity in Agricultural Technology Fields

	Harvesting Equipment				Animal Related Fields					Tillage Equipment		
	Hay Handling	Grain Reaping Threshing	Corn Hvsting	Mech. & Cotton Hvsting	Dairy Equips.	Lvstk. Housing	Poultry Equip.	Animal Harness	Crop Husbandry	Planters Drills	Cultivators	Plows
Pre-1830	2	13	0	0	0	0	0	1	0	0	0	91
1830–39	17	89	38	0	0	1	0	1	0	12	7	108
1840–49	22	74	32	0	0	5	3	10	2	52	18	97
1850–59	216	178	121	8	2	35	1	59	17	332	55	225
1860–69	903	401	143	30	10	292	11	226	84	997	691	934
1870–79	742	455	186	37	17	514	21	393	104	1172	640	660
1880–89	668	544	142	94	16	923	97	727	80	1661	656	438
1890–99	411	246	102	97	30	849	112	529	83	1263	489	341
1900–09	484	355	171	183	77	717	343	456	83	1131	470	392
1910–19	441	241	124	331	196	1100	385	302	225	875	381	339
1920–29	213	182	128	387	139	808	367	91	156	274	242	228
1930–39	147	162	97	622	62	425	282	28	239	421	112	125
1940–49	181	137	23	340	91	212	79	10	86	202	65	122
1950–59	379	118	45	901	244	399	96	9	209	179	69	135
1960–69	315	103	61	753	260	530	70	9	455	236	47	98
1970–79	257	185	79	638	218	582	34	11	457	227	56	75

was an active part of political life and served not only to
inform farmers of technical improvements emerging from the
incipient agricultural industries but to articulate farmer
interests to the public sector as well.

The Land Grant College Act of 1864, which established
funding for state universities oriented toward the agricultural
and mechanical arts, did not emerge as a result of bureaucratic planning in the federal government. It was passed because
several states had earlier established state universities with
colleges of agriculture and because these colleges had demonstrated that they could deal with real agricultural problems in
a far more effective way than could arts and science colleges
with their inherent academic elitism.

The Hatch Act of 1887, which provided for federal funding of agricultural research in each state, was also not the
result of USDA planning. It too gained political support on
the basis of earlier successful state - established experiment
stations.(6) As the system has evolved, the strong political
support base in certain states has been very important not
only in terms of funding but also in terms of research programs and station organization.

The public system today includes both federal (USDA)
and state (SAES) units. Most state units are integrated with
State Land Grant University teaching units and with State
Extension Programs. SAES programs tend to be departmental-
ized along academic lines and most researchers also hold
university teaching positions. In some SAES units, graduate-
student research is a large part of the research output. The
institutional structure of the SAESs includes very applied
disciplines, such as plant breeding and agronomy, and more
basic disciplines, such as genetics.

The argument that the character of the SAES organization
and the research programs have been importantly influenced
by farmer interest groups is presented in Evenson, Ruttan,
and Waggoner (1979). In some ways, the interest-group influence or articulation of demand produces certain inefficiencies.
Some duplication of programs and field trial and testing
programs has occurred. Many current critics of the system
(mostly from the federal government) note this duplication and
lack of coordination and also what they perceive to be low-
quality research. They do not generally note that the system
actually has some means by which its clients can articulate
their interests. This feature may well be worth (and I would
argue that it is) paying the price of some duplication and of a
fair amount of field testing that does not qualify as high-
quality research.(7)

The articulation of farmer interests works very differently
at the state and federal levels. At the federal level, research
interests tend to be of minor importance compared to the more
direct issues of price supports, tariffs and other farm policy

issues. In fact, they are sometimes seen to be in conflict with
the broader farm-income interests.(8)   At the state level,
price and income policy interests are not important since these
policies must be dealt with at the federal level.   The research,
extension, and teaching programs are important, however,
since they serve state residents directly.   States do not
generally take into account the effects of their programs
outside the state.(9)

Table 5.9 summarizes public sector investment in production oriented agricultural research and extension since 1890.
The expenditure data are in constant 1980 dollars to enable
comparisons over time,(10) and all expenditure data refer to
research and extension oriented to agricultural production
only.   Here we note that the system was relatively small prior
to 1910.   Most of the funding on research in the State Agricultural Experiment Stations (SAES) was from federal Hatch
Act funds.   The United States Department of Agriculture
(USDA) had developed the Beltsville, Maryland, station and
several others, and was investing almost as much on research
in these stations as were the states.

The 1910 to 1925 period exhibits a significant expansion
in both SAES and USDA research as well as the development of
the Federal Extension Service.   In contrast to the earlier
period, the contribution from state governments then became
significant, in support of both research and extension.   The
Granges and the Farm Bureaus were also instrumental in developing both state and federal support for research and, to
an even greater extent, for extension.

After 1925, a further major expansion of the research
system took place, again with significant state support.   Data
for 1935 indicate a significant new pattern of investment.   The
federal government, in expanding the USDA research program,
now began to locate a significant amount of its research
activity in the states, often locating scientists directly in the
state experiment stations.   Much of this expansion took place
in the southern states.

The post-World War II expansion of the research system
was most rapid from 1945 to 1960, and virtually all of this
expansion took place in the state experiment stations.   The
USDA investment outside the state experiment stations has
changed little since 1930.   Since we are considering only
production-oriented research in this table, we should note that
the USDA has expanded its research programs in the utilization and marketing of farm products very significantly since
1945.   It is interesting to note, in addition, that the federal
government through its investment decisions has been very
influential in changing the research system, even though state
governments have provided the majority of the funds.   In the
1930s and 1940s it located much of its investment in the
"lagging" regions, chiefly the South.   In this way it had a

Table 5.9. Expenditures by the Public Sector on
Research and Extension Oriented to Improved
Agricultural Production Technology 1890 to 1970
(in millions of Constant 1980 dollars).

| | Expenditures on Research State Agri. Exp. Stations | | | | |
Year	Total	State Funded %	Federal Funded %	USDA %	USDA outside state	Expenditures on Public Extension Service
1890	9.7	.22	.78		2.6	.3
1900	12.2	.34	.66		10.4	1.3
1910	37.0	.39	.61		47.4	2.4
1915	34.2	.72	.28		62.5	18.8
1920	28.7	.77	.23		49.0	46.4
1925	42.5	.85	.15		59.2	61.6
1930	75.6	.73	.27		96.5	77.2
1935	79.3	.57	.27	.16	66.3	70.2
1940	113.2	.54	.28	.18	119.9	107.7
1945	114.2	.56	.23	.20	97.8	102.0
1950	194.3	.63	.17	.20	83.4	140.8
1955	251.4	.63	.17	.20	89.2	152.1
1960	344.8	.55	.15	.30	87.6	169.5
1965	385.5	.58	.16	.26	67.8	179.7
1970	414.5	.66	.16	.18	109.5	221.0
1975	420.0	na	na	na	110.0	264.0
1980	428.0	na	na	na	110.0	314.0

Sources: Latimer (1962); Evenson (1968); USDA work sheets.

major impact on the regional nature of productivity. In the
1950s and 1960s it shifted emphasis to marketing and utilization
research, to a much greater extent than would have occurred
if the states were determining the investment pattern.(11)
   The data presented to this point do not adequately indicate how much research effort is being devoted to the solution
of particular problems. It is difficult to obtain a measure of
research "intensity" or research expenditures per "problem."
Later I will use a measure based on geoclimate region and on
commodity complexity, but here I want to use a simple summary measure. The research intensity measure that is presented in Table 5.10 is research expenditures per thousand
dollars of commodity value for each of the ten USDA regions.
   By this measure, the Southern regions, even in 1951,
were not lagging behind the rest of the country. In 1951, the
Southeast region had the highest livestock research intensity
and ranked fifth in crop research intensity. The Delta region
also had relatively high research intensities. The Corn Belt,
on the other hand, ranked low.
   This measure is imperfect for several reasons. First, the
intensities are not corrected for crops fed to livestock. The
value of forage and pasture crops not marketed should be
subtracted from the livestock intensity deflator and added to
the crop intensity deflator. This would bring the intensities
more closely in line with one another. Of more importance,
the dollar value of production in a region is not necessarily an
indicator of the difficulty of producing new technology of
value. The Corn Belt, for example, may have a more homogeneous set of geoclimate factors within it than the Southeast.
If so, a research finding in the Corn Belt will be adopted over
more units of production. Hence, the research activity per
economic problem may well be higher in the Corn Belt.
   In addition to research directly oriented to livestock and
crop production, two additional categories are shown. The
economic and engineering research includes only production-
oriented research, but basic research includes phytopathology,
soil science, botany, zoology, genetics and plant and animal
physiology in agricultural research institutions. Regional
differences in the shares of economic and engineering research
are somewhat greater than in the shares of basic research, as
the Southern regions have relatively high shares of economic
and engineering research and low shares of the more basic
research.
   The issue of research "deflation" to obtain a measure of
research effort per research problem is a difficult one.
Research per state, research per farm, and research per unit
of commodity value all have imperfections. A measure based
on geoclimate zones or regions that is closer to a meaningful
measure than the more conventional measures is introduced
here. I will report results based on this definition in later

Table 5.10. Research Orientation by Region:
U.S. Agriculture 1951 and 1963 Expenditures
(in millions of 1959 dollars) on Research by Orientation.

Region	Livestock		Crops		Economic & Engineering		Basic	
	Expenditure	Exp/Commodity[a] Value	Expenditure	Exp/Commodity Value	Expenditure	Share of Research Expenditure	Expenditure	Share of Research Expenditures
1. Northeast								
1951	3.66	1.79	5.86	5.96	.54	.047	1.27	.112
1963	6.03	2.65	7.42	7.47	1.06	.062	2.51	.147
2. Lake States								
1951	2.48	1.12	2.68	3.62	.48	.074	.84	.130
1963	3.91	1.56	4.10	4.38	.78	.076	1.59	.154
3. Corn Belt								
1951	4.48	.88	3.21	1.71	.77	.078	1.41	.143
1963	6.47	1.16	4.04	1.40	1.19	.084	2.44	.172
4. No. Plains								
1951	2.24	1.14	1.55	1.51	.27	.059	.54	.118
1963	4.47	1.85	3.14	2.26	.70	.075	.97	.104
5. Appalachia								
1951	2.19	1.81	2.63	1.41	.49	.082	.66	.110
1963	4.48	3.07	3.95	2.15	.81	.076	1.40	.131
6. Southeast								
1951	2.22	3.22	3.89	2.37	.69	.087	1.06	.134
1963	5.67	4.33	7.24	4.45	.91	.060	1.38	.891
7. Delta								
1951	1.22	2.32	2.70	2.64	.68	.135	.46	.091
1963	3.73	2.41	4.22	2.60	.55	.057	1.23	.126
8. So. Plains								
1951	2.32	1.79	2.24	1.90	.40	.074	.43	.080
1963	3.72	2.40	3.89	2.59	.65	.067	1.38	.143
9. Mountain								
1951	2.84	2.21	2.38	2.60	.61	.088	1.06	.153
1963	5.21	3.30	4.74	4.07	1.01	.092	1.67	.132
10. Pacific								
1951	3.93	3.00	4.91	2.18	1.75	.067	1.47	.132
1963	6.77	3.70	9.53	3.59	1.54	.073	3.07	.146

[a]Dollars research per thousand dollars of commodity value.

econometric specifications that relate research effort to productivity. In that work I deflate research by the "adjusted" number of commodities and the number of geoclimate zones within a state. I also use the regional research classifications to define the research activity relevant to the producers in each state.(12)

It is not possible, unfortunately, to obtain from the geography literature a standardized set of homogeneous crop production regions for the United States. It is not an easy task since several climate factors and a large set of soil and topography characteristics are important to crop production, and any attempt to define regions involves the explicit or implicit weighting of these factors. Of course, a number of them are reasonably highly correlated and this simplifies the task. Soil characteristics are determined to a large extent by climate factors, for example, and the definition of a geoclimate zone may not require a decision as to whether climate factors or soil characteristics are more important.

The extent or level of detail to incorporate into the definition of regions or zones is also arbitrary. It could be fine enough to distinguish between very small differences in soil texture, for example, and the soil surveys prepared for many counties in the United States by the Soil Conservation Service have such detail. Unfortunately, I am dealing with more aggregate economic units and require a broader definition. In particular, I want a region to be defined in terms that are meaningful to the transfer of technology between states.

I concluded that the regions and subregions defined by the researchers in preparing the 1957 Yearbook of Agriculture were best suited to this purpose. With some minor modifications to the regions presented in that report, I developed the regional configuration shown in Figure 5.1. In all, there are 16 regions, each defined on the basis of relative homogeneity of soil and climate factors. Each region has from one to five subregions (40 in all), and most subregions and all regions extend across state boundaries.

Table 5.11 presents research expenditures in constant dollars by region for selected years. The allocation of research expenditures to regions was done on a commodity basis. For each of 21 commodities, state research was allotted to each subregion according to the share of that commodity produced in the region. The regional totals then are the sum of commodity research plus a proportional allocation of the noncommodity-oriented research.

In Table 5.12, I present a summary of research expenditures by commodity in the State Agricultural Experiment Station system in 1966. It was possible to divide the production-oriented research into two subcategories, production-increasing, and "maintenance" research. Production-increasing

Fig. 5.1. U.S. Agricultural Geo-Climate Regions and Sub-Regions. (1 dot = 25,000 Acres Crop-land, 1964)

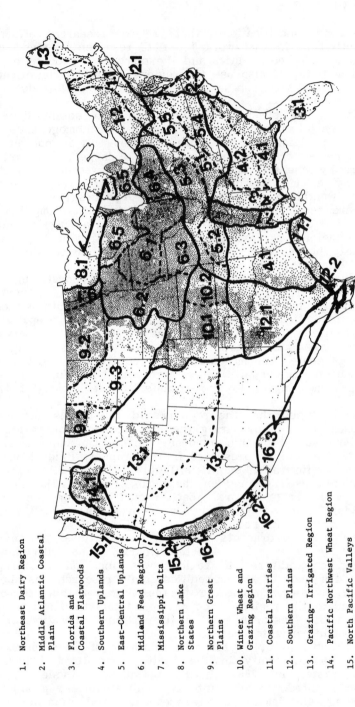

1. Northeast Dairy Region
2. Middle Atlantic Coastal Plain
3. Florida and Coastal Flatwoods
4. Southern Uplands
5. East-Central Uplands
6. Midland Feed Region
7. Mississippi Delta
8. Northern Lake States
9. Northern Great Plains
10. Winter Wheat and Grazing Region
11. Coastal Prairies
12. Southern Plains
13. Grazing- Irrigated Region
14. Pacific Northwest Wheat Region
15. North Pacific Valleys
16. Dry Western Mild-Winter Region

Table 5.11.   Research Expenditures by Geoclimate Region
(in millions of 1959 dollars).

Region	1915	1935	1950	1965	1969 Expenditures per Subregion
1. Northeast Dairy Region	2.09	3.84	8.29	13.35	4.45
2. Middle Atlantic Coastal Plain	.53	1.43	3.28	4.75	2.38
3. Florida and Coastal Flatwoods	.13	.94	2.68	4.63	4.63
4. Southern Uplands	.95	2.86	9.60	19.42	3.88
5. East-Central Uplands	1.42	2.39	6.28	10.84	2.17
6. Midland Feed Region	3.45	6.50	15.85	24.15	4.83
7. Mississippi Delta	.19	.45	1.55	3.17	3.17
8. Northern Lake States	.03	.01	.23	.37	.37
9. Northern Great Plains	1.17	1.76	3.99	6.55	2.18
10. Winter Wheat and Grazing Region	.61	1.50	4.26	7.15	3.57
11. Coastal Prairies (Texas-La.)	.01	.01	.02	.33	.33
12. Southern Plains	.18	.47	1.42	2.46	1.23
13. Mountain States Grazing-Irrigated Region	.85	2.26	5.42	8.95	4.48
14. Pacific Northwest Wheat Region	.34	.80	2.79	4.82	4.82
15. North Pacific Valleys	.01	.01	.35	.56	.56
16. Dry Western Mild-Winter Region	.76	3.23	7.61	16.98	5.66

Table 5.12.   Commodity Orientation of State Agricultural
Experiment Station Research, 1951, 1961.

Commodity	Research Expenditures in millions of 1959 dollars		1966 Research Expenditures			
	1951	1961	Millions of 1959 dollars	Expenditure per $1,000 of product	Share of "Maintenance Research"	Expenditures per Scientist Man-year
Livestock (total)	29.60	42.93	67.40	2.72	.40	53,534
Beef	14.33	18.13	17.48	1.67	.38	56,475
Dairy	4.37	7.11	15.99	2.91	.36	55,971
Swine	2.29	2.90	8.28	2.01	.45	60.272
Sheep & Lambs	1.25	2.22	5.52	16.33	.37	48,733
Poultry	3.87	5.88	14.36	3.47	.37	49,362
Other	3.49	6.69	5.77	-	.59	53,729
Crops Total	19.19	27.88	83.56	4.45	.43	36,567
Cereals	4.03	5.60	14.07	2.13	.40	34,340
Corn	-	-	5.65	2.23	.38	34,484
Sorghum	-	-	1.11	1.92	.18	30,248
Wheat	-	-	3.67	1.81	.52	35,475
Rice	-	-	.66	1.63	.36	32,031
Other small grains	-	-	2.98	5.63	.38	34,799
Cotton	1.16	1.42	9.69	6.14	.52	40,103
Oil seeds	.56	.70	4.67	1.64	.35	38,052
Soybeans	-	-	2.53	1.01	.31	36,544
Peanuts	-	-	1.21	4.48	.47	41,436
Other	-	-	.93	11.62	.33	37,556
Tobacco	.73	.81	3.51	2.90	.49	39,723
Sugar Crops	.28	.38	2.65	4.38	.53	37,656
Pasture & Forage	3.47	5.31	10.57	-	.22	36,972
Horticultural Crops	7.46	10.22	26.86	6.25	.50	35,596
Citrus Fruits	1.14	2.19	3.80	7.60	.51	38,122
Decid. Fruits & Nuts	2.47	3.15	10.71	8.86	.49	36,711
Vegetables	3.03	4.20	10.25	5.07	.49	33,586
Potatoes	.82	.68	2.10	3.57	.57	36,208
Miscellaneous Crops	.75	1.72	11.54	-	.33	32,714

research includes the improvement of biological efficiency, mechanization of cultivation and harvesting, improvement of crops' reproductive performance, and development of feed-efficient livestock. The concept of maintenance of technical gains is very important in agriculture because, in contrast to most mechanical technology, biochemical technology is subject to real loss or depreciation from diseases, insect pests, and internal parasites. The reader should be cautioned that the 1966 data are not strictly comparable to the 1951 and 1961 data. They include USDA research located in the states and because of a more detailed breakdown of the research program, the 1966 data are more accurately production oriented. The 1951 and 1961 data are comparable, however, and indicate that research expenditures on beef, dairy, sheep, and lambs, poultry pasture and forage, and citrus crops were increased by more than 50 percent over the decade.

The 1966 data enable more accurate comparative statistics and three are provided. The first, research expenditures per thousand dollars of commodity value, indicates relative research emphasis. This measure shows that crops receive more emphasis than livestock. It might be argued that research on pasture and forage should be allocated to the livestock sector, but even if this were done, crops would still be more research intense. Within the livestock group, sheep and lambs are very research intense. Within the crop sector one finds that the cereal grains and soybeans have low research intensities while cotton and the horticultural crops are quite research intense.

The second measure offered in the table is the share of maintenance research by commodity. It may be noted that wheat, sugar, cotton, and the horticultural crops are quite research intense.

The final computation presented in Table 5.12 measures a characteristic of the research system itself. The 1966 data allow a calculation of expenditures per scientist man-year by research program area. This gives some indication of the scientific equipment and related technical staff associated with different research programs. The average spending per scientist man-year by commodity is clearly highest for livestock research. Relatively little variation in the averages within the livestock and crop groups is apparent.(13)

The public research system exhibits a fair amount of variability in regional research intensities and in commodity intensities. Some of this variation is due to perceived differences in the range of problems to be solved. The fact that wheat, rice, and corn have low research expenditures per dollar of product ratios reflects the fact that they are produced in high volume in relatively homogenous areas. On the other hand, much of this variation is due to the political influence of producer interest groups. The role of political groups is considered in Section V of this chapter.

IV. THE ECONOMIC EFFECTS OF RESEARCH
AND EXTENSION

In this section I will summarize the conclusions of a number of studies that have addressed the question: does research (and extension) investment have quantifiable economic effects? This is a particularly important question with respect to public-sector investment in research and extension where one cannot rely on the monitoring and evaluation activities of private firms to determine whether valuable new technology has actually been produced. In fact, given the nature of the state support system for this research, there has been a considerable amount of informal evaluation of research effectiveness for a good many years. Farmer interest groups ask questions about the research program in each state and compare the release of varieties and other indicators of research output with those of neighboring states. Experiment station directors have been defending budgets on the grounds that research is productive since the stations were funded.

The earliest formal study of the economic effect of research was by T.W. Schultz (1953). this was followed by Zvi Griliches' classic study of hybrid-corn technology in 1958. Agricultural historians had, however, documented a large number of cases of improved technology and its adoption in earlier years. The reaper, for example, was studied in detail and calculations of cost savings were made. Many other examples of this type could be documented.

The methodology and formal quantification of the added production per unit of input (or savings in input costs per unit of output) added a new dimension to these studies and inspired a large number of such studies both in the United States and abroad. A recent review by Norton and Davis (1981) covered 180 or so such studies. A summary of the major studies reporting an internal rate of return measure for each is provided in Table 5.13. A number of these studies deal with U.S. agriculture. I will draw some generalizations from this large set of studies before discussing in more detail a study of three historical periods in U.S. agriculture.

First let us consider the studies summarized in Table 5.13. They are classified as imputation studies and statistical studies. The distinction between them is that the imputation studies rely on direct methods to associate economic effects with program costs. The Griliches hybrid-corn study, for example, used data on the relative yields of hybrid corn and the open-pollinated varieties that they replaced to measure the shift in the supply of corn due to hybrid-corn varieties. The statistical studies have estimated the economic effects of research and extension programs using regression methods. The typical study of this type specifies an aggregate production function (or an equivalent productivity decomposition model) which relates research and extension variables and conventionally measured factors of production, land, fertilizer, labor, and so on, to a measure of aggregate farm output. The research coefficient estimated in these studies allows one to compute the increment to output associated with an increment to research (extension) investment holding constant conventionally measured inputs or factors of production.

The statistical studies thus enable a calculation of marginal or incremental production owing to research (extension) investment. The imputation studies, on the other hand, enable one to compute the added production associated with an entire research program and hence provide an estimate of the average product of research. Perhaps the most important distinction between the two methodologies is that the data directly determine the estimates in the statistical studies. Imputations can be in substantial error and the reader of a given study may have little basis for judging the reliability of the calculations. In the statistical studies, the data can reject the a priori supposition that research is productive. The reader has statistical information (standard errors, R^2) with which to judge the result. All of the statistical estimates of research productivity in the table are "statistically significant" by conventional standards.

The internal rate of return is selected as the best common denominator to compare the conclusions of these studies. This rate should not be confused with other rates of return that treat investment capital as being available at a specific rate of interest. The internal rate of return is the rate of return realized on investments over the entire life span of the project. It should also be judged against other internal rates of return that are calculated for investments with long gestation periods and are of significant size.

The time dimension of research effects has been investigated in only two or three of the studies. These studies estimate that the effect of research on production begins to be observed about two years after the research investment is made. This effect then rises to a maximum approximately six to ten years after the investment. For most biological research, a real depreciation of the improved technology occurs several years after it is exposed to disease and pest environments. Because of this, part of the research effect is not permanent and disappears after several years (approximately 20 to 25, as estimated in these studies).

It is sometimes possible to measure very high rates of returns to small investments where there is a very short time lag between investment and effect. For investment with longer lags, where the life of the investment is 20 years or more, the internal rates of return measured in Table 5.13 are about as high as are generally observed. They are far above the range

Table 5.13. Summary Studies of Agricultural Research Productivity.

STUDY	COUNTRY	COMMODITY	TIME PERIOD	ANNUAL INTERNAL RATE OF RETURN (percentage)
IMPUTATION STUDIES				
Griliches, 1958	USA	Hybrid corn	1940-55	35-40
Griliches, 1958	USA	Hybrid sorghum	1940-57	20
Peterson, 1967	USA	Poultry	1915-60	21-25
Evenson, 1969	South Africa	Sugarcane	1945-62	40
Ardito Barletta, 1970	Mexico	Wheat	1943-63	90
Ardito Barletta, 1970	Mexico	Maize	1943-63	35
Ayer, 1970	Brazil	Cotton	1924-67	77+
Schmitz & Seckler, 1970	USA	Tomato harvester	1958-69	
		With no compensation to displaced workers		37-46
		Assuming compensation of displaced workers for 50 percent of earnings loss		16-28
Ayer & Schuh, 1972	Brazil	Cotton	1924-67	77-110
Hines, 1972	Peru	Maize	1954-67	35-40[a] 50-55[b]
Hayami & Akino, 1977	Japan	Rice	1915-50	25-27
Hayami & Akino, 1977	Japan	Rice	1930-61	73-75
Hertford, Ardila, Rocha & Trujillo, 1977	Colombia	Rice	1957-72	60-82
	Colombia	Soybeans	1960-71	79-96
	Colombia	Wheat	1953-73	11-12
	Colombia	Cotton	1953-72	None
Pee, 1977	Malaysia	Rubber	1932-73	24
Peterson & Fitzharris, 1977	USA	Aggregate	1937-42	50
			1947-52	51
			1957-62	49
			1957-72	34
Wennergren & Whitaker, 1977	Bolivia	Sheep	1966-75	44.1
		Wheat	1966-75	-47.5
Pray, 1978	Punjab (British India)	Agricultural research and extension	1906-56	34-44
	Punjab (Pakistan)	Agricultural research and extension	1948-63	23-37
Scobie & Posada, 1978	Bolivia	Rice	1957-64	79-96

Table 5.13. Summary of Studies of Agricultural
Research Productivity (continued)

STUDY	COUNTRY	COMMODITY	TIME PERIOD	ANNUAL INTERNAL RATE OF RETURN (percentage)
STATISTICAL INFERENCE				
Tang, 1963	Japan	Aggregate	1880-1938	35
Griliches, 1964	USA	Aggregate	1949-59	35-40
Latimer, 1964	USA	Aggregate	1949-59	Not significant
Peterson, 1967	USA	Poultry	1915-60	21
Evenson, 1968	USA	Aggregate	1949-59	47
Evenson, 1969	South Africa	Sugarcane	1945-58	40
Ardito Barletta, 1970	Mexico	Crops	1943-63	45-93
Duncan, 1972	Australia	Pasture improvement	1948-69	58-68
Evenson & Jha, 1973	India	Aggregate	1953-71	40
Kahlon, Bal, Saxena & Jha, 1977	India	Aggregate	1960-61	63
Lu & Cline, 1977	USA	Aggregate	1938-48	30.5
			1949-59	27.5
			1959-69	25.5
			1969-72	23.5
Bredahl & Peterson, 1976	USA	Cash grains	1969	36[c]
		Poultry	1969	37[c]
		Dairy	1969	43[c]
		Livestock	1969	47[c]
Evenson & Flores, 1978	Asia-- national	Rice	1950-65	32-39
			1966-75	73-78
	Asia-- international	Rice	1966-75	74-102
Flores, Evenson & Hayami, 1978	Tropics	Rice	1966-75	46-71
	Philippines	Rice	1966-75	75
Nagy & Furtan, 1978	Canada	Rapeseed	1960-75	95-110
Davis, 1979	USA	Aggregate	1949-59	66-100
			1964-74	37
Evenson, 1979	USA	Aggregate	1868-1926	65
	USA	Technology oriented	1927-50	95
	USA--South	Tech. oriented	1948-71	93
	USA--North	Tech. oriented	1948-71	95
	USA--West	Tech. oriented	1948-71	45
	USA	Science oriented	1927-50	110
			1948-71	45
	USA	Farm management research & agricultural extension	1948-71	110

[a]Returns to maize research only.

[b]Returns to maize research plus cultivation 'package'.

[c]Lagged marginal product of 1969 research on output discounted for an estimated mean lag of 5 years for cash grains, 6 years for poultry and dairy, and 7 years for livestock.

of returns estimated for most public-sector investment and well above the range of before-tax returns realized in the private sector as well.(14)

One might argue that those studies that have preselected obvious success cases such as hybrid corn present a biased picture regarding research productivity. the fact that those studies investigating aggregate research programs measure returns to investment of the same order of magnitude suggests that this bias is not great. It may also be noted from the table that the poorly developed research programs in developing countries have roughly the same rate of return as investment in the more sophisticated programs of the developed countries. This result is consistent with the interpretation that the potential for research discoveries is not being exploited at an optimal level and that this potential is high enough that even poorly organized research programs are highly productive.

Perhaps of more significance, however, is the conclusion that the U.S. agricultural research system has been highly productive over a long period of time. I will turn to the studies on which this conclusion is based shortly, but before doing so, two related topics deserve discussion. The first is the relationship between these estimates and the "growth dividend." The second is the implications these studies have for the transferability of research discoveries from one region to another.

First, regarding the growth dividend, Table 5.3 above reports measured productivity change which is interpretable as the additional product realized each year that is not accounted for by conventional resource use. It is a kind of growth dividend which, at the long-term 1940-1970 rate of productivity growth (1.9 percent), was worth roughly 2.7 billion dollars in 1980. A 2.7-billion-dollar income stream will support annual investment levels of several times this much. Public- and private-sector agricultural research and extension investment was approximately three billion dollars in 1980. The productivity growth that these studies attribute to public-sector research and extension is usually only one-fourth of the realized growth dividend. Thus, the high rates of return attributed to agricultural research and extension are not inconsistent with growth dividend data.

The data of Table 5.2 provide the basis for further illustrative calculations demonstrating large effects of research. The Griliches hybrid-corn study used data up to 1957 when national average yields were less than 50 bushels per acre. He reported very high returns to investment based on an 18 percent yield advantage for hybrids over open-pollinated varieties. Since 1957 corn yields have more than doubled as several generations of improved hybrids have been developed and diffused in the interior. Interestingly, in 1957, 95 percent of the hybrids sold and 75 percent of the inbred lines used in their production were developed by private companies. Initially, the hybrids were developed in the public sector but private firms proved to be superior to public experiment stations in their production and sale. Until the 1950s the private firms were dominating inbred line development as well. (The public sector was in direct competition in the sense that it made its lines freely available to private breeders.) This situation has sharply reversed itself in the past 20 years. Today the public stations supply more than 80 percent of the inbred lines used by private producers. The fact that public-sector stations have closer ties to more basic sciences appears to be the main factor in their ability to compete effectively with the private sector. They continue, however, to be unable to gain a significant share of the hybrid market.

The economic value of the improved corn technology (calculated after adjusting for increased fertilizer and related inputs) is huge. An update of the Griliches study would yield even higher returns for the 1957-1981 period. This study would also demonstrate some of the basis for sustained research productivity over long periods of time. The hybrid-corn story was far from over in 1957. The sequel demonstrates not only a public-private sector institutional compatability but also demonstrates the value of an institutional structure integrating science and technology.(15)

A second case that is only partially apparent in Table 5.2 is revealed in the soybean-yield data. The yield increase from the 1955-1959 period to the 1974-1978 period is a modest 22 percent. Almost all of this, however, is due to varietal improvement and 20 percent of a 15-billion-dollar crop provides a large annual benefit stream, indeed large enough to justify all public and private research expenditures on all commodities at normal returns on investment. Even with some conservative adjustments to the data, the income streams emerging in either corn or soybeans are large enough to justify the entire agricultural research and extension program. Reference to Table 5.2, however, will show large increases in yields of all other crops and indicators of major gains in productivity in livestock production as well.

It is not surprising then that the studies summarized in Table 5.13 find large effects. It may, however, be surprising that many of the statistical studies find these large effects in cross-section data. In other words, these studies find that regions with high levels of research intensity realize higher productivity gains. If research findings were easily transferred from one region to another, productivity growth would be similar among regions and not correlated with regional research intensities. Obviously, it is not.(16)

When cross-section data are used, some kind of spatial "spillover" specification must be made. It can be argued that schooling variables should be expressed on a per operator basis and that since much extension activity is location specific, expenditure or man-days variables should be measured on a per farm basis. This will not do, however, for the research variable where research findings from one state station clearly spillover into other states. It is also clear, however, that this spillover is incomplete. Environmental impediments such as soil and climate factors cause technology that is directed or targeted to a particular region to be less valuable to other regions. This incomplete transfer of technology is an important component of the public-sector motives for investment in research. (I will take this issue up in Section V.) It provides an incentive for state investment in research targeted to the state's farms.

A recent study of determinants of U.S. agricultural productivity proceeded to handle these issues in two stages. In the first stage, data on aggregate output, aggregate inputs, and cost shares were employed to produce a total factor productivity index. In the second state, this index was statistically decomposed by regressing it on its determinants, invention, schooling, extension and research.(17) The research variable, R_i, was first expressed in terms of the estimated time-shape weights. The research program that is influencing productivity in state or region i was then defined as:

$$R_i + \Theta_1 R_i^{ss} + \Theta_2 R_i^{sr}$$

where R_i is the research undertaken in the state; R_i^{ss} is research undertaken in similar subregions as described in Figure 5.1; and R_i^{sr} is research undertaken in similar regions. A procedure was devised for estimating Θ_1 and Θ_2. Knowing these parameters, we have an estimate of the transferability or spillover of research across regions. For example, if $\Theta_2 = 0$ and $\Theta_1 = 1$, research tends to be transferable only within the subregions depicted in Figure 5.2.

Table 5.14 summarizes the results of the study in terms of the estimated time between research spending and its maximum effect (section a in note 16 below) and the value of agricultural output in 1959 dollars of added investments in schooling, extension, and applied and scientific research. The internal rates of return based on these data are summarized in Table 5.13. Invention variables were incorporated in the 1868-1926 and 1927-1950 studies and these indicated that patented inventions were of high value.

These calculations indicate that for some regions and periods applied agricultural research (agronomy, plant breeding, etc.) can yield its maximum effect in as few as five years. For scientific research (genetics, plant and animal physiology, etc.), this time period is longer, approximately 11

Table 5.14.   Summary of Findings:
Productivity Decomposition Analysis:
U.S. Agriculture 1868-1926, 1927-1950, and 1948-1971.

	1868-1926	1927-1950	1948-1971
A. Timing			
No. of years between investment and maximum economic effect			
Applied Agricultural Research	18	5	North 7 South 5 West 7
Scientific Agricultural Research	18	15	North 15 South 11 West 15
Agricultural Extension (assumed)	n.c.	n.c.	1.5
Invention (assumed)	10	10	n.c.
B. Total Economic Effect			
Value of added farm production holding inputs constant (1959 dollars)			
From one year of primary schooling	n.c.	n.c.	260
From $1,000 added extension and applied economic spending	n.c.	n.c.	2173
From $1,000 added scientific agricultural research spending	n.c.	53000	4500
Proportion spilling into other states		.67	.68
From $1,000 added applied agricultural research spending	12,500	11,400	12,000
Southern states	n.c.		21,000
Northern states	n.c.		11,600
Western states	n.c.		12,200
Proportion spilling into other states	n.c.	.55	.43

to 15 years. It appears that roughly two-thirds of the economic effects of scientific research and 40 to 50 percent of the economic effects of applied research are realized in states other than the one where the research is carried out.

The data show that applied agricultural research has had large economic impacts in all periods. This is one of the more extraordinary facets of these data. They show that these research institutions have been productive over a long period of time. They have had the capacity to be oriented to clientele groups to produce research findings of value, and, at the same time by investing in related scientific research in the institutional framework of the experiment station, they have managed to replenish invention and discovery potential. I would not suggest that they have done this perfectly, nor that it was done smoothly. A constant state of tension between applied and practical research and extension workers and more scientifically oriented researchers has existed throughout the history of these institutions. Nonetheless, they have managed to be effective.

A final point regarding these estimates. For various reasons, investment in research in the South lagged behind the rest of the nation prior to World War II. During the 1930s the USDA located a disproportionate share of its research units in the South. After World War II, the Southern states embarked on an aggressive program to expand agricultural research both qualitatively and quantitatively. The data on productivity change combined with the research study indicate that this paid off.(18)

V. INSTITUTIONAL AND RESEARCH POLICY ISSUES

Previous sections of this chapter have provided a description of the development of agricultural research and extension institutions and reviewed a body of statistical evidence that attributes part of the productivity performance of the sector to their activities. It was further noted that these research and extension institutions were productive over the whole of the past century. This sustained productivity was attributed to a capacity for institutional change which in turn was associated with a capacity to respond to clientele interests.

In this final section I will elaborate further on the institutional forms of the agricultural research system and the issue of research quality. I will also discuss the political interests of farm and consumer groups and the distributional effects of research. Finally, I will discuss public-private policy issues.

The State Agricultural Experiment Station is an unusual institution. It has some of the characteristics of the conventional scientific research center and some of the characteristics of the applied industrial research organization, but is distinct from both. It generally has a disciplinary structure, as does a science center, but it also has a strong mission or applied orientation toward the resolution of real problems. The key to its institutional structure has been the development of agricultural science disciplines.

Agricultural science disciplines might be termed "applied" or "near" science. They bear a special relationship to one or more "mother" sciences, but are not governed by the same peer review quality standards or by the same pressure to work on particular problems. Their institutional separation from the mother science allows them to promote and reward work with a strong applied mission-oriented content while still realizing the advantages of a disciplinary structure. They can also accommodate a broader range of activities within departmental structures than can the mother science.

The strict peer review system by which research quality is judged and on which promotions are based has, in some ways, served certain fields of science well. Nonetheless, it has been a practical reality that such systems do not allow much weight to be placed on the economic value or relevance of the research contribution. Most scientific institutions do not have economic clients except very indirectly through federal public-support mechanisms. They tend to serve themselves.

The Land Grant College System was established to create new institutions that would recognize economic relevance in research work. The two areas emphasized were agriculture and engineering. Later, medicine was a third major applied science field to be given emphasis. The agricultural colleges and their associated agricultural experiment stations developed stronger political support bases than did any of the engineering fields. Experiment station directors regularly defend their budgets and programs in state legislatures as providing research findings of direct benefits to interest groups in the state. (They also link their contribution to the education services provided by colleges of agriculture.) They could not do this if they were not willing to fund projects and promote scientists for providing those benefits.

The departmental structures in agricultural research institutions have institutionalized a good deal of the tension between scientific elitists and practical researchers by encouraging both. This institutionalization is partly reflected in the peer review processes that emerge in each of the applied sciences. The agricultural sciences (e.g., agronomy) quickly developed their own scientific structure with journals, professional associations, and degree programs. Their peer review processes explicitly recognized economic relevance as a quality criterion. The emergence of the new applied sciences has tended to create unfortunate differentiation of products from the mother sciences. The mother sciences often become more elitist than before (particularly in the Land Grant universities).

Assessed by the criteria of the mother science, agricultural research is often judged to be of low scientific quality. A certain amount of such criticism is the natural consequence of the different criteria employed for judging quality and relevance. Even in a well-organized research discipline, research quality as measured in the discipline itself will vary by researcher and by research institution. It will vary by age of researcher and by type of graduate training. Furthermore, since many research disciplines are specialized in applied problems, they may have little control over the "research potential" with which they work and even high-quality researchers may find it increasingly difficult to produce research and output because of limited scientific potential.(19)

In some fundamental sense, the more important questions regarding research quality have to do with both the organization and design of research disciplines or areas within large research programs and with the consistency of the research objectives and quality criteria of each discipline to the objectives of the overall research program.

Crop improvement work, for example, is primarily done in plant breeding and agronomy disciplines. The closely related disciplines of plant pathology, soils science, and entomology, however, are very much a part of the crop-improvement system. Within each of these disciplines, subdisciplines with somewhat different quality criteria exist. Further, experiment stations have also incorporated plant physiology and other biological sciences in their institutional structure in an effort to produce new discovery potential. These disciplines exist outside the USDA-SAES system as well, and different quality standards may be applied by the two groups.

It is almost certainly the case that low-quality research is being undertaken in many agricultural research organizations. It is also the case that duplication of effort is a problem and that many pedantic and unimaginative field trials are undertaken year after year. These problems, however, may be unavoidable consequences of the mechanism by which clientele interests are expressed or articulated in the system.

The State Experiment Stations have a strong state political base while research and extension are not given high priority at the federal level. Producers rather than consumers form the interest groups supporting these activities. Given the importance of these activities in determining productivity growth, it is also important that we have a better idea of their political support base. To that end I find it useful to engage first in a nontechnical analysis of the gains and losses associated with new agricultural technology. I then turn to a fuller discussion of political interest.

Basically, research and extension programs can have a number of possible effects.

1. Research produces new technology. Extension facilitates its adoption and encourages further development of minor technological improvements and management technology. This technology can be:

   a. factor biased (e.g., labor using, etc.)
   b. scale biased (e.g., more profitable for large farms)
   c. region biased (e.g., not equally available to all farmers in different regions).

2. Research and extension may change the demand for farm products (i.e., introduce new products, encourage consumption via nutrition educations, etc.).

3. Research, especially private research and extension, may lower the cost of purchased inputs (such as fertilizer and the like).

4. Research, but particularly extension, may lower the cost of labor mobility between regions and sectors of the economy.

From these possible effects, we can focus the general question regarding the overall effects of agricultural research and extension on the distribution of incomes on four more particular issues:

1. The effects of agricultural research and extension on the distribution of incomes between consumers and producers;
2. The effects of agricultural research and extension on the distribution of income among agricultural factors of production;
3. The regional income effects of agricultural research and extension services; and
4. The impact of agricultural research and extension on the distribution of income among different size farms.

The analysis of the first issue is rather straightforward. Agricultural research and extension investment, insofar as it results in any rightward shift in the agricultural output supply functions, leads to consumer gains (lower agricultural output prices) as long as the demand function for agricultural goods is downward sloping. The size of these consumer gains depends on the relative elasticities of supply and demand.

The final distribution of consumer gains among all consumers (and producers insofar as they too are consumers) would depend on their expenditure patterns. Consumers who spend a high proportion of their budgets on agricultural products will benefit proportionally more from a decrease in food prices. It is important to bear this in mind because the poor generally do spend the highest proportion of their budget on food. Agricultural research and extension thus creates a progressive (i.e., more egalitarian) distributional effect for that proportion of benefits passed on to consumers in the form of lower agricultural output prices.

The second dimension of the distribution question regarding the distribution among factors of production has been the subject of a few theoretical studies (Evenson and Welch, 1974; Evenson, 1978; and Binswanger, 1978).

The simplest case of this distributional dimension is where there are only two factors of production, say land and labor. In an initial equilibrium, farmers demand a certain quantity of each factor at different prices of land and labor. The equilibrium price of each is determined by both this demand and the supply of the factor. Agricultural research that produces new technology causes a shift in the demand for each factor. If the technology is neutral, the demand curves for each factor will shift by the same percentage. Whether the shift is negative or positive depends on the demand for agricultural output from the producers affected. If output demand is relatively elastic, technical change will result in an increase in demand for both factors and both will gain. The extent of their gain depends on the elasticity of supply. The factor that is relatively inelastic in supply (land) will gain most. The reverse is true in cases where output demand is inelastic; both factors will lose.

During the 1950s, U.S. agriculture realized rapid gains in productivity and the demand for most agricultural products was relatively inelastic. This resulted in depressed farm prices and reduced demand for factors. During the 1970s, U.S. agriculture enjoyed expanding international demand and faced a more elastic demand curve. The demand for factors increased and land owners were the big gainers. In Section I of this chapter, I noted that capital gains from land-price appreciation have exceeded net farm income for almost every year in the past 15. This has produced a new wealth class in the U.S. economy. Farm labor, on the other hand, has benefited little because of its elastic supply. When wages are bid up, laborers move from nonagricultural sectors into agriculture.

The third and fourth distributional dimensions are important for understanding the basis for political support by farmers, chiefly landowners, at the state as opposed to the federal level. The essence of the analysis here is that when there are two groups of producers in the same market (e.g., large farms and small farms or farms in region 1 and farms in region 2), technical change that is available to only one group can have very strong effects for both groups. It benefits the group receiving the technology more than it would had both groups received it. This is because output prices fall less than they would if both groups had cost decreases. However, prices do fall because of increased output from the receiving group. This imposes real losses on the group not receiving the technology because their prices fall but their costs do not.

We thus have a basis for a political rivalry model between regions. Landowners in one state can expect other states to produce technology better suited to their own environments and hence to impose losses on them. They have a strong incentive to invest in extension to "pirate" as much of the other states' technology as possible. They also will support research in their own state to enable them to adopt technology from another state. A recent study (Rose-Ackerman and Evenson, 1981) of the political support base indicates that the "net" response to a strong research program in a neighboring state is negative, however. The "free-rider" effect is strong enough across regions (as defined in this chapter) to cause states to reduce spending on research when close neighbors are undertaking it.

What has been said about farmers in different states applies also to farms of different sizes. New technology is often differentially accessible to different groups of farmers. Within the same region, large farmers have more incentives to search and to experiment than small farmers since the benefits from search are proportional to farm size while the costs are not. This naturally leads to early adoption of new technology by large farms, providing them with innovator rents. These innovator rents to large farms may be transitory unless new technology itself has a scale bias, i.e., the new technology reduces costs for large scale farms much more than for smaller ones, or unless input and credit markets remain accessible only to large farms. These rents provide incentives for large farmers to perform experiments in a given year. This also lowers the cost of learning and experimenting for the smaller-scale producers who would have access to and benefit from them in the immediate future. In the case where innovator rents tend to be more permanent, institutional changes that facilitate access to new technology become necessary. Agricultural extension services then become an important feature of any institutional package designed to eliminate the permanent nature of some innovator rents. Extension activities lower the cost of learning and experimenting and thus lower the levels of innovator rents. Reducing rents to innovation via extension does not necessarily produce too little innovative activity since extension can also reduce the real cost of innovation. Again, however, the payoff to such activities depends on the capacity of small-scale farmers to process and use new and cheaper information to their advantage.

As noted earlier, agriculturally related industries have grown rapidly relative to the farm production sector. The agricultural supply industries (fertilizer, farm chemicals, farm machinery, and animal health) sell inputs to agriculture that now account for a large part of the total cost of production. Value added in the processing and marketing industries using agricultural products as raw materials far exceeds the value of agricultural production.   Furthermore, the large-scale corporate farm production sector is growing and now accounts for more than 40 percent of farm production.

The supply industries have incentives to undertake R&D because they can appropriate, through patent protection and their own sales, the benefits of this research.   The processing and marketing industries have similar incentives.   These firms do undertake a considerable amount of R&D.   In 1972, patent protection (or patentlike protection) was extended to sexually reproduced plants, thereby encouraging an expansion in private plant-breeding activity.

How does one organize a public-sector program in the presence of these private interests?   Is it possible to contribute to the public welfare without producing rents to particular private firms and industries?

These questions have been partially addressed by the agricultural experiment stations in certain fields of technology. Perhaps the most interesting is the relationship between public research on hybrid corn and private research and production of hybrid corn.   As the study by Griliches (1958) noted, the early work on hybridization was undertaken in the public sector.   At a certain stage, a number of private firms began to use inbred lines from the public sector in crosses that produced the "double-cross" seed that was sold to farmers. Several experiment stations competed directly with private firms in the production and sale of hybrids.   In general none of the stations selling hybrid corn was highly successful in the large markets, but they did provide seeds for specialized climate zones and for specialized purposes where the private sector was not producing.   They produced varieties both for northern and southern parts of the corn belt before private breeders designed breeding programs for these areas.

As the hybrid-corn seed industry grew, the major firms (Pioneer, DeKalb, Funks, etc.) invested not only in the applied breeding work that produces hybrids, but in the development of inbred lines as well.   By the 1950s or 1960s, almost all of the final hybrid lines used to produce them were products of the private sector.   Interestingly, the trend has reversed in recent years and the public sector is now producing 80 percent or so of the inbred lines.

Here we have, then, a natural comparative advantage situation between public and private sectors.   The public sector is producing basic breeding material, the private sector is providing the refined final development and serving specialized markets.   The public sector provides R&D for small markets and, by engaging directly in hybrid-corn production, helps to keep the industry competitive.   With competition, the products of the public sector do not produce rents captured by private firms. New public inbred lines, for example, are freely available to all firms. With competition, private firms capture rents only on the increments contributed by their own research.

Engineering research in the experiment stations has a similar institutional setting. When the public-sector enterprise competes directly with private engineering firms it comes off second best. Engineering departments do make contributions in terms of building prototype machines where the private sector firms do not invest. Most new farm machines have been brought through the early stages of development in recent years by wildcat inventors with some assistance from the public sector. On the whole, agricultural engineering research does not enjoy a strong reputation for being highly inventive, nor does it enjoy a reputation for contributing to the high-level process engineering important to manufacturing firms.

Research in chemical herbicides, insecticides, and animal pharmaceuticals is somewhat different from engineering because there is not much wildcat invention by individuals in these fields. The public sector develops generic chemicals and engages in considerable testing and related work. Some work of a highly basic nature on growth regulators, etc., is undertaken in the experiment stations.

Overall, the experiment stations have generally moved their work into areas where they have a comparative advantage vis-a-vis the private sector. In direct competition with market-oriented private firms, the public sector does poorly and generally does not invest heavily in research of that type. It tends to be pressed into a good deal of work of a testing and certifying nature, designed to help farmers make choices among suppliers of inputs. In recent years it has played a major role in facilitating adjustment to regulations both in the chemical inputs fields and in food technology.

The system has attempted to respond to changes in rural communities by expanding social-science work in recent years. Sometimes this has been in response to problems in rural industries such as local dairy manufacturing plants. In many cases the interests of farm producers and agricultural interests coincide (as with policy toward grains exports, for example) and experiment stations have responded to these interests.

There has been growing concern in recent years that the experiment stations are losing their public character and serving the interests of select interest groups, chiefly agribusiness interests. The tomato harvester, partly developed by the California Experiment Station, is put forth as a case in which public funds have been used to support the interests of a particular group, in this case tomato growers. The rapid expansion of the corporate sector and the increasing wealth of farmers generally are seen as factors removing the

public system from its traditional role of serving the farming
and rural sectors generally.

I have argued in this chapter that the system has long
served fairly specific interest groups and that this has been
an important factor in its success. The fact that the interest
groups are changing may require modification, but the elimination of the political support system is not the answer. The
state system is currently concerned with its relationship with
its clients. That in itself is a sign that the system continues
to have the capacity for change that has served it well in the
past.

NOTES

1. Michigan (1837), New York (1853) and Maryland (1856)
were the earliest.

2. The political implications of this change are discussed
later. It should be noted that the sector is currently (1982)
experiencing relatively low prices and current farm incomes are
low.

3. Landau and Evenson (1973) discuss the details of the
calculations involved in these indexes and report a comparison
between indexes that use Laspeyres factor weights and the
Divisia-type index that shifts weights each year.

4. Landan and Evenson (1973) provide details of calculations
and reports state indexes as well.

5. This is discussed in Evenson (1978).

6. Connecticut established the first State Experiment Station
in New Haven in 1857. This station was later to be instrumental in the development of hybrid corn.

7. It is also the case that much duplication is more apparent
than real because of the genotype environment intervention
and other factors that make the development of technology
sensitive to climate and soil conditions. (This issue is
considered in more detail below.)

8. See Willard Cochrane (1970), A City Man's Guide to the
Farm Problem.

9. See Rose-Ackerman and Evenson (1981) for a fuller discussion. Also see the discussion in Section V below.

10. The price index used to deflate current expenditures is
taken from Evenson (1968). It is constructed by deflating
separately the expenditures on professional staff by an index
of university professors' salaries, technical and clerical staff
(skilled labor), equipment (metal and metal equipment), and

building investment (building materials). The 1970 and 1980 deflation is based on the general GNP deflator.

11.   The data in Table 5.9 and later tables referring to the public sector include some production-oriented research undertaken outside the state universities.   For example, if any USDA funding is involved, research conducted in a private university is included.   Nonetheless, a great deal of agriculturally related research is missed.   Research in plant and animal physiology, in plant and animal genetics, in cytology, in experimental design, and in a number of other scientific fields is of direct importance to applied agricultural research. In 1965, a USDA study group estimated that expenditures for agriculturally related research were approximately 70 percent of public-sector spending on agricultural research.

12.   Alternatively, this can be viewed as the spillover of research results from one state to another.

13.   Statistical analysis did not reveal significant differences in these figures by state or region.   Most of the state variance in this measure is associated with the commodity mix in the states.

14.   Again, the reader has to judge these estimated returns against the proper standard.   A number of writers on agricultural technology cite the Griliches hybrid-corn study as reporting a 700 percent rate of return.   Table 5.13 reports the internal rate of return to be in the 35-40 percent range. The 700 percent figure is nonsensical.   It is really a benefit-cost ratio based on the supposition that a government could borrow money at 6 percent and invest it in hybrid-corn research.   The ratio of the present value of the benefits to the present value of the costs in 1957 was 7.   Other studies report very high rates of return for short-term investments; again, these are not comparable to those reported here.   The reader should bear in mind that a one-dollar investment yielding a real internal rate of return of 700 percent would, compounded annually, grow to the size of the world's GNP in only 20 years.

15.   See the concluding section for a discussion of private-public compatability.

16.   The time dimension or shape of the relation between R&D and impact was investigated first by Evenson (1968).   Efforts to use standard distributed lag procedures to estimate the time shape have not been very successful in these studies.   A simpler approach in which alternate time lags of the form (a, b, c) are constructed and a nonlinear least-squares procedure employed to estimate a, b and c has been more successful.   In regression specifications a research stock variable is constructed from lagged expenditures as follows:

$$R_t = \sum_t^{t-a} W_t R_t + \sum_{t-a}^{t-a-b} R_t + \sum_{t-a-b}^{t-a-b-c} W_t R_t$$

The weights W_t rise linearly to one over a periods, equal one for b periods and fall linearly to zero over c periods. (An alternative to this specification is to specify a depreciation rate to approximate the c and b parts of the construction.)

17. A number of productivity studies conclude that residual productivity measures are a "measure of ignorance." This approach treats these measures as composed of several parts. Some part of them is systematically related to public programs and technology-producing investment. They still contain parts about which we are ignorant. The statistical regression model employed in these studies is designed to identify the systematic components. It makes no pretense of eliminating all of our ignorance about the growth process.

18. The measures of benefits used in these studies do not consider who benefits. Indeed this technical change, like almost all other types of change, has its gains and its losses. The final section of the chapter considers this issue to some degree.

19. Simple measures of research output per unit of input such as publications, patents, new varieties, etc., per SMY, are indicators of quality only in a restricted sense. An expansion in the number of scientists in a discipline may create an increase in the demand for journals and related publications, but not necessarily in the real products of the system. Patents and citations of publications, as output measures, avoid some of the worst problems. Patents are granted by examiners from outside the discipline and are more reliable indexes of certain types of research products. Citations are also determined partially outside the discipline.

REFERENCES

Ayer, H.W., "The Costs, Returns and Effects of Agricultural Research in Sao Paulo, Brazil," (Ph.D. dissertation, Purdue University, Lafayette, 1970).

Ayer, H.W. and G.E. Schuh, "Social Rates of Return and Other Aspects of Agricultural Research: The Case of Cotton Research in Sao Paulo, Brazil." American Journal of Agricultural Economics 54 (November 1972), 557-569.

Barletta, N. Ardito, "Costs and Social Benefits of Agricultural Research in Mexico." (Ph.D. dissertation, University of Chicago, 1970).

Binswanger, H.P., "Income Distribution Effects of Neutral and
Non-Neutral Technical Changes." Discussion Paper No.
281, New Haven: Yale University, Economic Growth Center, May 1978.

Bredahl, M. and W. Peterson, "The Productivity and Allocation
of Research: U.S. Agricultural Experiment Stations."
American Journal of Agricultural Economics 58 (November
1976), 684-692.

Danhof, J.C., "Agriculture" in H.F. Williamson (ed.) The
Growth of the American Economy, New York, N.Y., Pren-
tice Hall, Inc., 1951.

Denison, E.F., The Source of Economic Growth in the United
States and the Alternative Before Us. Washington, D.C.:
Committee for Economic Development, 1962.

Duncan, R.C., "Evaluating Returns to Research in Pasture
Improvement." Australian Journal of Agricultural Economics 16 (December 1972), 153-168.

Evenson, R.E., "A Century of Agricultural Research and
Productivity Change in U.S. Agriculture: An Historical
Decomposition Analysis." Discussion Paper No. 296, New
Haven: Yale University, Economic Growth Center, 1978.

_____. "The Contribution of Agricultural Research and
Extension to Agricultural Productivity." Ph.D. Dissertation, University of Chicago, 1968.

_____. "International Transmission of Technology in
Sugarcane Production." Yale University, New Haven,
1969. (Mimeographed).

_____. J.C. O'Toole, R.W. Herdt, W.R. Coffman, and
H.E. Kaufman, "Risk and Uncertainty as Factors in Crop
Improvement Research." Risk in Agriculture, SEARCA
1979.

_____. and D. Jha, "The Contribution of Agricultural
Research Systems to Agricultural Production in India."
Indian Journal of Agricultural Economics 28 (1973),
212-230.

_____. and H. P. Binswanger, "Technology Transfer and
Research Resource Allocation." in Induced Innovation:
Technology, Institutions and Development, ed. H.P. Binswanger and V.W. Ruttan, Baltimore: Johns Hopkins University Press, 1978 pp. 164-221.

_____. V.W. Ruttan and P. E. Waggoner, "Economic
Benefits from Research: An Example from Agriculture."
Science 205 (1979): 1101-1107.

_____. and F. Welch, "U.S. Agricultural Productivity: Studies in Technical Change and Allocative Efficiency." Unpublished manuscript, Yale University, Economic Growth Center, 1974.

Finlay, K.W. and B.M. Wilkinson, "The Analysis of Adaptation in a Plant Breeding Programme." Australian Journal of Agricultural Research 14, 6 (1963).

Griliches, Z., "Research Costs and Social Returns: Hybrid Corn and Related Innovation." Journal of Political Economy 66 (1958): 419-321.

_____, "Research Expenditures, Education, and the Aggregate Agricultural Production Function." American Economic Review 54 (December 1964).

_____, "The Source of Measured Productivity Growth, U.S. Agriculture, 1940-1960." Journal of Political Economy 71 (1962): 321-346.

_____, and D.W. Jorgenson, "The Explanation of Productivity Change." Review of Economics and Statistics 34 (1967).

Hayami, Y. and M. Akino, "Organization and Productivity of Agricultural Research Systems in Japan." In Resource Allocation and Productivity, ed. Thomas M. Arndt, Dana G. Dalrymple and Vernon W. Ruttan (Minneapolis: University of Minnesota Press, 1977) pp. 29-59.

Hertford, R., J. Ardila, A. Rocha and G. Trujillo, "Productivity of Agricultural Research in Columbia." In Resource Allocation and Productivity, ed. Thomas M. Arndt, Dana G. Dalrymple and Vernon W. Ruttan (Minneapolis: University of Minnesota Press, 1977) pp. 86-123.

Hines, J. "The Utilization of Research for Development: Two Case Studies in Rural Modernization and Agriculture in Peru." (Ph.D. dissertation, Princeton University, Princeton, N.J. 1972).

Huffman, W.E. and J.A. Miranowski, "An Economic Analysis of Expenditures on Agricultural Experiment Station Research." American Journal of Agricultural Economics 63 (1981): 104-118.

Kahlon, A.S., H.K. Ball, P.N. Saxena and D. Jha, "Returns to Investment in Research in India." In Resource Allocation and Productivity, ed. by Thomas M. Arndt, Dana G. Dalrymple and Vernon W. Ruttan, Minneapolis: University of Minnesota Press, 1977: 124-147.

Kendrick, J.W., Productivity Trends in the United States Princeton: Princeton University Press, 1961.

Kislev, Y. and R.E. Evenson, "Research and Productivity in Wheat and Maize." Journal of Political Economy June 1974.

Laing, D.R. and R.A. Fischer, "Preliminary Study of Adaptation of Entries in 6th I.S.W.N. to Non-Irrigated Conditions." Unpublished draft, CIMMYT, 1973.

Landau, D. and R.E. Evenson, "Productivity Change in U.S. Agriculture: Some Further Computations." Mimeo, Economic Growth Center, Yale University, 1973.

Latimer, R.G., "Some Economic Aspects of Agricultural Research and Education in the United States." Ph.D. dissertation, Purdue University, 1964.

Loomis, R.A. and G.T. Barton, "Productivity of Agriculture: U.S. 1870-1959." Agricultural Research Service, Technical Bulletin No. 1238, 1956.

Lu, Y. and P.L. Cline, "The Contribution of Research and Extension to Agricultural Productivity Growth." Paper presented at summer meetings of American Agricultural Economies Association, San Diego, 1977.

MacEarkean, G.A., "Regional Prosections of Technological Change in American Agriculture to 1980." Ph.D. dissertation, Purdue University, 1964.

Nadiri, M. Ishaq, "Some Approaches to the Theory and the Measurement of Total Factor Productivity: A Survey." J.E.L. 8 (December 1970): 1137-1177.

Nagy, J.G. and W.H. Furtan, "Economic Costs and Returns from Crop Development Research: The Case of Rapeseed Breeding in Canada." Canadian Journal of Agricultural Economics 26, February 1978: 1-14.

Nelson, R.R., "Aggregate Production Functions and Median Range Growth Projections." American Economic Review 54 (September 1964).

Norton, G.W. and J.S. Davis, "Review of Methods Used to Evaluate Returns to Agricultural Research." Department of Agricultural and Applied Economics, Staff Paper P79-16, University of Minnesota, 1981.

Pee, T.Y., "Social Returns from Rubber Research on Peninsular Malaysia." Ph.D. dissertation, Michigan State University, 1977.

Peterson, W.L., "Returns to Poultry Research in the U.S." Ph.D. dissertation, University of Chicago, 1966.

_____, and J.C. Fitzharris. "The Organisation and Productivity of the Federal-State Research System in the United States." In Resource Allocation and Productivity, ed. by Thomas M. Arndt, Dana G. Dalrymple and Vernon W. Ruttan, Minneapolis: University of Minnesota Press, 1977: 60-85.

Pray, C.E., "The Economics of Agricultural Research in British Punjab and Pakistani Punjab, 1905-1975." Ph.D. dissertation, University of Pennsylvania, 1978.

Richter, M.K., "Invariance Axioms and Economic Indexes." Econometric 34: 739-55. 1971.

Rose-Ackerman, S. and R.E. Evenson, "Public Support for Agricultural Research and Extension: A Political-Economic Analysis." Mimeographed, Economic Growth Center, Yale University, 1982.

Schultz, T.W., The Economic Organization of Agriculture New York: McGraw-Hill, 1953.

Schmitz, A. and D. Seckler, "Mechanized Agriculture and Social Welfare: The Case of the Tomato Harvester." American Journal of Agricultural Economics 52, November 1970: 569-577.

Scobie, G.M. and R. Posada, T., "The Impact of Technical Change on Income Distribution: The Case of Rice in Colombia." American Journal of Agricultural Economics 60, February 1978: 85-92.

Solow, R., "Technical Change and the Aggregate Production Function." Review of Economics and Statistics 39 1957, 312-20.

Tang, A., "Research and Education in Japanese Agricultural Development." Economic Studies Quarterly 13, February-May 1963: 27-41 and 91-99.

Wennergren, E.B. and M.D. Whitaker, "Social Return to U.S. Technical Assistance in Bolivian Agriculture: The Case of Sheep and Wheat." American Journal of Agricultural Economics 59 August 1977: 565-569.

6
The Pharmaceutical Industry
Henry G. Grabowski
John M. Vernon

I. INTRODUCTION

Although the pharmaceutical industry dates back to the last century, its development into a major industry with its current characteristics began about forty years ago. Prior to the 1930s, the industry was largely a commodity-based industry producing a relatively small number of chemical compounds and engaging in little research or development of new pharmaceuticals.

The present era of the research-oriented pharmaceutical industry had its origins in the mid-1930s when the first important group of antiinfective drugs were introduced. In particular, sulfanilamide was introduced in 1936 after it was discovered to be effective against streptococci bacteria without having toxic effects on human cells. This development stimulated considerable interest in research on other potential drug therapies. Several important drugs, most notably penicillin, and the other "magic bullet" antibiotics were introduced over the next decade and a half. After World War II, pharmaceutical research broadened to cover several different therapeutic areas. A number of new drugs were introduced to deal with cardiovascular, respiratory, neurological, and other disease categories.

The development of an R&D-oriented pharmaceutical industry has been closely intertwined with the rise of scientific understanding of conditions of health and sickness, and of the nature and causes of various diseases more generally. The post-World War II era has seen an enormous increase in research in the biomedical science, undertaken largely by scientists and physicians at medical schools, and funded largely by government.

This development of the pharmaceutical industry into a research-based industry competing in terms of new drug innovation also has been accompanied by the evolution of extensive government regulations of new drug innovations. Government regulation in this industry in fact dates back to the Pure Food and Drug Act of 1906.(1)   Early drug regulation, however, was directed primarily at patent medicine abuses.   In 1938, following a drug disaster that killed over 100 children, the Food, Drug and Cosmetic Act was passed by Congress.   This law required new drugs to be approved as safe by the Food and Drug Administration (FDA) before they could be introduced into interstate commerce.   It also provided the basis for the separation of pharmaceuticals into ethical drugs, that may be purchased only with a doctor's prescription, and proprietary drugs, that may be generally sold over the counter.   In 1962 Congress further passed the Kefauver-Harris Amendments to the Food, Drug and Cosmetic Act.   These amendments required that a new drug's efficacy, as well as its safety, be demonstrated on the basis of well-controlled scientific tests prior to marketing approval by the FDA.   Furthermore, they extended FDA regulatory controls to the clinical development process in order to protect human subjects involved in new drug testing.

In addition to these FDA regulatory controls, numerous other public policies impact the innovational process in the pharmaceutical industry.   The opportunities for new drug discoveries are enhanced by government support of basic research in the biomedical sciences.   The economic incentives for undertaking drug research and development are affected by federal patent and tax policies.   In addition, there are a number of federal and state programs that are directed at the marketing and distribution of drugs that also can have potentially significant effects on the economic returns to new drug innovation (e.g., state substitution laws, product formularies, the Maximum Allowable Cost program of Medicare and Medicaid reimbursements, and so on).

While the pharmaceutical industry has been one of our most innovative industries, the level of new drug introductions appears to have declined significantly from the earlier post-World War II period.   The reasons for and social significance of this decline have been the subject of considerable attention by both policy makers and academicians.   At the same time, there is cautious optimism in some circles at present about the future prospects for the industry in the next few decades, given the possibility of several important drugs now in the pipeline (especially in the emerging biomolecular research area).

Several important changes in government policies toward the industry, especially in the regulatory and patent areas, have been recently proposed and and are now under active debate. This is therefore a particularly apt time to examine the effects of government policies on innovation in the ethical drug industry. The first sections of this chapter provide an overview of industry structure and the character of technical progress in ethical drugs. The last half of the chapter then turns to an analysis of public policy impacts on drug innovation and also discusses the policy changes currently under active discussion by Congress and other related parties.

II. INDUSTRIAL ORGANIZATION

Industry Demand and Growth

Table 6.1 presents some historical data on the value of shipments for the Bureau of Census pharmaceutical preparations industry Standard Industrial Classification-a system of Census Bureau for defining industries - (SIC) 2834. These data are further disaggregated into domestic ethical drug and proprietary drug sales and overall exports to other countries. The rapid rate of growth in ethical drug industry sales since 1939 is clearly evident from the data in Table 6.1. In the period between 1939 and the early 1960s, growth occurred at a truly explosive pace with the value of shipments increasing more than an order of magnitude in nominal terms. Over the last two decades, the rate of growth in value of shipments has slowed significantly, but still remains above the average for all manufacturing.

Table 6.2 presents a breakdown of ethical drug sales for 1978 into broadly defined therapeutic categories. The two leading categories are central-nervous-system drugs (i.e., antiarthritics, tranquilizers, antidepressants, analgesics, drugs for epilepsy and stroke, etc.) and antiinfectives. These two categories collectively account for almost 40 percent of total sales. The remaining sales are divided rather evenly among the other therapeutic categories.

Table 6.3 presents information on the buyer side of the market for ethical drugs. This table shows that approximately 75 percent of ethical drug sales are made through retail pharmacies. Retail prescription sales currently account for about 5 percent of total national expenditures for health services and supplies.

The concentration of buyers in the retail market is very low. There are approximately 60,000 retail pharmacy outlets in the United States and perhaps 200,000 to 300,000 physicians who prescribe drugs on a regular basis. The individual doctor is an important decision maker, although he does not pay the price for the product. It is the doctor, and not the patient, who decides which product and which brand will be

Table 6.1.   Pharmaceutical Preparations, except
Biologicals, for Human Use[a]
(Value of Product Shipments in Millions of Dollars)

Domestic Sales

Year	Ethical	Proprietary	Exports	Total
1939	148.5	152.4	*	301.0
1947	520.7	317.6	*	838.3
1954	1088.9	368.3	*	1457.2
1963	2001.6	787.1	99.3	2888.1
1967	2885.8	999.5	112.7	3998.0
1972	4286.8	1427.8	125.0	5839.6
1977	6607.9[b,c]	2221.2[c]	260.3[d]	8829.1

(*) not reported separately

Definition of Terms

Ethical: products primarily advertised or otherwise promoted
to or prescribed by the health professionals

Prescription legend: a drug product which by federal law is
available only by prescription by a
licensed physician

Over-the-counter Professional: a drug product sold over-the-
counter and primarily promoted
to the professions

Proprietary: a drug product primarily advertised or otherwise
promoted to the general public.

(a) Includes pharmaceutical preparations of industries not
classified as SIC 2834.

(b) In 1977, the ethical category was split up into prescription
legend and over-the-counter professional.

(c) Includes exports.

(d) Figure obtained from Current Industrial Report 1977; MA
28G (77)-1 Table 5.

Source: Bureau of Census, Census of Manufacture Industry
Statistics, Group 28C

Table 6.2. Manufacturers' Domestic Sales of Ethical
Drugs for Human Use, by Product Class, 1978.

Class	Relative Share of Sales (in percent)
Central Nervous System	23.6
Antiinfectives	15.0
Gastrointestinal and Genitourinary	11.8
Neoplasms and Endocrine	9.7
Vitamins and Nutrients	9.6
Cardiovasculars	9.4
Respiratory System	7.8
Dermatologicals	. 2.9
Other	10.2
Totals	100.0

Source: Annual Survey Report, (Pharmaceutical Manufacturers
Association Washington, D.C; 1978).

prescribed. It is generally maintained that doctors decisions
in this regard are primarily influenced by considerations of
product quality and reputation of the manufacturer and only
secondarily by a product's price. Consequently, demand for
ethical drugs is often taken to be relatively inelastic over
broad ranges in price - in large part because of the quality
orientation of doctors' prescribing decisions. In recent years,
however, state substitution laws have given pharmacists more
scope for discretion in product selection for multisource products. The exact effects of these laws on retail dispensing
patterns still remains to be seen.
        The institutional sector - hospitals and various govern-
ment purchasing agencies - accounts for about 25 percent of
total sales and is considerably more concentrated than the
retail drug area. Drugs purchased by these institutions tend
to be bought in large quantity lots, often using competitive
bidding procedures. Consequently, demand in this market is
generally assumed to be more price elastic than it is in the re-

Table 6.3. Percentage Distribution of Manufacturers'
Domestic Sales among Retail Pharmacies,
Hospitals, Government Agencies, 1970.

	Percentage	($ millions)
Retail	74.5	4296.9
Hospital	14.4	831.8
Government	11.1	639.4[*]
Total	100.	5768.1

*Prescription drugs only.

Source: David Schwartzman. Innovation in the Pharmaceutical
Industry (Baltimore: Johns Hopkins University Press,
1976), p. 25, as compiled from the following original
sources: for retail and hospital sales, IMS America,
Ltd., U.S. Pharmaceutical Market, Drug Stores and
Hospitals (Ambler, Pa: IMS America Ltd., 1970); Data
summary, U.S. Dept. HEW, Social Security Admin.,
Office of Research and Statistics, SS Pub. 59-71
(5-71), 1971.

tail sector and generic product sales are more concentrated in
this section.

Supply-Side Structure

Three distinct segments or subgroups of competitors can be
identified in the ethical drug industry. The first and by far
the most important groups from the standpoint of industrial
innovation consists of the large, research-intensive multinational firms. These firms account for the major share of both
new product introductions and total ethical drug sales. At the
other end of the competitive spectrum is a large number of
generic manufacturers that specialize in producing unbranded
products at low prices after the originating firm's patent has
expired. In-between these extremes is a third group of primarily domestic firms that have research programs to develop
new drug products under their own brand names, but on a
much smaller scale than the multinationals.

There are perhaps 12 to 15 U.S. firms that can be placed in the research-intensive multinational group. These firms together with their foreign multinational counterparts compete in a worldwide market. Competition among the multinationals centers in the discovery and promotion of new drugs that are capable of winning significant market shares in the international market. These drug products are typically protected by product patents (and perhaps also process patents) and marketed under copyrighted brand names. Although many of the U.S. multinationals produce an extensive line of both brand-name and generic products for the domestic market, their profits and sales tend to be disproportionately tied to a handful of single-source products developed by the company and promoted under brand names.

Table 6.4 presents the ethical drug sales ranking for 24 pharmaceutical firms with U.S. hospital and pharmacy sales in excess of 100 million dollars in 1978. The U.S. pharmaceutical market is not dominated by a few firms. Instead, sales are distributed rather evenly across many major firms. This is reflected by the fact that the top four and eight leading firms account for only 24 and 42 percent, respectively, of ethical drug sales. Nevertheless, the 24 leading firms listed in Table 6.4 collectively account for nearly 80 percent of total sales and the multinational firms predominate among this group. In addition to several U.S. multinational firms, there are six foreign multinational firms among those leading firms (three headquartered in Switzerland, and one each from Germany, the United Kingdom, and Mexico).

Table 6.5 presents worldwide sales data for the U.S. human ethical pharmaceutical industry for the period 1965 to 1978. These data show the importance of foreign sales in the growth of the industry over recent periods. In 1978, foreign sales represented 41 percent of total sales compared with only 25 percent in 1965. Foregin sales have been growing at twice the rate of domestic sales for U.S. firms in recent years. In addition, a list of estimated sales for the top-ranked multinational firms in 1977 compiled by the United Nations Center on Transnational Corporations indicates 10 of the largest 20 pharmaceutical firms are U.S. firms, although the number-one-ranked firm is German (Hoechst).

In contrast to the competitive orientation of the multinationals around new product development and introduction in worldwide markets, the generic firms specialize in producing low-cost multisource products after patent rights have expired. There are at present several hundred manufacturers specializing in generic products but their collective market share is less than 10 percent of the ethical drug market. Their sales are concentrated in certain products with above-average tendencies for generic prescribing and for certain institutional buyers that are particularly price sensitive.

Table 6.4.   Leading Firms in U.S. Ethical
Drug Sales in 1978.

A.   Sales[*] Ranking of Manufacturers with Sales in Excess of
$100 Million in 1978

1.	Eli Lilly Co.	13.	Ciba Geigy
2.	American Home Products	14.	Searle
3.	Merck and Co.	15.	Squibb
4.	Roche	16.	Burroughs Welcome
5.	Smithkline	17.	American Cyanamid
6.	Johnson and Johnson	18.	Wander
7.	Warner Lambert	19.	Robins
8.	Bristol Myers	20.	Revlon Health Group
9.	Upjohn	21.	Sterling
10.	Pfizer	22.	Hoechst Roussel
11.	Abbott	23.	Richardson Merrell
12.	Schering	24.	Syntex

B.   Percentage of U.S. Ethical Drug Sales Accounted for by:

Leading 4 firms - 25.7
"   8 firms - 41.7
"   12 firms - 55.1
"   16 firms - 65.3
"   20 firms - 72.9
"   24 firms - 78.3

---

*Sales of ethical pharmaceuticals plus ethical OTC in all drug stores, discount houses, and hospitals.

Source: IMS America, Ltd., U.S. Pharmaceutical Market, Drug Stores and Hospital (Ambler, Pa: IMS America 1978).

As discussed above, there are at present some important policy developments and structural trends that may enhance the competitive position of generic products in future periods. In particular, several states have passed liberal substitution laws that encourage pharmacists to substitute low-cost products for the brands prescribed by physicians.   While the amount of substitution that has occurred to date has been minimal, some of the large chain stores have recently begun to promote and implement drug-substitution programs.   These developments, together with the tendency for the average effective patent life on new drugs to decline in recent years, may result in en-

Table 6.5. U.S. Human Use Ethical
Pharmaceutical Sales,
1965-1978 ($ millions).

Year	Domestic Sales	Foreign Sales (including exports)	Total (domestic and foreign)
1965	2,940	999	3,939
1966	3,178	1,162	4,340
1967	3,393	1,351	4,744
1968	3,808	1,494	5,302
1969	4,135	1,702	5,837
1970	4,444	1,981	6,425
1971	4,796	2,213	7,009
1972	5,136	2,603	7,739
1973	5,644	3,078	8,722
1974	6,273	3,683	9,956
1975	7,806	4,468	11,554
1976	7,867	4,908	12,775
1977	8,434	5,404	13,838
1978	9,411	6,567	15,978

Source: Pharmaceutical Manufacturers Association, Annual
Survey (years 1965-1978).

hanced competitive opportunities for generic firms in the
future. This issue will be discussed further later in the
chapter.

Concentration and Market Share Stability

While the data in Table 6.4 show that the pharmaceutical industry is not dominated by a few firms with large market
shares, most economists would still consider the industry to be
oligopolistic. In this regard, it is reasonable to argue that
the relevent markets should be defined in terms of therapeutic
categories rather than total ethical drug industry. This is

because drugs oriented to one therapeutic use (e.g., vitamins) are generally not substitutes for those in other categories (e.g., antibiotics or antidepressants). Although no classification scheme of "therapeutic markets" is likely to satisfy everyone, a prior attempt to define such markets by one of the authors yielded four-firm concentration ratios that averaged 68 (Vernon, 1971). These data are presented in Table 6.6 and cover a selected grouped of 19 therapeutic markets. In another study, Cocks and Virts (1974) constructed therapeutic markets by systematically evaluating physicians' prescribing habits. Their scheme yields markets that are generally more broadly defined than those in Table 6.6, and as a consequence, had somewhat lower concentration ratios. Nevertheless, however one defines therapeutic categories, one tends to observe much higher levels of concentration for these markets than for the industry as a whole.

Some analysis has been undertaken in recent years of the dynamic "instability" of the market shares of ethical drug sales as well as within particular therapeutic markets. Although drug markets are subject to considerable concentration at any given point in time, one might also expect to observe a high rate of turnover in firm market shares over time as a consequence of the rapid flow of new product introductions in this industry.

Douglas Cocks (1975) has undertaken an analysis of this issue. Specifically, he first computed an instability index for the ethical drug industry and compared it with similar indices computed for 20 industries by Hymer and Pashigian (1962) in a prior analysis. Only one industry was found to have a higher "instability index" than pharmaceuticals. In addition, he found a higher degree of volatility of firm market shares within particular therapeutic classes associated with rival new product introductions displacing established market leaders over the ten-year period examined in his analysis.

Conditions of Entry

Three major factors have been cited in the literature as important sources of entry barriers in the ethical drug industry: patents, brand differentiation, and scale advantages in research and development.

Patents play a significant role in the innovative process for the ethical drug industry. This is in apparent contrast with many other technologically progressive industries (Taylor and Silberston, 1973, chap. 10).

In the case of pharmaceuticals, the main output from the R&D process is the knowledge and evidence that a particular chemical entity is a safe and effective therapy in the treatment of a particular disease, plus the FDA certification of this evi-

Table 6.6. Concentration of Sales in the United States
Ethical Drug Industry, by Therapeutic Markets, 1968.

Therapeutic Market	4-Firm Ratio
Anesthetics	69
Antiarthritics	95
Antibiotics-Penicillin	55
Antispasmodics	59
Ataractics	79
Bronchial Dilators	61
Cardiovascular Hypotensives	79
Coronary-Peripheral Vasodilators	70
Diabetic Therapy	93
Diuretics	64
Enzymes-Digestants	46
Hematinic Preparations	52
Sex Hormones	67
Corticoids	55
Muscle Relaxants	59
Psychostimulants	78
Sulfonamides	79
Thyroid Therapy	69
Unweighted Average	68

Source: John M. Vernon, "Concentration, Promotion and Market Share Stability in Pharmaceutical Industry,"
Journal of Industrial Economics (July 1971).

dence in terms of marketing approval.   Once this knowledge becomes publicly available, however, the costs of imitation by rival producers are usually low.   Hence, there is little to stop rival firms from producing this compound on similar terms as the innovator in the absence of legal barriers such as those afforded by the patent system.

An important ruling by the Patent Office in 1948 concerning the antibiotic streptomycin opened the door to the patenting of new drugs.   That is, new drugs would not be patentable if they were simply natural substances.   In the case of streptomycin, the Patent Office ruled that the natural materials found by Waksman, streptomycin's discoverer, were not in suitable form for medical use.   By making chemical modifications to streptomycin so that it could be purified, however, the Patent Office ruled that Merck had created a "new composition of matter," and therefore should be awarded a patent on this new product.   Without this ruling, R&D on antibiotics would not have been an attractive, or necessary, competitive instrument for the pharmaceutical companies.

Firm R&D strategies are oriented around developing products that are patentable.   Over 90 percent of the new chemical entities coming to the U.S. market in recent years have involved drugs protected by product patents.   Furthermore, approximately half of all prescription drug sales at the present time involve single-source products protected by patents.

A patent barrier, of course, can be overcome by the development of chemically distinct substitutes for the established market leader's product.   As discussed above, there is in fact considerable market share turnover in this industry associated with the introduction of new chemical entities.   One strategy for inventing around an existing firm's patent that has received considerable attention in the literature is "molecular modification."   This refers to the development of a similar compound so as to retain a rival product's main therapeutic effects (or hopefully to improve them), but at the same time possesses a chemically distinct structure that can be patented.   Our discussion in Section III on the character of technical progress indicates that many such "families" of drugs with similar chemical structures and therapeutic properties have been developed in just this manner.

Nevertheless, the strategy of developing "me too" products through molecular modification is neither costless nor always guaranteed to produce an effective substitute for an established product.   In contrast to imitative products involving already approved substances by the FDA (i.e., generic equivalents) chemically differentiated products must undergo full-scale reviews of safety and efficacy by the FDA. Hence, these drugs must be tested on the same scale as all previously approved products.   Therefore, under current regulatory conditions, the imitating firm is faced with several million dollars in development costs and several years in lag time before chemically distinct follow-on drugs can be marketed as approved new drugs.

Data from trade sources indicate that firms in the drug industry as a whole spend a little over 10 percent of their sales on research and development expenditures and at least a comparable percentage on promotional outlays. Promotional outlays per dollar of sales tend to be greatest in the early stages of product life cycle when information on a new drug is being initially diffused. As noted above, products are promoted under brand names with the objective of building up a stock of good will or specific preference in the minds of physicians for the innovating firm's product. This has historically provided an important source of product differentiation advantages vis-a-vis new entrants after patents expire (i.e., competitively advertised brands and generic products).

There is also evidence that the firm that introduces the first product of a new "family" of drug therapies can obtain important product differentiation advantages relative to follow-on imitative products. Bond and Lean (1977) have examined this issue in a recent FTC study. In the case of the oral diuretic market, for example, they found that the first drug on the market, Merck's Diuril, introduced in 1958, enjoyed substantial competitive advantages over a number of therapeutically similar (but chemically distinct) drugs that quickly followed it on the market. These data indicate that Merck spent less than half as much per sales dollar on promotion for Diuril than follow-on products and also charged a significantly higher price than competitors. Despite these policies, in 1971, 13 years after the original introduction of Diuril, it was still the market leader with a 33 percent share of the oral diuretic market. Bond and Lean further found that those follow-on products that were most successful in capturing market shares were those that offered significant therapeutic gain over established diuretics, rather than merely relying on high promotion levels or price discounts.

At the present time there is evidence to suggest that the major drug firms are concentrating more of their R&D efforts on developing drugs that embody new approaches to disease treatment and less on development of "me too" products. This reflects, at least in part, the strong upward trends in the costs and times for developing and obtaining FDA approval of a new drug entity compared with a few decades ago. Data discussed in the next section indicate that R&D costs have escalated sharply relative to overall returns for new drugs and hence there is less economic incentive to develop "me too" products. Of course, firms are still motivated to explore compounds with chemically related structures to those of existing products in hopes of developing products with improved therapeutic properties. There does appear, however, to be an increased emphasis on drug candidates with significant market share potential to compensate for increased development costs.

It also appears that, as a result of the sharp increase in the costs and riskiness of developing new drugs, economies-of-scale considerations in drug R&D are a much more important factor than was the case a few decades ago. This is consistent with the findings of recent studies that drug innovation is now much more concentrated in the large drug firms than was the case in the early 1960s (Grabowski and Vernon, 1976).

Industry Profitability and Pricing Trends

The pharmaceutical industry has ranked near the top of the manufacturing sector in terms of overall profit rates for most of the post-World War II period. This aspect of industry performance, together with the high-price cost margins on particular products, has received considerable attention from congressional committees beginning with the highly publicized Kefauver Hearings (2) in the late 1950s and 1960s.

In recent years, however, there has been a noticeable tendency for industry profit rates to begin converging toward the average obtained by the entire manufacturing sector. In Table 6.7, earnings data based on FTC's Quarterly Financial Reports are presented for the pharmaceutical industry and the overall manufacturing sector for the period 1956-1979. These data indicate that pharmaceutical earnings as a percentage of net stockholder equity have consistently been above the average for all manufacturing over this period. At the same time there is a clear trend evident in these data for the difference in the profit rates to narrow over time. This is especially true in the pre-tax profit series. Among other things, this apparently reflects, with some response lags, the lower rates of industry growth and slower rates of new product introductions in recent years compared with the early postwar period. These trends (together with recent technological developments that have produced more optimistic assessments of industry's prospects over the immediate future) will be discussed in detail in the next part of the chapter.

Another issue that has received considerable attention, primarily in academic studies, is the potentially significant bias present in reported profit rates for the drug industry (and other industries with similar characteristics) that results from the expensing rather than capitalizing of so-called intangible capital outlays - i.e., R&D and advertising investment expenditures. It is standard accounting practice to expense these intangible capital outlays even though conceptually they are in fact investment expenditures with expected returns distributed over future periods. Recent academic analyses by Clarkson (1977) and Grabowski and Mueller (1978) have adjusted report-

Table 6.7. Rates of Return on Average Stockholders Equity, 1956-1979.

Year	Before Taxes		After Taxes	
	Pharmaceutical Industry	All Manufacturing	Pharmaceutical Industry	All Manufacturing
1956	34.6%	22.6%	17.6%	12.3%
1957	37.4	20.0	18.6	11.0
1958	34.5	15.4	17.7	8.6
1959	34.1	18.9	17.8	10.4
1960	32.5	16.6	16.8	9.2
1961	32.4	15.9	16.7	8.8
1962	32.4	17.6	16.8	9.8
1963	32.8	18.4	16.8	10.5
1964	34.1	19.8	18.2	11.6
1965	37.1	20.0	20.3	13.0
1966	37.0	22.5	20.3	13.4
1967	33.7	19.3	18.7	11.7
1968	35.1	20.8	18.3	12.1
1969	35.4	20.1	18.4	11.5
1970	32.3	15.7	17.6	9.3
1971	31.9	16.5	17.8	9.7
1972	32.7	18.4	18.6	10.6
1973	33.1	21.8	18.9	12.8
1974	29.7	23.3	18.7	14.9
1975	27.7	18.9	17.7	11.6
1976	28.1	22.7	18.0	13.9
1977	28.9	23.2	18.2	14.2
1978	28.5	24.5	18.8	15.0
1979	28.8[*]	25.7	19.3[*]	16.4

---

*A considerable number of companies were reclassified by industry. The percentage of companies reclassified in the drug industry is unknown.

Note: for purpose of this table the pharmaceutical industry is defined as corporations primarily engaged in manufacturing biologicals, inorganic and organic medicinal chemicals, and pharmaceutical preparations; and grading and grinding botanicals.

Source: Quaterly Financial Reports (for manufacturing, mining and trade corporations) 1957-1979, Federal Trade Commission.

ed profit rates in several industries including drugs and have found that profit rates in ethical drugs do have a significant upward bias on this account. In the Grabowski and Mueller study, for example, more than half the reported differences in profit rates between the drug industry and the overall sample mean were eliminated when R&D and advertising outlays were capitalized rather than expensed. While there is room for disagreement on the appropriate assumptions for making such profit rate adjustments, it is clear that these adjustments do tend to reduce further the difference between drug industry and overall manufacturing accounting profit rates observed in Table 6.7

Another issue that has received considerable public policy attention is the high rate of price inflation in health services sector. The price performance of prescription drugs, however, has been in marked contrast with other sectors of the health services industry. In Figure 6.1, we present trends in the consumer price index for prescription drugs, for medical care (excluding drugs), and for all items over the period 1965 through 1979. While overall health-sector prices have increased at a much more rapid rate than the CPI index over this period, relative prices for prescription drugs have declined over time. The decline is especially pervasive during the 1960s. It has continued at a diminished rate over more recent periods. It should also be noted that current government price indices tend to adjust for product-quality improvements inadequately so that they tend to overestimate the degree of inflation in technologically progressive sectors vis-a-vis nonprogressive ones.

In summary, the ethical drug industry, in common with many other technologically progressive industries, has experienced above-average profit rates and declining relative prices over time. Accounting measures further tend to overstate both profitability and price inflation in research-intensive industries with high rates of product innovation. Nevertheless, given these measurement error problems, there is also a definite tendency in recent periods for some convergence toward the average for all manufacturing evident in the time trends on both these variables. The possible reasons for this are discussed further in our analysis in the following sections on technical progress in this industry.

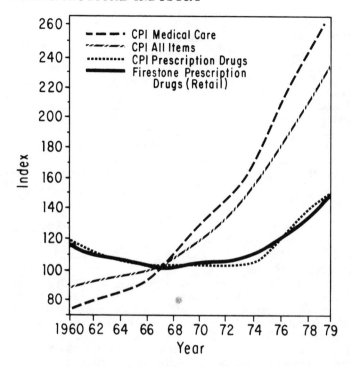

Fig. 6.1. Comparison of selected price indices, 1960-1979 (1967=100).

Source: Pharmaceutical Industry Fact Book as constructed from the following original sources - CPI indices; Consumer Price Index Detailed Reports, various issues; Firestone index - Firestone, various issues.

III. BASIC CHARACTERISTICS AND SOURCES OF TECHNICAL PROGRESS

Social and Economic Effects of New Drug Discoveries

As we noted earlier, the modern drug industry began in the mid-1930s with the introduction of the first sulfonamide drug. Since that time hundreds of new drugs have been introduced in the United States. Some of the major discoveries that have been introduced over the post-World War II period includes:

- Synthetic penicillins
- Tetracyclines

- Cortisone
- Chlorpromazine (major tranquilizer)
- Meprobamate (minor tranquilizer)
- Antihypertensives
- Antiinflammatories
- Oral contraceptives
- Diuretics
- Antidiabetics

Technical progress in the pharmaceutical industry has particular significance, of course, because of its key role in improving the quality of human life and health. In his well-known book on health economics, Who Shall Live, Victor Fuchs (1974) p. 105 stated:

> Drugs are the key to modern medicine. Surgery, radiotherapy, and diagnostic tests are all important, but the ability of health care providers to alter health outcomes - Dr. Walsh McDermott's "decisive technology" - depends primarily on drugs. Six dollars are spent on hospitals and physicians for every dollar spent on drugs, but without drugs the effectiveness of hospitals and physicians would be enormously diminished.
>
> The great power of drugs is a development of the twentieth century - many would say of the past forty years. Our age has been given many names - atomic, electronic, space, and the like - but measured by impact on people's lives it might just as well be called the "drug age."

Table 6.8 shows some major changes in mortality rates for selected diseases that have occurred since 1960. Drugs have had an important effect in explaining these declining mortality rates. The numbers of cases of many diseases have also declined because of improved drug therapies. Table 6.9 shows, for example, that new measles vaccines have reduced the number of measles cases by 46 percent between 1969 and 1978. Similarly, new antiinfectives have produced a 27 percent decline in the number of tuberculosis cases over the same period.

The introduction of new pharmaceutical agents has also resulted in significant benefits in the form of reductions in the need for and extent of hospitalization for many diseases. For example, the introduction of tranquilizers and antidepressants was instrumental in reducing the populations in mental hospitals from 565,486 patients in 1956 to 202,971 patients in 1975.

In addition, the cumulative advance in drug therapy has provided a relatively low-cost means of treating disease and producing good health. This is important because the health

Table 6.8. Mortality; Reductions in U.S. Deaths per
100,000 Population from Selected Diseases, 1960 and 1977.

Disease	1960	1977	% Reduction
Active Rheumatic Fever and Chronic Rheumatic Heart Disease	10.3	5.9	43
Hypertensive Heart Disease	37.0	4.8	87
Hypertension	7.1	2.6	63
Cerebrovascular Diseases	108.0	84.1	22
Arteriosclerosis	20.0	13.3	34
Pneumonia	32.9	23.1	30
Asthma	3.0	.8	73
Peptic Ulcer	6.3	2.7	57
Nephritis and Nephrosis	7.6	3.9	49
Infections of Kidney	4.3	1.7	60
Tuberculosis (all forms)	6.1	1.4	77
Meningitis	1.3	.7	46
Infectious Hepatitis	.5	.2	60

Source: Statistical Abstracts of the United States (Washington,
D.C., U.S. Bureau of the Census, 1979)

sector is characterized by scarce and expensive professional
manpower, labor-intensive activities, and complex technical
equipment - all contributing to a very high rate of cost
inflation in health services over recent years. By contrast,
the costs of ethical drugs have accounted for a relatively small
percentage of total health costs and have been a relatively
stable element in the presence of rapidly rising costs elsewhere
in the health sector.

Characteristics of the Drug R&D Process

In this section we shall examine the characteristics of drug
R&D at the level of the individual firm. First, we describe
the nature of drug discovery and review how a number of

Table 6.9. Reductions Reported in U.S. Cases
of Selected Diseases, 1969 and 1978.

Disease	1969	1978	% Reduction	Form of Treatment or Prevention
Diphtheria	241	76	68	Vaccines
Encephalitis	1,917	1,183	38	Antibiotics
Measles (all types)	83,542	45,170	46	Vaccines
Meningococcal Infections	2,951	2,505	15	Antibiotics
Whooping Cough	3,285	2,063	37	Vaccines
Acute Rheumatic Fever	3,229	851	74	Antibiotics and Steroids
Tuberculosis	39,120	28,521	27	Antiinfectives

Source: Reported Morbidity and Mortality in the United States, Annual Summary, 1978. (CDC) 79-8241, Center
for Disease Control, U.S. Dept. of Health, Education,
and Welfare, 1979.

important drugs have been discovered. The complex system of
drug development and FDA involvement is the next topic. A
well-known study of the cost of developing a marketable NCE
is also reviewed.

Drug Discovery

The process of drug discovery involves a multidisciplinary
research-team approach that is generally characterized by
considerable trial-and-error search effort. Serendipity has
also played an important role in many major discoveries.
Some examples of how drugs have been discovered will provide
further insight.(3)
     The original sulfa drug, sulfanilamide, was a life-saving
drug in many severe human infections. It was discovered in
1935 by Domagk who observed that the red dye sulfamidochry-
soidine was effective against streptococcal infections in mice.
It had, however, several serious side effects, including kidney
damage. Medicinal chemists therefore synthesized almost 5000
derivatives of sulfanilamide in search of compounds that were

free of the serious side effects. Two of the most successful drugs from this group have been sulfathiazole and sulfadiazine.

A chance clinical observation that patients taking sulfanilamide often excreted a larger than usual volume of urine led to the development of a whole new class of diuretic drugs. Again, testing of many closely related chemical compounds was necessary to discover the most effective diuretics. Similarly, serendipitous clinical observation of patients on sulfanilamide therapy led to antithyroid and oral hypoglycemic drugs.

There are many additional examples of drugs that were discovered by chance observation. The most famous is, of course, Fleming's discovery of penicillin. The important major tranquilizer, chlorpromazine, was the result of the unexpected discovery that certain of the anithistamines are potent depressants of the central nervous system. Others include the antihypertensive actions of the Beta-blockers, the antiinflammatory effects of the steroids, and the antigout action of allopurinol.

Random screening is another technique of drug discovery. Schwartzman (1976) has described one especially interesting example. In search of a drug to combat tuberculosis, Lederle Laboratories systematically tested a file of 103,000 chemical compounds that had been developed by its parent company for a variety of purposes. Eventually, after many years, a compound originally developed for use as an antioxidant additive for rubber was found to be effective against tuberculosis. Six hundred similar compounds were synthesized and the important antitubercular drug ethambutol was discovered.

These examples suggest that drug discovery is largely an empirical, trial-and-error process. However, this situation is changing dramatically. For example, a 1979 article in Business Week observes that:

> More and more, the development job is done today by setting forth in advance very specifically the characteristics desired in a new drug. The molecules of the chemical compound are designed, atom by atom, to affect a pretargeted physiological process in the body - inhibiting or stimulating, for instance, the flow of a specific enzyme. Examples of drugs developed in this fashion are Smith-Kline's Tagamet; Squibb's Capoten; and Lilly's Dobutrex.

We will return to "discovery-by-design" in the discussion below of sources of pharmaceutical innovation.

Of course the actual discovery of a new drug is only the first step in the lengthy process of drug innovation. In the following section we turn to the development of a drug once it has been synthesized and thought to possess potential therapeutic benefits.

Drug Development

Figure 6.2, reproduced from an article by William M. Wardell (1979) p. 10, provides a good overview of the present system of drug development and FDA regulation. As explained earlier, once a new chemical compound has been tested in animals and found to be worthy of human testing, the developer must file an IND (Investigational New Drug application) with the FDA

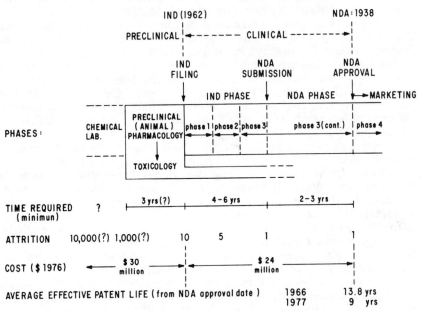

Fig. 6.2. Drug development (U.S.A.)

Source: Wardell (1979).

If approved by the FDA, the drug proceeds through three phases of clinical testing. The first phase is directed toward examining a drug's possible toxic effects and is performed on healthy individuals under highly controlled situations. If a drug successfully completes this stage, it is then tested on a relatively small number of patients to examine its effectiveness. It is then carefully evaluated from a therapeutic and marketing standpoint before the decision to begin phase three is made. Phase three involves expanded studies in large patient populations with a substantial escalation in development expenditures. If a drug passes these three phases of testing and is considered to have sufficient market value to warrant commercial introduction, an NDA (New Drug Application) is submitted to the FDA. Marketing can commence upon receiving an approved NDA.

Several further points should be made with reference to Figure 6.2. The time required to pass through the three testing phases is shown to be four to six years with an additional two to three years for NDA approval. The attrition rates show that for every ten drugs entering the IND stage, only one will have an NDA submission. Notice that Figure 6.2 shows no further attrition. According to Wardell, "the one survivor that reaches an NDA submission has a ninety percent chance of being approved by the FDA, given five years' for review at FDA."

The cost figures shown in Figure 6.2 are based on a study by Ronald W. Hansen. Hansen (1979) obtained survey data from 14 pharmaceutical firms on the R&D costs for a sample of NCEs first tested in man from 1963 to 1975. As shown, the discovery cost per NCE was estimated at $30 million and the development cost at $24 million, or a total cost of $54 million. This $54 million figure represents the capitalized value (at 8 percent interest and in 1976 dollars) at the date of marketing approval.

It should be pointed out that the $54 million includes the cost of NCE's that enter clinical testing but are not carried to the point of NDA approval. For example, Hansen found that by the end of 15 months of clinical testing, testing had ended on over 50 percent of the NCEs that had entered human trials. Hence, the $54 million figure should be interpreted as the average expected cost of discovering and developing a marketable NCE.

The Sources of Pharmaceutical Innovation

We begin by considering some data concerning the expenditures for health R&D in the United States. Table 6.10 shows total health R&D expenditures (not just drug-related), the portion of that total accounted for by the federal government, and the privately financed drug R&D outlays by the pharmaceutical industry.

Of the federal health R&D figure of $3.8 billion, $2.6 billion was health R&D support accounted for by the National Institutes of Health. While we do not know the total amount of federal support for drug R&D, there are several formal programs concerned with drug development. The largest is the National Cancer Institute Drug Development Program with an annual budget of over $200 million. This program is discussed further in the next section.

Table 6.10. Expenditures on Health R&D
(Billions of Dollars).

Year	Total Health R&D	Federal Health R&D	Private Drug R&D
1960	.9	.4	.2
1965	1.9	1.2	.3
1970	2.8	1.7	.5
1975	4.6	2.8	.8
1978	6.2	3.8	1.1

Source: U.S. Department of Health, Education and Welfare,
1979 NIH Almanac; PMA Factbook.

The relative importance of private versus public institutions (government, universities, and nonprofit foundations) in the discovery and development of new drugs has been a controversial issue. For example, the famous Kefauver Committee hearings on the pharmaceutical industry which began in December 1959, dealt extensively with the medical value of the R&D effort of the industry.

Comanor (1966) has referred to that controversy as the "battle of the lists." That is, the committee staff and the industry prepared competing lists of new drugs. The committee list tended to concentrate on drugs that embodied what it considered to be major therapeutic advances, and emphasized the role of public institutions. The industry list, on the other hand, "included new drugs that may not have embodied large steps forward but that are in frequent use and thereby seem to have the confidence of the country's physicians. A large majority of the drugs on this list were discovered and developed within industry laboratories."

Schwartzman (1976) p. 13 has assembled some more recent data on this question. As shown in Table 6.11, close to 90 percent of the NCEs introduced over the 1950 to 1969 period were discovered by private ethical drug firms (U.S. and foreign). Furthermore, this percentage exhibits a tendency to increase over time as evidenced by the 5 percentage-point increase in 1960 to 1969 over the earlier ten-year period. His analysis also reveals that the industry accounted over the 1960-1969 period for 86 percent of the therapeutically most important drugs, as classified by Martin Seife of the FDA. The result is consistent with similar analyses of this question by Schnee (1971) and Deutsch (1973).

Table 6.11. Percentage of New Chemical Entities
Discovered and Introduced by the Pharmaceutical Industry
1950-1959, 1960-1969, and 1950-1969.

	Periods in Which Drugs Were Introduced		
Source	1950-1959	1960-1969	1950-1969
Industry	86	91	88
Other	14	9	12
Total	100	100	100

Notes: List of NCEs. Selected from Paul de Haen, New Pro-
        duct Survey and Nonproprietary Name Index. Codis-
        covers are each given half credit where the source of
        discovery could not be determined, it was assigned to
        other.

Source: David Schwartzman, Innovation in the Pharmaceutical
         Industry, (Baltimore: Johns   Hopkins   University
         Press, 1976), p. 74.

        Such exercises as these, though useful, tend to deempha-
size a basic point. The roles of private and public institutions
are largely complementary rather than competitive. Professor
Ernst B. Chain, a Nobel laureate for his work in penicillin
development, has made this point well. Chain p. 450 (1963)
observed that large industrial laboratories are ideal for
"large-scale screening for new antibiotics, large-scale pharmacological testing, and the synthesis of a vast number of
analogous or related substances with the aim of improving one
or the other property of a drug." The academic laboratory,
on the other hand, is designed "to break fundamentally new
ground towards a better understanding of the laws of Nature,
and in this way to lay the basis for eventual industrial
exploitation of the scientific discoveries emanating from its
work."
        A more recent description of the complementary nature of
private and public R&D was given by Richard D. Wood, chief
executive of Eli Lilly, in a 1979 interview:

The industry depends on the productivity of research, and research goes in cycles. Some of it is serendipity, but progress depends mostly on what comes out of basic medical research and the knowledge it produces. Then industry can take hold of this knowledge and develop new drugs. Sometimes this occurs in stair-step fashion, and you reach a new plateau of medical knowledge that gives further impetus to new drugs. [Business Week p. 135, 1979].

On the other hand, David Schwartzman (1976, p. 30) has argued that, in pharmaceutical R&D "there exists no simple flow-through from basic to applied R&D. Basic research advances relevant to drug discovery, in contrast to the role of basic research in other fields, do not lead in any direct way to new drugs. New drugs cannot be designed by logical deductions from valid general principals; chemical theory alone is not enough and biological theory is woefully inadequate." Schwartzman goes on to observe that the majority of discoveries can be traced to one of three sources: naturally occurring compounds, accidental discoveries, and modifications of previously known drugs. In his view, this explains the relatively high proportion of drug discoveries made by the industry.

In this connection, we should emphasize the trend noted earlier that "discovery-by-design" appears to be replacing the more inductive trial-and-error methods emphasized by Schwartzman. One inference from this trend is the strengthening of the linkage between basic biomedical research and drug innovation.

According to Dr. William I. H. Shedden (Business Week, p. 137), vice president in charge of clinical evaluation at Eli Lilly, scientists at Lilly are now taking a "very fundamental biological approach" in some of their research.(4) Dr. Shedden observed that in the old days the chemists would make a batch of compounds and send them over to the biologists to put into animals to see what would happen. In contrast, today the biologists ask the chemists to design molecules to accomplish particular effects.

One highly successful example of drug design is the antiulcer drug, Tagamet, which was introduced by Smith Kline in 1977 and is already the second largest-selling U.S. drug. Knowing that the hormone histamine is a potent stimulant of the gastric secretions that can lead to ulcers, Smith Kline scientists sought a compound that would inhibit the flow of histamine. They finally succeeded in designing a molecule that would lock into a "receptor site," thereby blocking out the hormone and, in turn, the gastric secretions.

As Dr. P. Roy Vagelos, head of R&D at Merck, observed in a 1979 interview:

There has been a flowering of biomedical research. This is a fantastic time in biology. The companies with the right kinds of people and resources can capitalize on it and bring the new knowledge to bear on the right diseases and compounds [Business Week, 1979, p. 137].

The apparent trend toward closer ties between advances in scientific knowledge and new pharmaceutical products is well illustrated by recombinant DNA, or "gene splicing." This new process has been used to induce bacteria to produce human insulin and interferon, and has exciting possibilities in other areas. Several established drug firms now have research and development programs in this field and several small new firms have been founded to explore its commercial application. Even some universities are now considering the establishment of genetic engineering companies to develop the discoveries of its scientists (Time, November 10, 1980).

While, as stated, the major role of the universities in pharmaceutical R&D has been in basic research on the nature and causes of various bodily malfunctions, and on human biology in general, there often are more direct links between academic and corporate researchers. The interaction of university biomedical researchers and their counterparts at pharmaceutical companies can take various forms. We list below some of these relationships:

1. Most major pharmaceutical firms use university researchers on advisory boards or as consultants.
2. Many of the clinical studies required for FDA approval are conducted at university medical centers.
3. Pharmaceutical companies provide research grants to universities. Some companies award postdoctoral fellowships to encourage the supply of scientists for the pharmaceutical industry.
4. There are many scientific organizations which have publications authored by scientists from both industry and the universities. Also, there organizations sponsor conferences which promote industrial and academic interaction. A list of some of these organizations is given below:

   - Academy of Pharmaceutical Sciences
   - American Chemical Society
   - American Association for the Advancement of Science
   - American Medical Association
   - American Society for Pharmacology and Experimental Therapeutics
   - Drug Information Association
   - Industrial Research Institute
   - The American Association of Immunologists

- The American Physiological Society
- American Institute of Nutrition
- American Society of Biological Chemists
- American Association of Pathologists
- Society for Neoroscience
- American Academy of Allergy
- Biophysical Society
- New York Academy of Medicine
- American Society for Microbiology
- American Association for Cancer Research

5. Pharmaceutical companies' scientists and university scientists often collaborate in joint research projects. As an example, NIH has recently funded a three-year project to investigate the biosynthesis and mechanism of action of an anticancer agent, CC-10-65. The research will be carried out jointly at the University of Texas and at the Upjohn Company. According to F-D-C Reports, September 14, 1981:

> Upjohns joint research with U. Texas is apparently part of a collaborative research program the firm established over two years ago. The firm has set up a full-time scientific liaison office, headed by Douglas Shepherd, PhD, to serve as a basis for interaction and coordination of joint research projects. The company now has well over 200 interactions with universities outside of clinical studies, consultantships and contracted services.

> The scientific liaison office at Upjohn both funds projects through its own budget and seeks funding from other sources such as Natl. Science Foundation's Industry-University Cooperative Research Program and NIH. The U. Texas project is indicative of Upjohn's ability to extend its research activities and budget through the use of university talent and federal money.

6. A final form of interaction is where a drug is discovered by a university scientist and is then licensed to a pharmaceutical company for development and marketing. An example is the drug calcitriol, an agent used for the management of hypocalcemia in patients undergoing chronic renal dialysis.

> Dr. H.F. DeLuca of the University of Wisconsin was awarded a patent for the drug calcitriol. Dr. DeLuca then assigned the patent to the Wisconsin Alumni Research Fund (WARF). In 1975, WARF

licensed Roche to develop the drug and make it available to the medical community. It was recognized that substantial resources were needed to determine the efficacy, safety, and acceptability of the drug, and that an economic method for manufacturing the compound was needed. Roche was able to accomplish these objectives and gain marketing approval in 1978.

Government Drug-Development Programs

A clear exception to the tendency for government efforts to be concentrated in the basic end of the biomedical research spectrum is the National Cancer Institute's Drug Development Program. This program was established in 1955 with an initial annual budget of five million dollars.(5) Given the nation's overriding desire to cure cancer - as expressed in the National Cancer Act of 1971 - Congress has dramatically increased support for this program over the past decade. The annual budget in 1979 was 238 million dollars (Table 6.12).

The NCI's drug-development activities in the cancer area have several components. The program screens some 15,000 potential therapies in animals annually. The NCI also sponsors all of the major cancer clinical trial groups in the United States. All new cancer drugs, regardless of source, must be tested in these groups. NCI's involvement at the clinical trial stage range from complete support and sponsorship to various forms of collaboration (contractual as well as informal in character) with private firms. The NCI also attempts to evaluate on a continuing basis the optimal role of all marketed drugs in the management of neoplastic diseases. The government has not become involved in the marketing of any new drugs but has licensed its developed drugs to private firms for commercial production and marketing.

Since the NCI drug-development program originated in 1955, 20 new anticancer drugs have been developed with varying degrees of government collaboration and support. (In addition there were seven new anticancer drug applications on file with the FDA in 1980.) Of these 20 marketed drug introductions, the NCI was involved with 10 of these drugs on a preclinical basis (e.g., synthesis, screening, or toxicology testing). In addition, a favorable externality of the anticancer drug developments of government and industry has been a half-dozen or so additional introductions in other therapeutic areas.(6)

Table 6.12 indicates that the other drug-development programs of the government are very modest in size. These programs are conducted in other branches of the National Institutes of Health or the military and generally have annual

Table 6.12. Government Drug-Development Programs.

Year Began	Program Name	Annual Budget FY 1979 (in millions)	NDAs*	Target Condition or Process
1955	Drug Development and Cancer Therapy Evaluation Programs, NCI	$238	27	Cancers
1956	Psychopharmacology Research Branch, NIMH	17	1+	Methodology of Psychiatric Drug Development
1963	Army Drug Development Program, Department of the Army	4	1	Malaria, Leishmaniasis, Trypanosomsis, Schistosomiasis, Radiation Sickness, Chemical Warfare Injuries
1965	Vaccine Development Program, NIAID	8	2	Prophylaxis of Viral and Bacterial Infections
1968	Antiepileptic Drug Development Program, NINCDS	2.5	3	Epilepsies
1969	Antiviral Substances Program, NIAID	4.2	0	Viral Infections
1971	Contraceptive Development Program, NICHHD	3.5	0	Reproductive Process
1971	Narcotics Abuse Treatment Drugs, NIDA	2	2	Narcotics Abuse
1971	National Caries Program, NIDR	1.9	0	Dental Caries
1974	Iron Chelator Development Program NIAMDD-NHLBI	0.5	1	Hemosiderosis 2° to Thalassemia
1974	Blood Substitutes Program, NHLBI	1.3	0	Anemias
1978	Antisickling Agents Program, NHLBI	0.5	0	Sickle Cell Crises

*New drug applications for which the particular program has provided either scientific or financial support.

Source: Barrett Scoville, presentation at the Foresight Seminar on Pharmaceutical Research and Development, Washington, D.C., December 2, 1980.

budgets of less than five million dollars. They have resulted in only a handful of new drug applications in other therapeutic areas.

One noteworthy program in this regard is the Antiepileptic Drug Program of the National Institute of Neurological and Communicative Disorders and Stroke (NINCDS). This program played a significant role in bringing an important anticonvulsant drug, Depakene, or valproic acid, to the market. This drug entity was first synthesized in 1881 but its anticonvulsant activity was discovered serendipitously in 1961. The clinical tests on valproic acid were first performed in Europe and the drug was widely marketed in several foreign countries. Since the drug compound had only limited patentability and the FDA accepts only a limited amount of foreign data in support of an NDA, there was insufficient commercial interest in this country to perform the large-scale phase III testing necessary to get the drug approved for marketing. NINCDS and the Epilepsy Foundation of America actively sought out a sponsor for clinical development for this drug. Abbott Laboratories eventually agreed to develop it using in part data from clinical tests conducted by the Epilepsy Branch of NINCDS. Valproic acid, which represents a significant advance over existing therapies, was approved by the FDA in 1978.

The valproic acid case provides an example of a more general policy issue, the "orphan drug" question. This is an issue that has been receiving considerable attention in recent years. In particular, orphan drugs are therapies for which there is limited commercial interest, either because the drug treats a relatively rare disease (i.e., it has insufficient market size) or because it is not patentable (i.e., it involves inadequate property rights). Although there are clearly weak economic incentives to undertake large-scale targeted R&D activities toward diseases of very limited incidence (or where patent rights don't exist) a number of such drugs in fact have been introduced into the market (or alternatively have been made available through open INDs at major medical centers). These drug introductions can be rationalized as providing "good will" to the sponsoring firm. They often result from a serendipitous discovery late in the clinical trials process or are clinically investigated and marketed because the firm is more generally interested in the knowledge associated with a particular class of pharmacologic entities.

The general policy question of whether the government should do more to support drug development on disease areas of limited commercial opportunities, either at institutions like NIH or through targeted R&D subsidies to private firms, is currently an active area of congressional interest. We shall consider the orphan drug question again in the final section dealing with current policy issues.

IV. ADVERSE TRENDS IN THE DRUG INDUSTRY
DURING THE 1960s AND 1970s:
THE ROLE OF REGULATORY AND NEW REGULATORY FACTORS

Annual Levels of New Product Introductions

Table 6.13 provides a list of the annual number of new chemical entities (NCEs) introduced in the United States between 1940 and 1978. (New chemical entities are new compounds not previously marketed and include nearly all major therapeutic advances. New products that are not NCEs include combinations of existing drugs and new dosage forms.)

The rate of introduction of NCEs has clearly declined since the late 1950s. For example, from 1955 to 1960, an average of about 50 NCEs per year were introduced. The corresponding number for the 1965 to 1970 period is only 17 NCEs, and for the most recent six-year period the average is 17 also.

Table 6.13. New Single Entity Drug Introductions
in United States.

Year	Number	Year	Number
1940	14	1960	50
1941	17	1961	45
1942	13	1962	24
1943	10	1963	16
1944	13	1964	17
1945	13	1965	25
1946	19	1966	13
1947	26	1967	25
1948	29	1968	12
1949	38	1969	9
1950	32	1970	16
1951	38	1971	14
1952	40	1972	10
1953	53	1973	17
1954	42	1974	18
1955	36	1975	15
1956	48	1976	14
1957	52	1977	16
1958	47	1978	23
1959	65		

Source: Pharmaceutical Manufacturers Association, Prescription Drug Industry Factbook, 1980.

This decline in new product introductions has been accompanied by corresponding structural trends on the input side of the innovational process. As discussed above, Hansen estimates that the current costs of developing and marketing an NCE are on the order of 24 million dollars. We may compare this finding to prior studies by Clymer (1970), and Mund (1970), and Sarett (1974) that put the uncapitalized development cost of a new NCE in the one-to-two-million-dollar range in the early 1960s. Moreover, Clymer estimated that the attrition rate for drugs undergoing clinical tests was two out of three in the pre-1962 period. Current data analyzed by William Wardell and reported earlier in Figure 6.2 suggest that only one in ten clinically tested drug entities becomes a new drug introduction. Finally, the average gestation period for a successful new drug has also increased significantly from four to six years in the early 1960s to the current ten years or more.

Thus, there has been a decline in annual new drug introductions accompanied by strong upward trends in the costs, time, and risks associated with discovering and developing new drugs. In economists' terminology, there has been a shift in the "production function" for new drug innovation in the direction of lower R&D productivity - that is to say, fewer new drug introductions are emanating from larger resource committments by the industry.

The causes and importance of this decline in new drug introductions has been the subject of considerable controversy. This debate has centered around the effects of increased regulation resulting from the 1962 Kefauver-Harris amendments as a major cause of this decline in innovation.

An initial response by the FDA was to argue that the observed decline in pharmaceutical innovation was in fact actually compositional rather than real in character:

> The relevant question is not and never has been how many new drugs are marketed each year, but rather how many significant, useful and unique therapeutic entities are developed. . . . The rate of development and marketing of truly important, significant, and unique therapeutic entities in this country has remained relatively stable for the past 22 years [Alexander Schmidt, 1974a].

It is difficult, however, to substantiate the FDA claim that the observed decline in new drug introductions has been largely confined to marginal-type drugs. As discussed above, it is true that the much higher costs and risks of developing new drugs have caused firms to focus less in their research programs on imitative "me too" drugs. These drugs do appear to have declined disproportionately over time. Nevertheless,

there is also evidence that suggests a decline in therapeutically significant drugs as well. Most classifications of important
therapeutic advances by academic analysts show such a decline, as does at least one prior FDA ranking of important
drugs.(7)

Furthermore, measures of pharmaceutical innovation based
on economic criteria also suggest a real decline has occurred.
For example, if we examine a "market share" type measure(8)
that indicates the relative importance of NCE sales to total
ethical drug sales, we find that the share of NCEs has fallen
from 20 percent in 1957-1961 to 8.6 percent in 1962-1966, and
to 6.2 percent in 1972-1976 (Grabowski and Vernon, 1976). Of
course, these economic measures will tend to give little weight
to major therapeutic advances for relatively rare diseases. It
is unlikely, however, that the downward trend can be primarily explained by an increasing proportion of such innovations
over time given the adverse economic shifts in the costs of
discovering and developing new drugs that occurred over this
period.

Sam Peltzman (1973) has analyzed a related drug quality
issue as to whether the large decline in NCE introductions
could be explained by fewer ineffective drugs entering the
marketplace after the 1962 amendments were passed. His analysis of data from three groups of experts - hospitals, panels
employed by state public-assistance agencies, and the American
Medical Association's Council on Drugs - does not support this
view. These data suggest that only a small fraction of the
pre-1962 and post-1962 NCE introductions could be classified
as ineffective.(9)

In sum, the hypothesis that the observed decline in new
product introductions has largely been concentrated in marginal or ineffective drugs is not generally supported by empirical
analyses. If one accepts that a significant decline in drug
innovation occurred in the 1960s and 1970s, the question still
remains as to the role of regulatory versus nonregulatory
factors in explaining this decline. In the remainder of this
section we consider various possibilities in this regard and the
evidence from various aggregative analyses of this issue.

Regulatory Developments in the 1960s

As noted above, a major legislative change occurred in 1962
with the passage of the Kefauver-Harris amendments to the
Food, Drug and Cosmetic Act. This law was passed following
the well-known and tragic events associated with the drug
thalidomide (a drug introduced in several foreign countries but
not the U.S.). The 1962 amendments had two basic provisions
that directly affected the drug innovational process - a proof
of efficacy requirement for new drug approval and establish-

ment of FDA regulatory controls over the clinical (human) testing of new drug candidates.(10)

With regard to the efficacy requirement, the amendments required firms to provide substantial evidence of a new drug's efficacy based on "adequate and well controlled trials." Subsequent FDA regulations interpreted this provision to mean using experimental and control-group samples to demonstrate a drug's efficacy as statistically significant. The preferred mode of study was "double blind" control where neither patient nor physician was aware whether he was receiving the experimental drug or a standard therapy of placebo. According to industry sources, these substantial evidence criteria led to large increases in the amount of resources necessary to obtain an NDA approval, especially in therapeutic areas where subjective analyses of patient responses are necessarily involved (such as analgesics or antidepressants).

The second major change in the 1962 amendments influencing the drug innovational process were the Investigational New Drug (IND) requirements on clinical testing. Prior to any tests on human subjects, firms were now required to submit a new drug investigational plan giving the results of animal tests and research protocols for human tests. Based on its evaluation of the IND and subsequent reports of research findings, the FDA may prohibit, delay, or halt clinical research that poses excessive risks to volunteer subjects or does not follow sound scientific procedures. Hence, as a result of the IND procedures the FDA shifted in the post-1962 period from essentially an evaluator of evidence and research findings at the end of the R&D process to an active participant in the process itself. This is another potentially important factor leading to the higher development costs and times observed over more recent times.

In addition to these two major changes in the 1962 legislation, the external environment surrounding FDA decisions on new drug approval also changed significantly. The thalidomide disaster received wide publicity in the popular press. This in turn galvanized congressional and media attention on new drug approvals.

Former FDA Commissioner Schmidt has emphasized the problem these external pressures create for the maintenance of a balanced and rational decision-making structure. He notes:

> For example, in all of FDA's history, I am unable to find a single instance where a Congressional committee investigated the failure of FDA to approve a new drug. But, the times when hearings have been held to criticize our approval of new drugs have been so frequent that we aren't able to count them. . . .
> The message of FDA staff could not be clearer. Whenever a controversy over a new drug is resolved

by its approval, the Agency and the individuals involved likely will be investigated. Whenever such a
drug is disapproved, no inquiry will be made. The
Congressional pressure for our negative action on
new drug applications is, therefore, intense. And it
seems to be increasing, as everyone is becoming a
self-acclaimed expert on carcinogenesis and drug
testing [Alexander Schmidt, 1974a].

The expanded attention from Congress and the media thus
tended to reinforce the natural incentives of FDA officials to
err on the side of caution or delay rather than the alternative
type of error of allowing a drug with excessive risks into the
marketplace.

A final set of factors influencing R&D costs and regulatory
delays relates to FDA resource capabilities and its management
procedures. The FDA's regulatory responsibilities expanded
dramatically after the 1962 amendments. Little thought was
apparently given, however, to the resource and management
problems that might arise in implementing the new law. This
point has come up repeatedly in outside and intraagency reviews of the FDA over the past two decades.

The most recent analysis of this question was a recent
General Accounting Office study(11) that focused on the NDA
approval process. Despite the fact that over 90 percent of all
NDA's are eventually approved, the FDA now takes between
two to three years on the average to approve an NDA. The
GAO cited the following problems in FDA procedural reviews:

1. FDA guidelines are imprecise.
2. Reviewers of the NDA change, slowing the process.
3. Scientific and professional disagreements between FDA and
    industry are slow to be resolved.
4. FDA feedback to industry about deficiencies is slow.
5. Chemistry and manufacturing control reviews are especially
    slow.
6. Industry submits incomplete NDAs.

In responding to the GAO report, the FDA has indicated
the goal of reducing over a three-year period the processing
time on NDAs by 25 percent for drugs that represent important or modest gains and 15 percent for all other drugs.

To sum up, over the post-1962 period, there has been a
substantial increase in both the scope and the intensity of
regulatory controls on ethical drugs. As a consequence, it
has been postulated that the costs of discovering and developing a new drug, along with the risk and uncertainty of drug
innovation, have increased and that this in turn has been a
major factor underlying the observed decline in new drug
innovation in the United States.

Alternative Hypotheses for Explaining Declining Innovation Levels

Several other factors have been advanced in the literature as explanations for the decline in drug innovation over the past few decades.

Depletion of Research Opportunities

This hypothesis has been given the most attention in the literature as an alternative to increased regulation. Adherents of the research depletion hypothesis argue that major drug innovations tend to occur in waves or cycles and that in many major therapeutic areas we have currently reached a point where the probability that a new discovery will be an advance over existing therapies is quite low. They further argue that we are on a research plateau because the major diseases areas left to conquer are the ones where we have the least adequate scientific understanding of the underlying biological processes.

Former FDA Commissioner Schmidt has expressed the research depletion hypothesis in the following terms:

> Today's world includes a great number of important therapeutic agents unknown a generation ago. These include antibiotics, antihypertensive drugs, diuretics, antipsychotic drugs, tranquilizers, cancer chemotherapeutic agents, and a host of others. . . . In many of these important drug groups there are already a large number of fairly similar drugs. As the gaps in biomedical knowledge decrease, so do the opportunities for the development of new or useful related drugs. As shown by the declining number of new single entity drugs approved in the U.S., England, France and Germany, this is an international phenomenon. This does not reflect a loss of innovative capacity, but rather reflects the normal course of a growth industry as it becomes technologically more mature [Schmidt, 1974b, p. 272].

This hypothesis, advanced by the FDA and others, has been received with considerable skepticism in many scientific quarters. Some have challenged the hypothesis on conceptual grounds. Others have pointed to the vast expenditures on basic biomedical research by the National Institutes of Health and other organizations as creating a renewed pool of basic knowledge which should offset any tendency toward a depletion of opportunities from prior drug discoveries (Bloom, 1976).

Changing Expectations

In addition to the factors of increased regulation and research depletion, Lebergott (1973) p. 9843 has pointed to the effects of the thalidomide tragedy on the behavior and expectations of physicians and drug firms as further confounding factors. In particular, he argues:

> Do any of us believe that after that catastrophe, consumers were quite as likely as before to prefer new drugs to ones tested by experience? Were physicians henceforth quite as likely to prescribe new drugs - with the prospect of acute toxicity (and malpractice suits) when the one chance of 10,000 ran against them? Which of our leading pharmaceutical firms would henceforth endanger its reputation (and its entire existing product line) on behalf of a new drug on quite the same terms as it did in the days when biochemists could do no wrong? . . . Such massive changes in the U.S. perspective on drugs - we may call them shifts in both supply and demand curves - had to cut the number of more venturesome drugs put under investigation since 1962. It would have done so if the entire FDA staff had gone fishing for the next couple of years.

Thus, Lebergott argues that strong shifts in the incentive structure facing physicians and manufacturers occurred after thalidomide and that this would independently operate to increase R&D costs and lower the number of new drug introductions. His analysis points up the analytical difficulties in trying to identify the effects of regulatory and nonregulatory factors that changed simultaneously as a result of the thalidomide incident.

Advances in Pharmacological Science

Dr. Pettinga (1975) and others have also pointed to scientific advances in pharmacological science over the past few decades as another potentially important factor. In particular, he suggests that these advances, which have made teratology and toxicological studies much more sophisticated and costly in nature, would have been incorporated into drug-firm testing procedures even in the absence of regulatory requirements to do so. That is, drug firms would undertake many of these increased tests in their own self-interest, in order to reduce the likelihood of future losses in good will and potential legal liabilities.

Several plausible hypotheses have thus been advanced with respect to the observed downward trend in drug innovation. These hypotheses are not mutually exclusive and may all have contributed significantly to declining innovation in ethical drugs. In the next section we discuss the empirical evidence concerning the relative importance of increased government regulation versus these alternative explanations of declining drug innovation.

Aggregate Analytical Studies of Pharmaceutical Innovation

Time Series Studies by Peltzman and Baily

Sam Peltzman's 1973 study of the effect of the 1962 amendments has received considerable attention in both economic and policy circles. Peltzman (1973) employs a "demand pull" model in which the supply of new drugs in any period responds with a lag to shifts in demand-side factors. The model is estimated on preamendment data (1948-1962) and is then employed to forecast what the number of NCEs would have been in the post-1962 period in the absence of regulation. The effects of the 1962 amendments are computed as the residual difference between the predicted and actual flow of NCEs. Using this approach, Peltzman concludes that "all the difference between the pre-1962 and post-1962 new chemical entity flow can be attributed to the 1962 amendments" (Peltzman, 1973). However, his approach never formally includes or considers any of the supply-side factors in the hypotheses cited above. All of the observed residual difference after 1962 is attributed to increased regulation. Since this residual difference can plausibly reflect the effects of a number of the other factors cited above (i.e., research depletion, changing expectations, and scientific factors), it probably encompasses various nonregulatory phenomena as well.

Martin Baily (1972) employed a production-function model of drug development that does try explicitly to separate the effects of regulation from the depletion of scientific opportunities. He postulates that the number of new chemical entities introductions in any period will be a function of lagged industry R&D expenditures and that both regulation and research depletion effects operate to shift this R&D production function over time. Regulation is captured explicitly in Baily's model by a time intercept shift variable and depletion by a moving average of past introductions. Both variables were quantitatively and statistically significant when his model was estimated over the period 1954 to 1969. However, when the model was later estimated for the period extending through 1974, the research depletion variable became insignificant and unstable over time.

Thus, the early time series studies of this issue by Peltzman and Baily both found strong negative impacts of regulation on new drug innovation. Neither study, however, provided very satisfactory approaches for isolating the effects of regulation on innovation from other confounding effects discussed above. This is a difficult econometric problem to handle in the context of aggregate time series analysis of U.S. introductions.

Wardell's Drug Lag Analysis

In order to separate the effects of increased regulation from other hypothesized factors, one would ideally perform an "experiment" involving two different states of the world: one with the 1962 amendments in effect and one where they are not. Given the impossibility of this experiment, a second-best type of analysis may be to find another country that is as similar to the United States as possible, but which differs significantly in terms of regulatory controls and procedures.

With this kind of methodological approach in mind, William Wardell, a clinical pharmacologist, performed a series of comparative analyses of drug introduction in the United States and the United Kingdom in the post-1962 period. The United Kingdom is similar to the United States in terms of high standards of medical training and practice and also has a very research-intensive multinational drug industry. However, the regulatory systems in effect in the United Kingdom and United States have important differences in the post-1962 period. Premarket safety reviews of new drugs essentially began in 1963 in the United Kingdom as a response to the thalidomide tragedy. The safety reviews in the United Kingdom have been characterized as high quality in terms of the depth of review process and the type of evidence necessary to gain approval (FDA, 1975). At the same time, the United Kingdom did not require formal proof of efficacy until its Medicine Act was implemented in 1971; before this the task of evaluating a drug's efficacy was essentially left to the market mechanism. Furthermore, the U.K. IND procedure was on a voluntary basis until 1971. Third, the British system utilized the judgment of external committees of academic medical experts in making approval decisions and and emphasizes postmarket surveillance of new drugs to a much greater degree than the United States. As a result, the British system has been characterized as less adversarial and bureaucratic than the U.S. system, which relies to a greater extent on the decisions of career civil servants, congressional oversight hearings, and the judicial process.

Wardell's first comparative study(12) of new drug introductions in the United States and United Kingdom covered nine therapeutic classes for the period 1962-1971. For this period he finds that the number of new chemical entities introduced into the United Kingdom was roughly 50 percent higher than

the number introduced into the United States (159 NCEs compared to 103 for the United States). Moreover, for the drugs
that were mutually available in both countries by 1971, twice
as many were introduced first in the United Kingdom as were
introduced first in the United States. This "drug lag" was
found to be the greatest in the areas of cardiovascular,
diuretic, gastrointestinal, and respiratory medicine. On the
other hand, in cancer chemotherapy, Wardell found both countries had comparable availability of new therapies.

In a related paper, Wardell attempted to assess the therapeutic consequences of these different rates of introduction
through a detailed discussion of the individual drugs available
in the two countries. He concludes:

> From the present study, it is clear that each coun-
> try has gained in some ways and lost in others. On
> balance, however, it is difficult to argue that the
> United States has escaped an inordinate amount of
> new-drug toxicity by its conservative approach; it
> has gained little else in return. On the contrary, it
> is relatively easy to show that Britain has gained by
> having effective drugs available sooner. Further-
> more, the cost of this policy in terms of damage due
> to adverse drug reactions have been small compared
> with the existing levels of damage produced by older
> drugs. There appear to be no other therapeutic
> costs of any consequence to Britain. In view of the
> clear benefits demonstrable from some of the drugs
> introduced into Britain, it appears that the United
> States has, on balance, lost more than it has gained
> from adopting a more conservative approach than did
> Britain in the post-thalidomide era [Wardell, 1974 p.
> 90].

In a follow-up study(13) to his original drug lag study,
Wardell found comparable trends for the 1972-1976 period in
the aggregate numbers of exclusive introductions and comparable lags in mutually available drugs. However, he also noted
some tendency for the largest clinical differences to narrow
over time. He attributed this convergence in part to more
"realistic" regulatory standards in the United States in some
(but not all) areas and a trend to more conservative practices
abroad.

Further International Comparative Analyses

In a recent paper, Grabowski (1980) analyzes the time pattern
of all NCE introductions in the United States for the period
1963 to 1975 relative to three European countries - the United

Kingdom, Germany, and France. He finds a significant lag
has characterized NCE introductions in the United States relative to the United Kingdom and Germany in the post-1962
period. This is true for both NCEs discovered in this country
and those discovered abroad. For France, the data indicate
that the United States still generally leads that country in the
introduction of U.S.-discovered NCEs, but not foreign-discovered ones. Second, his analysis also indicates that the lag
with Europe is not confined to drugs with little or modest
medical gain, but also includes drugs ranked as significant
therapeutic advances (as classified by the FDA itself). Third,
there is evidence, from a regression analysis performed in the
paper, that regulatory approval lags have been an important
factor contributing to this introduction lag. Finally, the
analysis further indicates that regulation has had an especially
strong impact on the introduction lag for foreign-discovered
drugs over this period.

The recently released GAO study(14) of the FDA drug-
approval process discussed above also examined the availability
of 14 therapeutically important drugs in the United States and
five other countries (Canada, Norway, Sweden, Switzerland,
and the United Kingdom). This study focuses on drugs introduced in the United States between 1975 and 1978. They
found that all but 1 of these 14 drugs were available first
abroad with lags ranging from 2 months to 13 years in length.
Furthermore they found the average FDA approval time on
these drugs of 23 months was significantly greater than that
for all other countries except Sweden (with England and
Switzerland having average regulatory approval times of 5 and
12 months respectively).

While a pattern of lagging U.S. NCE introductions (including therapeutically important drugs) thus emerges from a
number of recent studies, a broader issue is the effect of
regulation on the level, rather than the timing, of introductions. This may be characterized as the issue of "drug
loss" rather than "drug lag." This is the issue addressed by
the earlier econometric analysis of Peltzman and Baily. As
noted above, however, these aggregate times-series studies
had substantial difficulties in separating the effects of regulation from other confounding factors such as research depletion.

One, of course, may view the drug lag findings as symp-
tomatic of broader impacts of regulation on the innovational
process; that is a scenario of regulation leading to greater
costs, development times, and commercial uncertainties for new
drugs and hence to fewer annual NCEs being developed and
introduced each year. The magnitude of these impacts, however, is arguable and remains an important issue for empirical
research.

In a study(15) that we performed jointly with Lacy Thomas, we have examined aggregate "R and D productivity" changes in the United States and the United Kingdom to gain some insights into the effects of regulation on the level of innovation. Our strategy in this analysis was to structure the analysis in terms of an econometric model and to use international data as a means of separating confounding regulatory from nonregulatory factors. We found in this analysis that American R&D "productivity" - defined as the number of new chemical entities discovered and introduced in the United States per dollar of R&D expenditure - declined to about 1/6 of its former level between 1960-1961 and 1966-1970. The corresponding decrease of R&D productivity in the United Kingdom was to about 1/3 of its former level. On the basis of a regression analysis that used these and other data, we concluded that increased regulation in the post-1962 period has probably at a minimum doubled the cost of obtaining an NCE. At the same time, nonregulatory factors (such as research depletion, scientific advances in detecting toxicology, changing expectations) also apparently have significantly increased costs here and in the United Kingdom. However, the specific mechanisms and magnitudes of these different regulatory and nonregulatory factors await a more extensive and disaggregative analysis.

Summary and Implications

The various empirical studies discussed above do not provide definitive conclusions on the exact role of regulatory versus nonregulatory factors in explaining the lower levels of drug innovation experienced in the 1960s and 1970s. On analytical grounds, it is difficult to separate the effects of these contemporary factors. Nevertheless, the studies do provide a number of different analytical approaches to the problem and a consistent finding is that increased regulation is one of the important explanatory factors in this regard.

From a policy standpoint, the evidence has been sufficient to shift the perception of law makers quite dramatically compared with the situation in the early 1960s. At the time of the passage of the 1962 amendments, apparently little thought or credence was given to the notion that increased regulation could have unintended or undesirable side effects on innovation. However, given the industry's experiences of the past two decades, and the evidence from various academic studies (especially the drug lag studies), even the proposed regulatory reform laws of liberal congressmen include at least provisions for improving regulatory performance so that useful new drug therapies can be obtained by patients on a speedier basis.

In the last section of this chapter, we provide a detailed analysis of current legislative proposals in this regard. Before doing so, however, we turn in the next two sections to some more microeconomic-oriented studies on the returns and determinants of pharmaceutical R&D investment. Using a more microeconomic framework, we also attempt to analyze the effects and interactions of other government policy variables on firm R&D investment behavior.

V.   STUDIES OF THE RETURNS TO AND DETERMINANTS OF PHARMACEUTICAL R&D

Rate of Return Studies

Several empirical studies of the rate of return to drug innovation have been performed in recent years. We will discuss two of the most influential studies first and then turn to our own recent work on this topic.

David Schwartzman 1975 Study

Schwartzman (1975) begins his analysis by computing the annual sales revenues generated by the new chemical entities introduced in the United States in the 1966-1972 period. In order to calculate an expected rate of return to discovering and developing these drugs, he further estimated (1) the level and time pattern of research and development costs incurred to obtain these NCEs, and (2) the current and expected future profits generated by these new product sales.

Schwartzman's estimates on the average cost and revenues streams over this period yielded an after-tax rate of return on pharmaceutical R&D of only 3.3 percent. Schwartzman's analysis clearly embodies a number of important assumptions. Perhaps the weakest link in his chain of assumptions concerns his procedures for estimating expected profit margins for new drugs and expected product lifetimes (see Grabowski, 1976). Schwartzman, however, does perform a sensitivity analysis to see how his rate of return results change with different assumptions on these parameters. Other things constant, a 40 percent profit margin (instead of 25.6 percent) and a 20-year product life (instead of 15 years) yields an after-tax return for this period studied by Schwartzman of 7.5 percent (instead of 3.3 percent). This is still a very low rate of return for what is generally considered to be a very risky activity.

Perhaps the most interesting finding of Schwartzman's analysis is not his absolute estimates on the rate of return to drug innovation but the rate of change that he observes in this measure when his methodology is used on data from an earlier period. Specifically, Schwartzman found an after-tax rate of return of 11.4 percent in 1960 (compared to 3.3 percent in 1966-1972) using conservative estimates of the model's parameters. This is generally consistent with findings of higher returns in prior analyses by Baily (1972) and Clymer (1970) for this earlier period. In contrast to Schwartzman's approach, Baily constructed a two-equation econometric model from which he calculated the rate of return. Hence these two studies, despite the use of quite different methodologies, seem to be in general agreement.

The Virts and Weston 1980 Study

Virts and Weston (1980) p. 107 were concerned with the expected rate of return for pharmaceutical R&D in 1976. While they did not explicitly calculate a rate of return estimate, they did provide estimates of the present value of net revenues. These estimates were then compared with the present value of R&D costs taken from the Hansen study (discussed above). According to Virts and Weston:

> The implications are that the average return on investment (ROI) to pharmaceutical R&D has been significantly less than 8% and that these results have varied markedly among individual NCE's. Clearly, the owners of those NCE's falling in the top 25% in terms of market performance have achieved much higher returns and the average ROI to R&D for those companies exceeds the average for the drug industry and for all manufacturing industries. On the other hand, most NCE's have apparently failed to generate cash flows sufficient to earn a conservative estimate of the cost of capital.

Grabowski and Vernon 1982 Analysis

Because studies of returns to R&D require numerous important assumptions about such variables as product life, profit margins, etc., we have performed a sensitivity analysis (Grabowski and Vernon, 1982) of expected profitability for 37 NCEs discovered and introduced in the United States over the 1970-1976 period. A brief description of our analysis is given below.

The primary data used in the analysis are U.S. sales and promotion expenses for each NCE and R&D costs by therapeutic class. Two additional important types of data were not available: the cost of producing the NCEs after FDA approval and the net revenues resulting from sales in foreign countries. In both cases we have relied on estimates made by Celia Thomas as part of her Ph.D. dissertation at Duke University.

For example, her best estimate for production costs (16) as a
fraction of sales is .30.   However, because of the uncertainty
about this estimate, we have also examined the effect of estimates of .20 and .40.   A similar approach was taken with
respect to Thomas' estimate of 1.75 as the ratio of worldwide
net revenues to U.S. net revenues.   That is, estimates of 1.5
and 2.0 were also used in a sensitivity analysis.

The R&D cost estimates are based on a study by Hansen.
As described earlier, Hansen (1979) obtained survey data from
14 pharmaceutical firms on the R&D costs for a sample of NCEs
first tested in humans from 1963 to 1975.   The average discovery cost was $19.6 million and the average development cost
was $14.1 million, for a total of $33.7 million.   The $33.7
million represents the capitalized value (at 10 percent interest
and in 1967 dollars) at the date of marketing approval.

At our request, Hansen (1980) estimated the costs per
NCE on a therapeutic class basis.   These are the cost estimates used in this analysis and as will be shown, reveal a
rather large variation across classes.   We should also note that
Hansen's estimates include the costs of NCEs that enter clinical
testing but are not carried to the point of NDA approval.
Hence, the estimates should be interpreted as the average
expected cost of discovering and developing a marketable NCE.

As observed above, Hansen's estimates are expressed as
capitalized values at the date of marketing.   For example, the
capitalized expected cost of discovering and developing a cardiovascular drug at the date of marketing is $30.6 million in
1967 dollars.   Because he worked with constant dollars,
Hansen used real interest rates; in the example above, the
interest rate is 10 percent.   The natural measure for comparison with Hansen's cost estimate is the present value of the
new revenue stream resulting from the NCE.   To be consistent, of course, the net revenue stream must be deflated to
1967 dollars and discounted to the date of marketing at the
same real interest rate.   The ratio of present value of net
revenue to capitalized R&D cost is termed the profitability
index (PI) in the finance literature, and it will be the measure
of expected returns used here.   Clearly, a PI = 1 implies a
project that just breaks even.

Before turning to the results of our analysis in Figures
6.3, 6.4, and 6.5, we provide some general information about
the data in Table 6.14.   What is particularly striking about
Table 6.14 is the relatively low estimate of the expected
present value of R&D costs for a new chemical entity in antiinfectives (19.1 million dollars) compared to therapeutic classes
such as psychopharmacology (70.0 million dollars), metabolic-
antifertility (65.3 million dollars) and antiinflamatory (68.3
million dollars).   This is consistent with a significant
regulatory effect on the cost of developing new entities since
antiinfectives is the easiest area to establish efficacy using the
"large and well controlled trials" criterion of the FDA.

Table 6.14.   Characteristics of Sample of
37 NCEs Used in Sensitivity Analysis.

Therapeutic Class	Hansen's R&D Cost (10%, 1967 dollars)	Number of US NCEs
A. Cardiovascular	30.6	4
B. Neurologic, Analgesic	36.3	6
C. Psychopharmacology	70.0	3
D. Metabolic, Antifertility	65.3	5
E. Antiinfective	19.1	12
F. Antiinflammatory	68.3	4
G. Gastrointestinal, Respiratory, Surgery	28.5	3
		—
Total		37

Figure 6.3 shows the PI versus Product Life relationship for four alternative real interest rates (cost of capital). As stated, the PI variable is a weighted average PI for the 37 NCEs, where the weights applied are the R&D costs. The fraction of production cost to sales is held at .30 and the ratio of world net revenues to U.S. net revenues is taken to be 1.75. If we assume that the appropriate real cost of capital (inclusive of a risk premium) is 10 percent, then the product life necessary to break even on average is 19 years. An 8 percent cost of capital reduces the break-even life to 12 years.

Since the assumptions about production costs and foreign sales are uncertain, Figure 6.4 was prepared to reflect this uncertainty. Given the subjective probability distributions shown in Figure 6.4, a band of one standard deviation in width about the weighted average PI is presented. The one standard deviation band brackets the break-even life between approximately 14 and 30 years.

Figure 6.5 focuses on a different type of uncertainty. It shows a frequency distribution of the PIs of the 37 NCEs. Clearly, the distribution is highly skewed - with only 13 of the 37 projects breaking even or better. The letters are codes for the innovating firms and indicates that firm "A" had three "winners," while the remaining ten were spread over ten different firms.

Of course, the 24 NCEs that have PIs of less than unity fail to break even only in the sense of not covering fully allocated discovery and development costs, including a share of

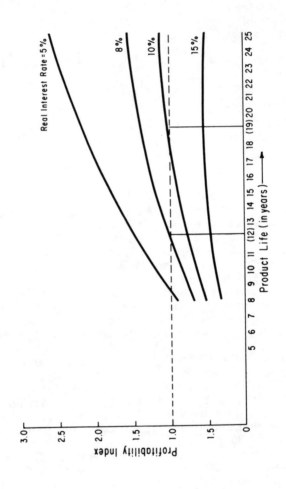

Fig. 6.3. Weighted average PI vs. life for various interest rates (weights are R&D costs).

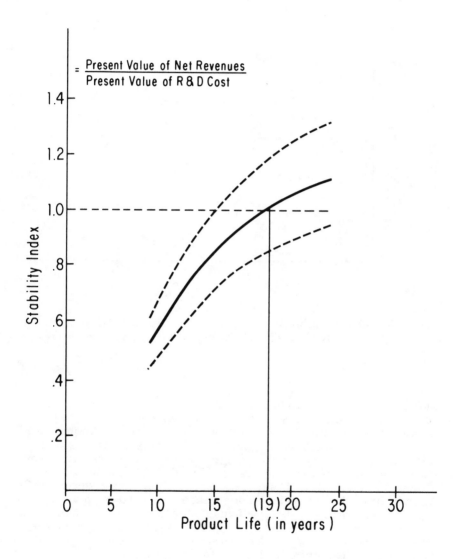

Fig. 6.4. Weighted Average P.I. vs. Life with Uncertainty
Bands (Uncertainty due to estimate of M.F.)

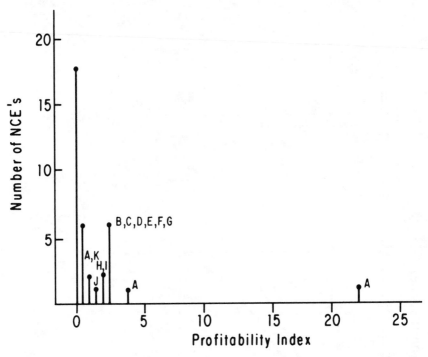

Fig. 6.5. Distribution of PI's of 37 NCEs 1970-1976 (letters indicate firms introducing the 13 NCEs with PIs > 1).

the costs of drugs that never make it to the point of NDA submission. This is the nature of Hansen's R&D cost estimate. If we consider only the development costs of a single NCE (neglecting discovery costs and attrition costs), the capitalized R&D costs decline substantially. For comparison with the values in Table 6.13, they range between $1 million and $2.3 million. As one would expect, substituting these lower R&D figures into the PI calculations leads to a large number of "break-even" NCEs. In particular, the number of NCEs that fail to cover their own development costs is only seven. Hence, in only 7 of 37 cases were firms worse off economically by carrying through the projects to marketing.

     Thus, the thrust of these three studies is that the <u>average</u> return to pharmaceutical R&D is relatively low, although there are a small number of big winners. Also, given that the average return has declined over time, one might expect to observe a decline in real resources devoted to drug R&D and a corresponding shift of these resources elsewhere to activities offering a higher return. We turn to this issue in the next section.

Trends in and Determinants of R&D Expenditures

In Table 6.15 we show R&D expenditures by the pharmaceutical industry for the 1965-1978 period. The first column shows that in absolute dollars, the amount of domestic R&D outlays increased in every year over the period. However, if one adjusts the R&D expenditures for inflation, the result is a growth rate of around 3 percent per year in constant dollars over the last four or five years of the period. We should note that the GNP price deflator probably understates the true rate of price change in R&D activity, so the true growth rate may in fact be zero or even negative.

Table 6.15. Domestic and Foreign R&D Expenditures of U.S. Ethical Drug Industry, 1961-1974[a].

Year	Domestic R&D Current Dollars (millions)	Domestic R&D Constant Dollars[b] (millions)	Foreign R&D Current Dollars (millions)	Ratio of Foreign R&D to Total R&D (percent)	Ratio of R&D to Sales to Current Dollars[c] (percent)
1965	304.1	304.1	24.5	7.5	8.3
1966	344.2	333.3	30.2	8.1	8.6
1967	377.9	355.4	34.5	8.4	8.7
1968	410.4	369.4	39.1	8.7	8.5
1969	464.1	397.8	41.7	8.2	8.7
1970	518.6	421.9	47.2	8.3	8.8
1971	576.5	446.2	52.3	8.3	8.6
1972	600.7	446.5	66.1	9.9	8.6
1973	643.8	452.2	108.7	14.4	8.6
1974	726.0	465.1	132.5	15.4	8.6
1975	828.6	484.3	144.9	14.9	8.4
1976	902.9	501.7	164.9	15.4	8.4
1977	984.1	516.5	197.7	16.7	8.5
1978	1089.2	532.3	222.0	16.9	8.2

---

[a]For human-use pharmaceutical research and development. (veterinary-use pharmaceutical R and D is excluded).

[b]Deflated by GNP price deflator converted to 1965 base.

[c]Global pharmaceutical R&D and sales of U.S. firms.

Source: Pharmaceutical Manufacturers Association, Factbook 1980 (Washington, D.C.: 1980).

Table 6.15 also presents the time pattern of foreign research and development expenditures for the period 1965–1978. While it is not clear how to deflate these outlays, it is clear from these data that slower growth rates in domestic R&D have been offset in part by faster growth in foreign R&D expenditures. The proportion of total R&D accounted for by foreign R&D roughly doubled from 7.5 percent in 1965 to 16.9 percent in the most recent year. This is consistent with the greater percentage of revenues from foreign markets and also the possibility of incurring less stringent regulatory controls in early clinical trials abroad. It may, of course, reflect other economic factors as well.

The final column in Table 6.15 gives the time trend in the ratio of global R&D expenditures to sales (i.e., including both domestic and foreign pharmaceutical activities) for U.S. firms. This ratio has been quite stable over the period, ranging between 8 and 9 percent.

One other trend in industry behavior is also worth noting at this point. Specifically, U.S. firms have been increasing their degree of participation in nonpharmaceutical activities in recent years. This is reflected in Virts and Weston's analysis(17) of changes in the aggregate percentage of pharmaceutical to nonpharmaceutical sales for eight leading firms over the period 1973-1978. This measure declined from 58.9 to 55.3 percent for this six-year period. Moreover, analyzing the longer period 1962-1975, we found a significantly declining trend in overall, corporate R&D to sales ratios (that is, including pharmaceutical and nonpharmaceutical corporate activities). This trend is also consistent with increased firm diversification into less research-intensive activities such as specialty chemicals and cosmetics.

In sum, the larger firms seem to be exhibiting a mixed strategy in their investment behavior in recent years - maintaining their R&D activity in pharmaceuticals with low rates of growth in real terms, while devoting somewhat more managerial and financial resources to nonpharmaceutical areas. While these trends may be viewed as providing some support for the findings of low rates of returns on pharmaceutical R&D by Schwartzman and others, they are much less than one might expect if firms really expected the low rates of returns observed by Schwartzman to hold on their current R&D activity. There is thus an apparent paradox. If current rates of return are so low, why do pharmaceutical firms continue to invest such substantial sums of money in R&D?

In our recent study(18) of the determinants of R&D expenditures, we attempted to answer this question.

Basically, we performed a multiple regression analysis on a sample of ten firms over the period 1962-1975. The dependent variable was the firm's R&D to sales ratio. The two primary explanatory variables were measures of past R&D success and cash flow.

The measure of past R&D success was essentially a moving average of firm's introductory sales of NCEs over a prior five-year period divided by its R&D expenditures over this period. It, of course, was intended to reflect the firm's expected rate of return from R&D investment. The cash flow measure, lagged profits plus depreciation, was included to test the hypothesis that firms impute a lower cost to internal funds, because of the lower transactions costs and risks, than they impute to external funds. Cash flow also seemed especially important for investment in activity characterized by such great uncertainty as pharmaceutical R&D.

Our regression results indicated that firms do react to lower realized returns on R&D activity in the expected manner, but the adjustment process is a very gradual one with relatively long lags. This is perhaps not surprising given that new product innovation historically has been a central and quite profitable mode of competition for the industry dating back to the pre-World II era. Moreover, the high degree of uncertainty and serendipity that characterizes discovery research and early clinical development trials in pharmaceuticals is also consistent with a cautious response to lower realized returns on past R&D efforts. Future returns may be very different from current or past returns, especially at the individual-firm level.

In this regard, it is worth mentioning once again that many firms apparently expect that industry returns from new drugs will be much greater in the coming decades as a result of several basic research "breakthroughs" discussed in Section III above. It is, of course, expectations about future rather than past returns that ultimately count in terms of firm investment behavior. It remains to be seen, however, whether these basic research advances can be translated into profitable new drugs in the forseeable future.

Our regression results also indicated that the general availability of internal funds or cash flow is another important factor that influenced R&D behavior over this period. We found a statistically significant stable positive relation between firm research intensities and their lagged cash flow margins. Moreover, these margins were relatively high over much of the period under study as a result of the record number of products introduced in the 1950s. These products remained under patent protection and generated high cash flows for the innovating firm well into the 1960s and even the 1970s in many cases.

Hence, we can infer from our analysis that the relatively high levels of internal cash flow over much of post-1962 period operated to moderate what would otherwise have been a more dramatic decline in R&D investment patterns.

In sum, our regression analysis indicates that both expected returns and cash flow are two major economic factors

influencing firm willingness and ability to invest in R&D outlays for new drug products.   From a policy standpoint, these
results therefore indicate that R&D expenditures will be sensitive to the spectrum of government policies that impact on
these variables.   The remainder of this chapter is concerned
with an analysis of various policy impacts in this regard.

VI.   GOVERNMENT'S IMPACTS ON INNOVATION

Many different government laws and regulations affect the
process of pharmaceutical innovation.   Some regulations directly
affect innovation, e.g., the FDA's regulations concerning
safety and efficacy testing increase the costs of developing
new drug compounds.   On the other hand, some laws are less
direct in their impact.   A good example is the current movement to repeal state antisubstitution laws.
      State antisubstitution laws prohibit pharmacists from sub-
stituting generic products for brand-name products prescribed
by physicians.   Repeal of these laws should lead to increased
competition for the innovator's drug by imitative drug products, thereby reducing expected returns to innovation.
      Important interdependence can exist among the various
laws and regulations.   For example, the effects of the new
substitution laws on innovation incentives must be considered
in light of government patent or regulatory policies.   Since
substitution laws alter the expected revenues of a new drug
only after the patent expires and alternative suppliers enter
the market, their impact on innovational returns depends on
the patent protection.   The effective patent life for new pharmaceuticals is typically much shorter than the legal life of 17
years owing to the long gestation period that is required to
develop and gain regulatory approval for a new drug entity.
Hence, drug substitution, patent, and regulatory policies have
potentially significant interactive effects on the incentives for
drug innovation investment.
      From a normative or policy perspective, these public
policies are also obviously interrelated.   If changes in drug
substitution laws were seen as leading to suboptimal incentives
for drug innovation, policy makers have the option of adjusting patent life to increase incentives.   It would not be necessary to maintain substitution restrictions on all pharmaceuticals
in order to maintain sufficient incentives with respect to drug
innovation.   This latter objective could be accomplished by
changing the patent life on new drugs.   This point is developed in more detail later.
      The objective in this section is to provide a comprehen-
sive discussion of how government laws and regulations affect
the expected return to R&D investment.   Ideally we would also

provide an assessment of the relative importance of these various laws and regulations in stimulating or retarding innovation. Unfortunately, adequate evidence does not exist for such an assessment in many cases. This is necessarily the case for policies, such as the new substitution laws, which are just now becoming operational. Hence, our assessments in such cases will necessarily be somewhat speculative.

It will be useful to organize our discussion around the standard investment model of the firm.

Suppose that an NCE is expected to be introduced in year t. It will involve R&D costs over m years and earn positive profits for n years after introduction, p of which are subject to patent protection. Then the rate of return, r, for this particular product introduction is found by solving the equation that equates the present value of investment outlays to net revenues generated after the product is commercially introduced.(19)

This expected rate of return calculation abstracts from potential differences in risk associated with specific development projects. The expected return from each project would have to be adjusted for such risk differentials across projects (unless the firm is risk neutral). The firm's decision to invest in a particular development project would depend on whether its adjusted rate of return exceeds or falls below the firm's capital cost, which reflects the opportunity cost of alternative investments for the firm and its shareholders.

The firm is assumed to make such calculations for all possible new drug-development opportunities. It then uses this information to construct a marginal rate of return (MRR) schedule by arranging projects in order of decreasing rates of return. The intersection of MRR and the marginal cost of capital schedule (MCC), which reflects the opportunity cost of alternative investments for the firm and its shareholders, determines the optimal level of R&D investment, R^*. This is shown graphically in Figure 6.6.

We now begin an analysis of how various government policies can be expected to affect R&D investment decisions.

Funding of Basic Biomedical Research

In Section III the large expenditures on basic biomedical research made by the federal government were shown. We concluded that while it is impossible to quantify precisely the impact of advances in basic science on pharmaceutical innovation, the impact is undoubtedly of great importance.

In terms of Figure 6.6, it is useful to view such advances as shifting the MRR schedule rightward as new opportunities for drug development are made possible.

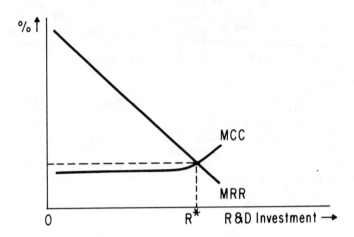

Fig. 6.6. Optimal level of R&D investment.

Given the lengthy discussion in Section III of the role of government-supported basic research in drug discovery and development, we shall not discuss it further here. It might be recalled, however, that many experts believe that a revolution is now taking place in molecular biology and this might make the social payoff to funding of basic research especially high at this time.

FDA Regulation

FDA regulation affects both the cost and revenue sides of the rate of return calculation. Earlier we gave a brief description of Hansen's estimates of R&D costs. FDA regulations exert important effects on these costs, e.g., by specifying the number of tests and the amount of evidence on safety and efficacy that must be accumulated. And, as described above, the two-to-three year period of FDA review of the NDA adds significantly to the cost. (Earlier we referred to our 1978 study which concluded that the increased FDA regulation resulting from the 1962 amendments more than doubled R&D costs.)(20)

FDA regulations also exert effects on the expected revenues from an NCE. There are several possibilities here, some of which have opposite implications for expected revenues.

Regulatory controls will reduce the probability of commercialization for many compounds and lower expected revenues.

One of the primary benefits of regulation is the extent to which the regulatory agency screens out and deters drug entities that present risks that the majority of consumers would not knowingly and willingly undertake. Evaluating whether the FDA has been too conservative in its risk/benefit decisions is one of the most difficult and controversial areas of regulatory analyses.

There are also several ways that regulation can operate to increase the expected revenues of drugs approved for marketing by the FDA. First, regulations serve a certification function. Stringent regulatory processes provide physicians and patients with confidence in a new drug's safety and efficacy, thereby facilitating rapid market diffusion and penetration for new drugs. Second, drugs that are approved in a stringent regulatory regime face less actual and potential competition than is the case in an unregulated market. This is true for two basic reasons. First, many marginal drugs will be undeveloped, given the greater costs of developing drugs under regulation. Second, the minimum scale at which R&D can be profitably undertaken will tend to increase under regulation, lowering the number of firms engaged in pharmaceutical innovation.

Regulation also affects the effective patent life for a new drug entity. Since the average time to develop an NCE and gain regulatory approval now far exceeds the time necessary to obtain a patent, regulatory-derived increase in development or approval times will operate to lower the effective life of a drug patent. While the length of patent protection has been of secondary import historically in the drug industry, this situation could change dramatically with the repeal of antisubstitution laws.

How do these regulatory effects balance out and what is their new impact on the rate of return to innovation? Of course, there is no definitive answer to this question, but the evidence surveyed earlier suggests that increased regulation has been at least one important factor underlying the declining trend in average innovation returns. We should emphasize that these studies all dealt with past time periods; the likely impact of FDA regulation in the future is less certain given various proposed legislative and administrative reforms currently under active consideration.

Substitution Laws

As noted earlier, the repeal of antisubstitution laws might result in increasing the importance of the length of patent protection. To see how this might come about, some background on the antisubstitution laws should be useful.

These laws were enacted in the early 1950s in response to the drug "counterfeiting" problem, i.e., the dispensing by pharmacists of drugs similar in size, color, and packaging to popular brand-name products, but of unknown quality or origin. Antisubstitution laws were adopted by all 50 states and generally prohibited any form of substitution for the brand written on the prescription.

The laws made it possible for innovating firms, through strong brand loyalties, to maintain dominant market positions for their products even after patent expiration. Hence, even though lower-cost generic products became available upon patent expiration, in many cases physicians have continued to prescribe the original brand-name product.

A major structural change taking place in the pharmaceutical industry today is the repeal of state antisubstitution laws. Over 40 states have passed product selection, or drug substitution laws. While the state-enacted laws have significant differences, essentially all enable pharmacists to substitute generic products (some mandate substitution) unless a physician prevents substitution by checking a preprinted box or writing "dispense as written" (DAW) on the prescription form.

If substitution laws foster increased competition for the innovator's product, then the degree of patent protection assumes a critical role in the appropriability of drug returns. A shorter effective patent life shifts the impact of drug substitution forward in time, amplifying the impact of revenue losses on the expected return to innovation. Table 6.15 shows that the effective patent life for pharmaceuticals has been declining and is currently in the range of nine to twelve years.

We can show the sensitivity of the expected profitability of R&D to changes in the effective patent life and the degree of substitution by using the profitability index (PI) analysis described in the first part of Section V. Recall that we calculated the average PI for 37 NCEs introduced over the period 1970 to 1976 for a number of alternative assumptions about product life, real interest rate, etc. The results were shown graphically in Figure 6.3. Here, using this same sample, we take as our benchmark case a product life of 20 years and a real interest rate of 10 percent. The PI corresponding to these assumptions is 1.029.

In order to study the sensitivity of this PI of 1.029 to changes in the effective patent life and the impact of substitution on net revenues, we imposed selected values of these parameters on our data and recalculated the PIs. One case examined was an effective patent life of eight years and a 50 percent reduction in U.S. net revenues after patent expiration.(21) The PI for this case was only .863 as compared with the benchmark of 1.029. The results for all cases are given in Table 6.17.

Table 6.16.   Average Effective Patent Life for New Chemical
Entities Introduced into the United States from
1966 to 1977*.

Year	Average Effective Patent Life (years)
1966	13.6
1967	14.4
1968	13.5
1969	12.7
1970	14.4
1971	12.2
1972	10.9
1973	12.1
1974	13.0
1975	11.4
1976	11.3
1977	9.6
1978	10.5
1979	9.5

*Effective patent life refers to the length of time from the date
of FDA approval until the date of patent expiration.

Source: University of Rochester, Center for the Study of
Drug Development, Department of Pharmacology and
Toxicology.   See Eisman and Wardell (1981).

As one would expect, the calculated PIs in Table 6.17 are
lower for shorter effective patent lives and for greater
percentage reductions due to substitution.   Under the most
unfavorable conditions for R&D activity considered here - an
eight year patent life and a 50 percent reduction in U.S. net
income - the rate of return is reduced to .863, or by about 16
percent from the 1.029 benchmark.   On the other hand, when
a 30 percent net income reduction and a 12-year patent life are
assumed, the PI is .974, or roughly a 5 percent reduction due
to substitution.   These estimated effects are not negligible
and, other things constant, may be expected to make some
R&D projects no longer attractive to pharmaceutical manufacturers.
The results in Table 6.17 underscore the fact that the
effects of substitution on R&D returns are highly sensitive to
the length of patent protection.   If the patent life for drugs

Table 6.17. Sensitivity Analysis Showing
Profitability Index for Alternative Assumptions
About the Impact of Substitution and the
Effective Patent Life

Percentage Reduction U.S. Net Income upon Patent Expiration	Effective Patent Life		
	8 Years	12 Years	17 Years
-10	.996 (-3.2)	1.011 (-1.7)	1.023 (-.6)
-30	.930 (-9.6)	.974 (-5.3)	1.011 (-1.7)
-50	.863 (-16.1)	.937 (-8.9)	.998 (-3.0)

Notes: 1.   The standard against which the above Profit-ability Indexes (PIs) should be compared is 1.029.   This is the PI for a 20-year commercial life with no reduction in U.S. net income.   It is also assumed that the ratio of production cost to sales is .3, the ratio of world net revenues to U.S. net revenues is 1.75, and the real interest rate is .10.

2.   It is assumed that at the end of the effective patent life, substitution will result in the alternative reductions in U.S. net income given above for the remaining years of the 20-year commercial life.

3.   The numbers in parentheses are the percentage re-ductions for each PI from the standard PI of 1.029.

actually equalled the legal life of 17 years, the effects of increased substitution on R&D returns would be quite modest. For example, with a 17-year life, a 50 percent reduction in U.S. net income from substitution causes R&D profitability to decrease by only 3 percent in the present example.   On the other hand, as patent lives decrease, the effects of drug substitution are magnified.

The results in Table 6.17 are somewhat tentative in char-acter.   The analysis is based in aggregative data sources and contain the simplifying assumptions discussed above.   Never-theless, the results suggest that the effects of substitution laws on innovation incentives are consequential in nature and are quite sensitive to the longevity of patent lives over the ranges considered (i.e., 8 to 17 years). These issues are considered further in the final section on current policy iniatives.

Other Laws and Regulations

To conclude this analysis of the effects of various government policies on the incentives to undertake R&D, we briefly discuss the Maximum Allowable Cost (MAC) program and the federal income tax code.

The MAC program(22) is somewhat similar to the new substitution laws in the way it affects the expected rate of return to pharmaceutical innovation. Specifically, MAC is a program designed to limit federal government third-party reimbursement, primarily under Medicaid, for prescription drugs. It limits reimbursement to the lowest price at which a particular multisource drug is generally available. Since MAC is only applicable to multisource drugs, it is similar to the substitution laws in acting to reduce an innovator's net revenues after patent expiration. On the other hand, MAC only applies to drug purchases that qualify for government reimbursement. Medicaid prescriptions, for example, account for only about 15 percent of all prescriptions.

Since the first maximum cost limit for a drug product was set for ampicillin in 1977, it is clear that this program is just getting started. Hence, it is too early to assess the overall impact of MAC on innovation incentives.

The U.S. Internal Revenue Code has been designed by Congress to assist U.S. possessions in obtaining employment-producing investments by U.S. corporations. Through Section 936, so-called "possessions corporations" can be exempt from federal tax on income from operations in Puerto Rico, American Samoa, Guam, and the Panama Canal Zone. As a result, many pharmaceutical firms have set up operations in Puerto Rico and thereby obtained large tax savings.

The tax savings are, in fact, quite substantial and are concentrated especially in the pharmaceutical industry. For example, the Treasury Department has estimated that in 1977, 45 percent of all tax savings to U.S. corporations accrued to the pharmaceutical industry. It also reported that 16 drug firms had a total of $344 million in tax savings in 1977 under Section 936.(23) This sum represents about 10 percent of the pretax income for these firms.

The sizable tax savings from Puerto Rican operations add significantly to industry cash flows. Given the importance of cash flows as a determinant of R&D expenditures as noted earlier, a change in tax policy to reduce this tax advantage could have a significant negative effect on R&D incentives. Just such a change is possible if the IRS successfully argues in a current court case that Lilly has allocated excessive profits to its Puerto Rican subsidiary. A ruling in favor of the IRS could possibly be applied to the other pharmaceutical firms. In addition, various legislative proposals to amend these tax provisions have been introduced into Congress recently.

To summarize briefly, we have examined how six government laws and regulations affect the expected rate of return to pharmaceutical innovation. These six policies are funding basic biomedical research, FDA regulation, patent policy, state substitution laws, the MAC program, and the corporate income tax. A key point that has been made throughout the discussion is that these policies are interdependent and that policy changes must be considered in light of that interdependence.

It is clear that public policies have had both significant positive and negative incentive effects on innovation. Historically, it appears that the main positive effects have been derived from government funding of biomedical research while the main negative effects have been associated with health and safety regulations. Other public policies, currently in an evolutionary state, such as MAC and state substitution laws, could also have significant negative impacts on the economic returns to innovation over future periods. This will be so if the effective patent lives for new drugs continues to trend downward over time and these evolving new laws and regulations cause a dramatic increase in generic drug usage after patents expire. These negative incentive impacts could be offset by various compensatory policy actions. These are discussed in our final section on current policy initiatives.

VII.   CURRENT PUBLIC POLICY INITIATIVES

In the final section of this chapter we consider current policy issues in the areas of regulatory reform, patent legislation, and government funding of research and development.

Regulatory Reform

In 1979, the Carter administration introduced parallel bills into the House of Representatives and Senate with several co-sponsors that would have comprehensively overhauled all stages of the drug regulatory process. This legislation came to be known as the Drug Regulatory Reform Act of 1978. In addition, bills with similar (but not identical provisions) were introduced into the Senate by Senator Kennedy and the House of Representatives by Congressman Rogers.

These regulatory reform measures were introduced during a period of changing attitudes toward drug regulation and attempted to balance a number of somewhat conflicting objectives. Among the apparent objectives of the bills were: (1) to speed up the approval of significant new drug therapies; (2) to increase the degree of public participation in the drug approval process and make it more open to outside scrutiny; (3) to facilitate the entry of generic producers into the market after patent expiration by removing duplicative testing requirements; and (4) to expand FDA regulatory controls over the postmarketing period.

As one might expect, none of the interested parties here - the drug manufacturers, consumer advocates, practicing physicians, pharmacists, and the academic medical community - was completely satisfied with all sections of these proposed drug regulatory reform bills and they worked vigorously to amend certain provisions. In September 1979, after extensive hearings, the Senate passed an amended drug regulatory reform act that contained some important compromise features and omitted some of the more controversial provisions of the original bill. However, the Drug Regulatory Reform Act was never reported out of committee in the House and thus the bill died when the 96th Congress expired.

With the inauguration of the Reagan administration in January 1982, primary attention has shifted from legislative to administrative reform of the drug regulatory process. The new administration announced its intention to examine all the major regulations that evolved from the broad statutory authority granted under the Food, Drug and Cosmetic Act and its amendments. In fact, many of the main features the Drug Regulatory Reform Act of 1978 would have codified FDA procedures already adopted through administrative rules or internal policy decisions. With a few exceptions (primarily involving postmarketing controls), administrative rule making could have been used to accomplish all of the major reforms embodied in the Drug Reform Act. Administrative rule-making changes of course do not necessarily have the same permanence as legislative changes. The new administration has chosen the former approach for its initial strategy in it pursuit of a speedier and more efficient regulatory process.

In the remainder of this section we consider some of the major proposed policy changes of the drug regulatory process that have been discussed in various recent academic and government studies and task forces. These involve both administrative and legislative reform proposals.

More Flexible Controls over Early Clinical Trials

Regulatory delays in the early stage of clinical research process can have an especially significant effect on resource costs and time because of the uncertain recursive nature of the research process. Generally, about ten substances are tested clinically for every one that is taken through full development to an NDA with the FDA. However, the information garnered from testing the unsuccessful compounds on a small number of individuals in phase I and II provides a cumulative feedback effect that is incorporated into successful drug therapies. Delays in the early stages of clinical process therefore have a compound effect on outcomes and tie up the most creative part of a firm's research organization.

Clinical trials are currently approved and supervised by institutional review boards at the medical centers where they are performed in addition to the controls exercised by the FDA in the IND process. The safety record in these early trials is very good. This is because of the intensive monitoring and highly controlled nature of early clinical trials. Cardon, Dommel and Trumble (1976) of the National Institutes of Health have reviewed the injury data to research subjects and concluded in this regard that "the data suggest that risks of participation in nontherapeutic research may be no greater than those of everyday life and in therapeutic research, no greater than those of treatment in any other setting."

Decentralizing primary responsibility for early clinical trials into the hands of institutional review boards is a recommendation of several recent studies of the drug process including the GAO and the Staff Report of House Subcommittee on Science Research and Technology. The Drug Reform Act (which passed the Senate but not the House in the last Congress) would have had the FDA issue general regulations and then certify certain delegated health institutions (such as research hospitals) to approve and supervise phase one and two clinical investigations. The FDA would still retain oversight authority, however, to revoke any drug investigations approved by these delegated institutions. Current regulations governing initial clinical trials are part of the general administrative review of the drug regulatory process being undertaken by the new administration.

More Reasonable Premarketing Standards Combined with Increased Postmarketing Surveillance

Drug regulation in the current system has an all or nothing character. Before approval, candidate drugs are restricted to small patient populations under highly controlled experimental conditions. After approval, usage often increases with minimal regulatory surveillance. Under such circumstances, it is not surprising that regulatory officials tend to err on the side of conservatism in new drug approvals. At the same time, despite very great preapproval conservatism, many of the adverse side effects of drugs, those that occur with frequencies of less than 1 in 1,000 or are longer term in nature, can realistically be discovered only after a drug has been consumed by large patient populations.

The FDA premarketing approval conservatism has manifested itself by an evolving expansion over time in its interpretation of what constitutes "adequate and well controlled" investigations of a drug's effectiveness. This frequently puts investigators at the FDA in the position of delaying a drug's entry into the market until the pivotal scientific studies are performed, even where there is little doubt about a drug's safety or effectiveness and where the new drug offers some very significant advances over existing therapies. Furthermore, low incidence risks (less than one per thousand) cannot be generally detected in these pivotal clinical studies. The best way to detect these is through more extensive and effective postmarketing surveillance.

An alternative approach to the current "all or nothing" system of drug approvals would be to allow new drugs on the market sooner based on significant evidence of safety and efficacy while expanding postmarketing surveillance on these drugs after they enter the marketplace. This is an approach favored by a large number of academic and medical experts as a principal means of getting new drugs to patients sooner consistent with present safety objectives. There is less than complete consensus, however, on how to change existing procedures to accomplish this general objective. Some specific policy changes in this regard are discussed further below.

Greater Acceptance of Foreign Clinical Data on Safety and Efficiency

The most serious drug lag cases generally occur for important new drug therapies discovered or developed in foreign countries. One of the basic reasons for this drug lag is the limited acceptability of foreign data trials as proof of safety and effectiveness for U.S. approval. Prior to 1975, the FDA did not accept foreign data at all as positive evidence in support of an NDA's approval. Since that time, the FDA has begun accepting foreign data. Nevertheless, since 1975, no drugs have been approved by the FDA on the basis of foreign data alone, irrespective of the amount of evidence in support of a drug's safety and efficacy from foreign experiences. The usual requirement is that at least two U.S. studies be conducted to supplement and verify foreign evidence.

A general recommendation, endorsed by a wide number of academic and government studies, is to have the FDA place greater reliance on foreign data as a way of speeding up the availability of new drug therapies to U.S. patients. This appears to be an important step for improved regulatory performance. Drug discoveries from foreign laboratories now account for approximately 40 percent of U.S. introductions and U.S. firms are now conducting an increasing percentage of their research and development abroad (Wardell, 1978). The GAO further found that in technically advanced countries, drug sponsors design clinical trials to meet the most stringent standards.

The documented case history on sodium valproate presented in hearings of the House Subcommittee on Science Research
and Technology (1980) p. 66 is instructive on this issue:

> There were over 200 favorable reports in the medical
> literature on the effectiveness of sodium valproate in
> the treatment of epilepsy. Thirty of these were
> clinical scientific studies involving a total of more
> than 1,300 patients. It is extremely unusual for
> even a U.S. drug to have undergone such wide-
> spread investigation before marketing. FDA's Neu-
> rological Drug Advisory Committee reviewed these
> studies, and the committee's chairman reviewed them.
> Both reviews found that at least two studies were
> adequate and well controlled and clearly demonstrat-
> ed valproic acid's effectiveness and safety in con-
> trolling absence seizures in epilepsy and unanimously
> recommended that the FDA immediately approve the
> drug.
> FDA staff questioned aspects of the studies and
> required that at least one U.S. study be conducted.
> When completed this study showed nothing signifi-
> cantly different from the foreign well controlled
> studies. The result was several more months delay
> in the approval of the drug in the United States.
> Sodium valproate is one of a few, if not the only
> drug which the FDA has approval on the basis of
> just one U.S. study accompanied by foreign studies.

The subcommittee concluded that a mechanism was needed
within the FDA to evaluate foreign clinical data to determine on
a drug-by-drug basis whether it meets the intent of U.S.
criteria and can be used in lieu of domestic tests.

Conditional Release of New Drugs

A concept that has been frequently advocated in the literature
is a conditional type of approval process for new drugs.
Under this concept, new drugs could be marketed when significant evidence of safety and efficacy become available, subject
to continued testing and monitoring and FDA authority to stop
product sales quickly if warranted. Some of the variants of
this proposal would also have initial marketing restricted to
particular institutions or particular medical specialties.
There have been precedents for this type of regulatory
approval. The FDA has granted some important drugs (for
example, L-DOPA) early release for marketing on the condition
that manufacturers monitor and report effects on a given number of patients. In addition, the NIH with its network of
clinical centers for patients trials in the cancer area has

something like a conditional availability of promising new anticancer drugs prior to formal approval by the FDA for marketing. This involves liberal use of the IND procedures to undertake treatment of patients in terminal situations. This is not, however, typical in other therapeutic areas.

The Drug Reform Act of 1978 contained a provision allowing the conditional release of "breakthrough" drugs for use in life-threatening, severely disabling, or severely debilitating situations. This provision would have relaxed the standard of evidence from "substantial" to "significant" to allow patients access to such breakthrough drugs while final testing on them was being performed. The statute was designed, however, to restrict the discriminatory use of this provision to only a small portion of new drug introductions.

The breakthrough drug approach for conditional approval was criticized in hearings for its narrowness of application and also because it relied heavily on FDA judgment to determine breakthrough status. The FDA, for example, initially classified valproic acid as a moderate therapeutic advance while the NIH Anti-Epileptic Drug Program was actively campaigning for its rapid approval as a significant advance over established therapies. It was also argued that scientific advances in the drug areas, as in other fields, are more often incremental in character and frequently cumulate only gradually over time to major gains in social welfare. This has been the case historically for example, in antihypertensive therapy and combination chemotherapy for cancer. Furthermore, the "breakthrough" status of a new drug sometimes becomes apparent only after a drug is in general use and often for a purpose different from what was originally intended (e.g., the diuretic qualities of the sulfa drugs). These factors often make forecasting of the significance of medical advances difficult.

It should be noted that the FDA has already begun implementing a program of "fast-tracking" certain drugs in the allocation of its resources during both the IND and NDA phases. In particular, all INDs are now classified at a fairly early stage in the development process into three basic categories - drugs likely to be (a) an important advance, (b) a modest advance, or (c) little or no advance. The intention is to have priority reviewing in accordance with how a drug is placed under this classification scheme. While this approach may get some important therapies into public hands sooner, it is also subject to similar problems as discussed above for the Breakthrough Drug Provision. In particular, if the FDA's judgment on a new drug's therapeutic value is in error, it may delay rather than speed up the time required for a drug to clear regulatory hurdles (i.e., by putting an important drug on a slower track). This is an area where further research on FDA performance seems desirable.

Regulatory Incentives and Accountability

As noted earlier in the chapter, FDA incentives are strongly skewed toward officials avoiding the acceptance of a "bad" drug while being much less concerned about rejection or delay of a "good" drug. The agency's authority evolved as a response to a few widely publicized drug tragedies. Its mandate is drawn in very narrow terms - i.e., to insure the safety and efficacy of new drug products. All of the burden of proof rests on the sponsoring firm to demonstrate this to the satisfaction of the regulatory authorities. There is also the absence of any effectively functioning appeals process in the case of scientific or other kinds of agency firm conflict short of a full-fledged suit in the judicial system.

In view of these characteristics, it is not surprising that the drug approval process is a long and costly affair or that the GAO and others have found a number of management deficiencies in the agency that contribute to this outcome. While these regulatory problems are receiving increasing attention by Congress and the general public recently, there seems to be little popular support for removing the basic system of premarket controls in pharmaceuticals in favor of other forms of government protection. Nevertheless, there is clearly a desire being manifested in many quarters for a more balanced regulation structure that will get approvable drugs into the hands of the public without long delays and one that also puts minimal restraints on the innovative process. The congressional sessions on drug lag and in related areas seems to be having a growing impact in this regard.

Obviously, the incentive structure at the FDA is not an easy matter to change through either legislative or administrative action. The Drug Reform Act attempted a beginning by declaring the encouragement of innovation to be an important objective of public policy along with basic safety protection. Beyond stating this objective, however, Congress might consider some specific institutional mechanisms for insuring that a more balanced perspective will in fact be reflected in regulatory decisions.

One idea that has been advanced is to create a distinguished panel of scientists and medical experts from elsewhere in the health community to review annually the FDA's progress on new medicines as well as to consider potentially valuable new drug therapies already in use abroad. This type of body would be a logical extension of the FDA advisory committees. In contrast to the latter, however, which become involved only in the later stages of the approval process for specific medicines, the proposed panel would have a broader oversight function and would be designed to bring the perspective of scientists and medical prescribers of drugs into the regulatory decision process in a more complete and systematic way. The

greater use of outside experts has been one of the more successful aspects of the British system of drug regulation.

A related measure is the development of an effective
appeals process for scientific disputes between the FDA reviewers and sponsoring firms. Such an appeals mechanism
exists in many other countries. For example, in the United
Kingdom, applicants can appeal an adverse decision of the
Licensing Committee to the Medicine Commission, a 14-member
body composed of scientists, physicians, veterinarians, and
representatives of the pharmaceutical industry (Dunlop, 1973).
A similar kind of appeals mechanism might be considered in the
United States as an additional check on regulatory decision
making.

While the effectiveness of such policies in the United
States is open to question, it would seem worth experimenting
with such measures is order to try to generate a more balanced decision-making environment.

Patent Protection

Patent protection in the pharmaceutical industry has been
another major area of interest for policy makers. As discussed
above, the period of patent protection in drugs now averages
about ten years in length and has been trending downward in
recent years as a result of the long development periods and
regulatory approval times for new drugs. Furthermore, there
is the prospect of increased substitution and market penetration by generic products after patents have expired in the
future periods as a result of the spread of state substitution
laws and the growth of programs like MAC. Given these
trends, a number of policymakers have advocated restoring
part or all of the effective patent term lost during the IND
and NDA regulatory periods.

Proposals embodying the basic concept of patent restoration in ethical drugs and other similarly affected industries
have been introduced recently into Congress. For example, a
bill (S255) introduced by Senator Mathias into the 97th Congress, and passed by the Senate, would add back to the patent life at the time of FDA approval, any time lost during the
clinical testing and FDA review period, up to a maximum of
seven years. The bill would also be applicable to other industries with premarket approval of new innovations (e.g., food
additives, pesticides, and medical devices). A similar bill
(HR1937) is being considered by the House of Representatives.

The selection of any specific number of years for patent
protection necessarily gives rise to difficult tradeoffs (i.e.,
the possibility of too little incentive for innovation versus the
encouragement of too much market power). These tradeoffs
must be evaluated under considerable uncertainty. Neverthe-

less, there appears to be growing concern among policy makers about the potential adverse implications of passively allowing the continued downward drift in effective patent lives for drugs.

In our sensitivity analysis of the mean profitability of new drugs introduced in the period 1970–1976 (Grabowski and Vernon, 1982) we estimated an average product life of 12 to 19 years was needed by firms to cover R&D costs and provide a real rate of return on investment of 8 to 10 percent. Average effective patent life is therefore currently considerably less than average product life necessary for profitable operation. In the emerging environment of increased competition from generic products after patent expiration, the length of patent protection is likely to become an increasingly important economic incentive factor influencing R&D investment decisions.

It is quite difficult to say, however, what type of innovational activity would be most stimulated by increased patent protection. On theoretical grounds, patent restoration can be expected to have a greater impact on the R&D incentives for products with longer expected product lives before being made obsolete by rival innovations. (If a drug had a very short expected life before technological obsolesence occurs, it would be essentially unaffected by patent restoration.) One class of drugs that should be positively affected by patent restoration are "breakthrough" type drugs with above-average riskiness but also longer expected product lives before technological obsolesence. Another class of drugs positively affected would be marginal type drugs with a small but stable market share niche. The exact size or quantitative impacts of patent restoration on specific kinds of innovational outputs is very uncertain, however.

Government Support of R&D

As we discussed earlier, the most relevant government-funded R&D activities for the pharmaceutical industry have been in the basic scientific research area. Several of the industry's most promising current R&D projects are the outgrowth of basic discoveries in such diverse scientific fields as immunology, enzyme chemistry, recombinant DNA, and prostaglandins. New pharmaceuticals has become one of the most important applications of the government's basic biomedical research activities.

Except in the cancer area, the government role in the direct development of new pharmaceuticals has been limited. As discussed above, Congress dramatically increased the budget of the National Cancer Institute for new drug development during the past decade. The annual budget in this respect, now over 200 million dollars annually, is larger than the total R&D budget of any of the major U.S. pharmaceutical firms.

The large-scale government funding to find a cure for cancer has been controversial from the start. In the case of applied programs like new drug development, scientists warned that targeted budgeted outlays might outpace promising scientific opportunities. In addition, some of the major areas of government funding, such as the work on interferon, are areas where commercial firms are also very active. Thus, there is always the danger of significant duplication of private development efforts.

The record to date of the anticancer development program has been somewhat mixed. There have been some notable successes in treating cancer through government-aided drug therapies, particularly in Hodgkin's disease and childhood leukemia. At the same time there have not been the dramatic advances or "magic bullet" cures that the most ardent congressional supporters of the "War on Cancer" anticipated (perhaps naively) when budget outlays were rapidly increased in the early 1970s. It is still too early, of course, to judge the overall success of the government anticancer efforts but there remain many skeptics in the scientific community who emphasize the heavy opportunity costs, in terms of the diversion of scarce scientific human resources, associated with this program.

Another issue that has recently been receiving considerable attention is government support for the development of so-called "orphan drugs." As discussed above, this label is variously used to describe drugs of limited commercial value, either because they treat rare diseases or because they are for some reason nonpatentable. The orphan drug problem is in considerable part an outgrowth of the large development costs and times necessary to satisfy all the regulatory hurdles for a new drug entity. This in turn significantly limits the willingness of firms to undertake development projects with small expected revenues.

To the extent that the problem results from the inability to patent a compound that otherwise would have a sufficiently large market to be commercially viable, there are a number of possible property-rights-type solutions that could be used to deal with the problem. For example, some countries such as Japan grant an exclusivity period to the firm that undertakes the initial development of the data necessary to gain regulatory approval. The Drug Reform Act also had a provision that would have prohibited imitators from relying on the innovator's scientific data on safety and efficacy for a five-year period. This was included specifically to provide some incentives for the development of drugs without patent rights. Another approach might be to strengthen the applicability of use patents in situations where a firm discovers a new use of an existing chemical compound. Currently, use patent exist but their degree of protection and enforceability is subject to a number of problems.

A more difficult category of orphan drugs involves the case of disease areas whose incidence is too small to warrant commercial development and regulatory approval. If such diseases are to be the subject of systematic targeted R&D, some kind of subsidy to producers appears necessary. While orphan drugs are not public goods in the normal use of this term, government subsidies for them can be rationalized as expenditures for "merit" wants or goods in common with many other expenditures in the social welfare area (e.g., food and medical treatment for low-income individuals, special provisions for the handicapped, etc.) However, orphan drug development must compete with other social welfare outlays for a limited (and contracting) budget for such expenditures.

As noted above, there are currently a few drug-development programs in the NIH directed at orphan-drug-type problems. There has been some success to date in facilitating development of antiepileptic drugs. The proposed Drug Reform Act would have expanded government R&D efforts in this regard by creating a Center for Drug Science with one of its main functions being targeted R&D on rare diseases. A recent government task force has recommended the creation of an independent government advisory board to coordinate and fund orphan drug research in both the government and the private sectors. Some congressmen in hearings have also mentioned requiring private firm R&D projects on rare diseases through regulation, although very obvious significant administrative and enforcement problems are associated with this.

The appropriate degree of government support for orphan drugs is obviously highly debatable. In the present period of contracting government size, the likelihood of any major new funding programs for orphan drugs (or for other types of merit wants) appears slight. At the same time, the orphan drug question remains an understandable area of intense concern for certain groups. Perhaps some novel solutions could be fashioned through the charitable giving process. Some of the relevant disease areas are the subject of sizeable charitable campaigns currently and some of these funds might be channeled toward the development of new therapies. This and related approaches appear to warrant further consideration in the current period of reduced government expenditures in the social welfare area.

NOTES

1. For a historical account of the regulation of drugs, see Wardell and Lasagna (1975), chap. 1, and Temin (1980).

2. See, for example, U.S. Senate (1960).

3. Two good sources on this topic are Clarke (1973) and Schwartzman (1976).

4. See Business Week (1979).

5. For a description of this program, see V.T. DeVita et al. (1979).

6. The antitumor drugs with clinical effectiveness in other therapeutic areas include Vidarbine (antiviral), Allopurinol (uricosuric), Flurocytosine (anti-fungal), Imuran (immunosuppressive) Iododeoxy-uridine (antiviral), Primethamine (antimalarial), and Trifluoromethyl (antiviral); DaVita et al. (1979), p. 205.

7. See Grabowski (1976), especially Table 2.

8. The market share equals the average annual sales of all NCEs introduced during the period as a percentage of total ethical drug sales in the last year of the period.

9. In particular, these data suggest that the incidence of ineffective new drugs was less than 10 percent in the pre- and post-1962 period. Peltzman also analyzes the growth rate patterns of NCEs in the pre- and post-1962 periods and argues that they also support the findings of expert evaluations in this regard.

10. Actually, the controls over clinical testing were the result of rules written by the FDA "in anticipation of the new law and in coordination with it" (Crout, 1981).

11. See U.S. General Accounting Office (1980).

12. See Wardell (1973).

13. Wardell (1978).

14. U.S. General Accounting Office (1980).

15. Grabowski, Vernon, and Thomas (1978).

16. We use the term production costs to include not only cost of goods sold, but also an allowance for general selling and administrative costs.

17. Virts and Weston (1980).

18. Grabowski and Vernon (1981).

19. In mathematical terms, the rate of return, r, is found by solving the equation:

$$(1) \quad \sum_{i=1}^{m} C_{t-i} (1 + r)^i = \sum_{j=0}^{p} \frac{R_{t+j}}{(1 + r)^j} + \sum_{j=p+1}^{n} \frac{R_{t+j}}{(1 + r)^j}$$

where

C_{t-1}, C_{t-2}, ..., C_{t-m} are R&D costs and other investment expenditures:

R_t ... R_{t+p} = net income stream before patent expiration;

R_{t+p+1} ... R_{t+n} = net income stream after patent expiration.

m = gestation period
n = product life
p = patent life.

20. Grabowski, Vernon, and Thomas (1978).

21. In terms of the present value, equation (1) in note 19, p was set equal to 8 and all the Rs in the second summation term were reduced to reflect a 50 percent reduction in U.S. net revenues.

22. A good description of MAC is contained in Jean Paul Gagnon and Raymond Jang, Federal Control of Pharmaceutical Costs: The MAC Experience, Roche Laboratories, 1979.

23. See U.S. Department of the Treasury, The Operation and Effect of the Possessions Corporation System of Taxation (Washington, D.C.: U.S. Government Printing Office, June 1979), Table 5, p. 59.

REFERENCES

Baily, Martin N., "Research and Development Costs and Returns: The U.S. Pharmaceutical Industry." Journal of Political Economy (January/February 1972).

Bloom, B.M., "Socially Optimal Results from Drug Research." In Impact of Public Policy on Drug Innovation and Pricing, ed. S.A. Mitchell and E.A. Link (Washington, D.C.: American University, 1976).

Bond, R. and D. Lean, Sales, Promotion and Product Differentiation in Two Prescription Drug Markets. Federal Trade Commission staff report (Washington, D.C.: Government Printing Office, February 1977).

Business Week, "Eli Lilly: New Life in the Drug Industry" (October 29, 1979).

Cardon, P.F., F.W. Dommel, Jr., and R.R. Trumble, "Injuries to Research Subjects: A Survey of Investigators." The

New England Journal of Medicine 295 (September 16, 1976): 650.

Chain, E.B., "Academic and Industrial Contributions to Drug Research." Nature (November 2, 1963).

Clarke, F.H. (Ed.), How Modern Medicines are Discovered (Mount Kisco, N.Y.: Futura Publishing Co., 1973).

Clarkson, K.W., Intangible Capital and Rates of Return, (Washington, D.C.: American Enterprise Institute, 1977).

Clymer, H., "The Changing Costs and Risks of Pharmaceutical Innovation." In The Economics of Drug Innovation, (Washington, D.C.: The American University, 1970).

Cocks, D.L., "Product Innovation and the Dynamic Elements of Competition in the Ethical Pharmaceutical Industry." In Drug Development and Marketing, ed. R.B. Helms (Washington, D.C.: American Enterprise Institute, 1975).

Cocks, D.L. and J.R. Virts, "Pricing Behavior of the Ethical Pharmaceutical Industry." Journal of Business (July 1974).

Comanor, W.S., "The Drug Industry and Medical Research: The Economics of the Kefauver Committee Investigations." Journal of Business (January 1966).

Crout, J.R., "Discussion." In Drugs and Health, ed. R.B. Helms (Washington, D.C.: American Enterprise Institute, 1981).

Deutsch, L.L., "Research Performance in the Ethical Drug Industry." Marquette Business Review (Fall 1973).

DeVita, V.T., et.al., "Clinical Trials Programs of the Division of Cancer Treatment, National Cancer Institute." Cancer Clinical Trials 2, 2 (1979).

Dunlop, Sir Derrick, "The British System of Drug Regulation." In Regulating New Drugs, ed. Richard Landau (Chicago: University of Chicago Center for Policy Study, 1973).

Eisman, M. and W. Wardell, "The Decline in Effective Patent Life of New Drugs." Research Management (January 1981)

Fuchs, V.R., Who Shall Live? (New York: Basic Books, 1974).

Grabowski, H.G., Drug Regulation and Innovation: Empirical Evidence and Policy Options (Washington, D.C.: American Enterprise Institute, 1976).

_____., "Regulation and the International Diffusion of Pharmaceuticals." In The International Supply of Medicines ed. R.B. Helms (Washington, D.C.: American Enterprise Institute, 1980).

Grabowski, H.G. and D.C. Mueller, "Industrial Research and Development, Intangible Capital Stocks, and Firm Profit Rates." Bell Journal of Economics (Autumn 1978).

Grabowski, H.G. and J.M. Vernon, "Structural Effects of Regulation on Innovation in the Ethical Drug Industry." In Essays on Industrial Organization in Honor of Joe S. Bain, ed. R.T. Masson and P. Qualls (Cambridge, Mass: Ballinger, 1976).

_____., "Consumer Protection Regulation in Ethical Drugs." American Economic Review (February 1977).

_____., "Consumer Product Safety Regulation." American Economic Review (May 1978).

_____., "Substitution Laws and Innovation in the Pharmaceutical Industry." Law and Contemporary Problems (Winter-Spring 1979).

_____., "A Sensitivity Analysis of Expected Profitability of Pharmaceutical R & D." Managerial and Decision Economics (March 1982).

_____., "The Determinants of Research and Development Expenditures in the Pharmaceutical Industry." In Drugs and Health, ed. R.B. Helms (Washington, D.C.: American Enterprise Institute, 1981). (b)

Grabowski, H.G., J.M. Vernon, and L.G. Thomas, "Estimating the Effects of Regulation on Innovation: An International Comparative Analysis of the Pharmaceutical Industry." Journal of Law and Economics (April 1978).

Hansen, R.W. "The Pharmaceutical Development Process: Estimates of Current Development Costs and Times and the Effects of Regulatory Changes." In Issues in Pharmaceutical Economics, ed. R.I. Chien (Cambridge, Mass: Lexington Books, 1979).

_____., "Pharmaceutical Development Cost by Therapeutic Categories." University of Rochester Graduate School of Management Working Paper No. GPB-80-6, March 1980.

Hymer, S. and P. Pashigian, "Turnover of Firms as a Measure of Market Behavior." Review of Economics and Statistics (February 1962).

Lebergott, Stanley. Statement before the Subcommittee on Monopoly of the Select Committee on Small Business, Competitive Problems in the Drug Industry Part 23: Development and Marketing of Prescription Drugs 93rd Congress, 1st Session, 1973.

Mund, Vernon A., "The Return on Investment of the Innovative Pharmaceutical Firm." In The Economics of Drug

Innovation, ed. J.D. Cooper (Washington, D.C.: The American University, 1970).

Peltzman, S., "An Evaluation of Consumer Protection Legislation: The 1962 Drug Amendments." Journal of Political Economy (September/October 1973).

Pettinga, C., "Discussion." In Regulation, Economics, and Pharmaceutical Innovation, ed. J.D. Cooper (Washington, D.C.: American University, 1975).

Sarett, L.H., "FDA Regulations and Their Influence on Future R&D." Research Management (March 1974).

Schmidt, A., "The FDA Today: Critics, Congress, and Consumerism." Speech to National Press Club, Washington, D.C., October 29, 1974.(a)

_____ . Statement before the Subcommittee on Health of the Committee on Labor and Public Welfare Examination of the Pharmaceutical Industry, 1973-74 Part I: Hearings on S.3441 and S.966 93rd Congress, 2nd Session, 1974.(b)

Schnee, J., "Innovation and Discovery in the Ethical Pharmaceutical Industry." In Research and Innovation in the Modern Corporation, ed. E. Mansfield et al. (New York: W.W. Norton, 1971).

Schwartzman, D., The Expected Return from Pharmaceutical Research (Washington, D.C.: American Enterprise Institute, 1975).

_____ ., Innovation in the Pharmaceutical Industry (Baltimore: Johns Hopkins University Press, 1976).

Taylor, C.T. and Z.A. Silberston, The Economic Impact of the Patent System (Cambridge, England: Cambridge University Press, 1973).

Temin, P.M., Taking Your Medicine: Drug Regulation in the United States (Cambridge, Mass.: Harvard University Press, 1980).

Thomas, Celia, "The Return To Research and Development in the Pharmaceutical Industry" Unpublished Ph.D. Dissertation, Duke University, 1981.

U.S. Food and Drug Administration, "The Analysis of Economic Impact of the Disclosure of Safety and Efficacy Data." Mimeo, 1978.

U.S. General Accounting Office, FDA Drug Approval - A Lengthy Process that Delays the Availability of Important New Drugs, HRD-80-64, May 28, 1980.

U.S. House of Representatives, Committee on Science, Research and Technology, Report on the Food and Drug

Administration's Process for Approving New Drugs, 96th Congress 1st Session (Washington, D.C.: U.S. Government Printing Office, 1980).

U.S. Senate, Committee on the Judiciary, Subcommittee on Antitrust and Monopoly, Hearings on Administered Prices, 86th Congress, 1st Session (Washington, D.C.: U.S. Government Printing Office, 1960).

Vernon, John M., "Concentration, Promotion, and Market Share Stability in the Pharmaceutical Industry." Journal of Industrial Economics (July 1971).

Virts, J.R. and J.F. Weston, "Returns to Research and Development in the U.S. Pharmaceutical Industry." Managerial and Decision Economics (September 1980).

Wardell, W.M., "Introduction of New Therapeutic Drugs in the United States and Great Britain: An International Comparison." Clinical Pharmacology and Therapeutics (September/October 1973).

_____., "Therapeutic Implications of the Drug Lag." Clinical Pharmacology and Therapeutics (January 1974).

_____., "The Drug Lag Revisited: Comparison by Therapeutic Area of Patterns of Drugs Marketed in the United States and Great Britain from 1972 through 1976." Clinical Pharmacology and Therapeutics (November 1978).

_____., "The History of Drug Discovery, Development, and Regulation." In Issues in Pharmaceutical Economics, ed. R.I. Chien (Cambridge, Mass: Lexington Books, 1979).

Wardell, W.M. and L. Lasagna, Regulation and Drug Development (Washington, D.C.: American Enterprise Institute, 1975).

Wardell, W.M. et al., "The Rate of Development of New Drugs in the United States." Clinical Pharmacology and Therapeutics (May 1978).

7
Residential Construction
John M. Quigley

I. INTRODUCTION

Public concern with housing in the United States has both efficiency and distributional bases. Because housing expenditures are such a large fraction of consumers' budgets, and because poor households have such small budgets, a series of federal programs to provide "adequate" housing for "poor" households has evolved, beginning with the Public Housing Act of 1937. On narrow efficiency grounds, however, there has also been increasing concern about public policy and its effect upon the production and distribution of housing services. It is alleged that residential construction is a "backward" industry, characterized by a low rate of technical progress and that supply prices for new construction are higher than would be indicated by efficiency in production. Concern is with the effect of existing policies upon the structure of the market and with the design of public policies to foster technical progress, reduced costs, and increased output of housing services.

In any practical context, of course, these distributional and efficiency considerations are hardly separable. Nevertheless, a reading of the reports of two presidential commissions established in response to inadequate living conditions of the urban poor (the Douglas and Kaiser Commissions) indicates widespread dissatisfaction with the economic health of the construction sector(1) as distinct from the delivery of basic

*This chapter benefited from extensive comments and suggestions by Eric Hanushek, Joseph Sherman, and Francis Ventre, and from the research assistance of Bruce Gillis.

services to the needy. The reports of these commissions, incorporated into the language of the Housing and Urban Development Act of 1968, indicated that the goal of "a decent home and a suitable environment for every American Family"(2) required two types of public policies: policies to increase the flow of newly constructed, unsubsidized dwellings at affordable prices, as well as subsidy policies to improve the quality of existing dwellings and to increase the housing consumption of the poor.

This chapter considers public policy and the efficiency of the residential construction sector. Section II below records basic facts about the industrial structure and relative performance of the house-building sector. It summarizes postwar empirical research about changes in productivity and the costs of construction, and describes briefly some of the more important innovations in materials and techniques. It also assesses, largely on the basis of interview data and expert opinion, the magnitude of cost savings attributable to some of these innovations. Section II also notes the relationship between reductions in labor and materials inputs and their effects upon the supply cost of housing services and costs of occupancy for housing consumers. Finally, limited and suggestive information is presented about the nature of private research and development activity.

Section III discusses three aspects of industry structure and its relationship to public policy. This section investigates the cyclical sensitivity of the house-building sector, the fragmented nature of the industry and its regulatory environment, and the federal role in supporting research and development and technical innovation.

Conclusions are presented in Section IV.

II. SECTORAL PERFORMANCE AND INDUSTRIAL STRUCTURE

Productivity Measures

Throughout the 1960s the conventional wisdom held that productivity trends in house-building lagged behind other sectors of the economy. Table 7.1, which provides a summary of postwar productivity studies, confirms this general impression. The estimates of average annual productivity growth in construction range between 0.5 percent per year and 2.8 percent, depending upon the methodology employed and the period of analysis. Productivity growth estimates for contract construction are uniformly lower than those estimated for the rest of the economy.

Underlying all comparisons of the rate of technical progress in this sector are at least four methodological and measurement problems: (1) appropriate adjustment for quality changes; (2) consistent definitions of inputs; (3) adjustments required by variations in the mix of site and off-site activity; and (4) disaggregation of construction activities into the residential and nonresidential sectors. Although analogous methodological problems are inherent in the measurement of technical progress in all sectors of the economy, there are indications that these issues present more difficulties in the analysis of the construction sector (Sims, 1968). For example, Rosefields and Mills (1979) have argued that more appropriate adjustments for quality improvements in contract construction would lead to substantially larger productivity estimates for the sector (presumably, however, a more refined measurement of quality change would lead to higher productivity estimates in other sectors as well).

Nevertheless, a consensus seems to exist that housing was a "backward" sector of the economy through most of this century. For example, by comparing independently derived indices of building costs and new home prices, Grebler, Blank, and Winnick (1956) concluded that productivity in residential construction had remained relatively constant from the turn of the century through the mid-1950s. Applying a similar methodology to nonresidential contract construction led the authors to conclude that "productivity has increased significantly in heavy construction, but much less so in residential building" (Grebler, Blank, and Winnick, 1956, p. 357). Dennison's (1962) analysis of economic growth found an absolute decline in input productivity in the construction sector during the 1930-1960 period. Dacy's analysis of price trends and productivity during the 1947-1960 period similarly concludes that contract "construction productivity lagged considerably behind the average for the economy and even behind total services" (Dacy, 1965, p. 406). Kendrick's exhaustive study of postwar productivity trends (1973) provides estimates of total factor productivity during the period 1948-1969. Of 34 industry groups considered by Kendrick, the average productivity change in contract construction ranks 31st.

Raw productivity change measures for the more recent period are presented in the bottom part of Table 7.1. During the 14-year period, 1966-1979, productivity increases in contract construction were smaller than increases observed in the overall economy or in the manufacturing sector in 12 of the years. Productivity changes in contract construction exceeded those elsewhere in the economy in two years. In 8 of the past 14 years, moreover, the raw productivity index (measured as constant dollar output per worker-hour) actually declined. The period as a whole indicates a modest decline in productivity in contract construction activity.

Table 7.1.   Postwar Productivity Trends in Construction.

A.   Estimates of Annual Growth in Productivity in Percent

Author	Time Period	Contract Construction	Residential Component	Manufacturing	Private Domestic Economy
Sims	1947-1966	2.3			
Gordon	1948-1965	1.4-2.8		3.4	
Dacy	1947-1963	3.0			
Domar	1948-1960	2.0		3.4	2.6
Ball	1962-1969		1.5^a		
U.N.	1953-1967	0.5			3.6
Alterman	1947-1955	2.5			3.6^d
Cassimatis	1947-1967	$1.6-2.8$			
	1952-1965	1.5^c			
C.E.A.	1947-1966	1.9			2.8
Clague	1948-1953	2.7			
Haber	postwar	1.5			
Kendrick	1948-1966	1.5^c		2.5^c	2.5^c
	1948-1953	3.6^c		2.9^c	2.8^c
	1953-1957	2.8^c		1.5^c	1.9^c
	1957-1960	1.1^c		2.0^c	2.3^c
	1960-1966	-1.0^c		3.2^c	2.9^c

B.   Average Annual Change in Productivity, 1966-1979.

Author	Time Period	Contract Construction	Manufacturing	Private Domestic Economy[b]
Chase	1966	-3.5	2.2	0.4
	1967	11.0	4.8	3.8
	1968	-7.1	3.2	1.2
	1969	-9.8	-0.3	-1.6
	1970	7.3	0.0	3.3
	1971	4.0	5.1	2.8
	1972	2.8	4.2	4.5
	1973	-16.2	1.1	-2.7
	1974	-4.5	-2.5	-2.2
	1975	9.3	5.6	5.6
	1976	1.0	3.4	2.4
	1977	-0.2	0.4	-0.3
	1978	-7.0	3.1	-0.1
	1979	-5.5	1.3	-1.5

(continued)

Table 7.1. Postwar Productivity
Trends in Construction (continued)

[a]Single family housing.
[b]Private nonfarm sector.
[c]Total factor productivity.
[d]Total private economy.

Sources: Evsey Domar et al., "Economic Growth and Production in the United States, Canada, United Kingdom, Germany, and Japan in the Post-war Period." Review of Economics and Statistics (February 1964), p. 36.

Douglas C. Dacy, "Productivity and Price Trends in Construction Since 1947." Review of Economics and Statistics (November 1965), pp. 406-411.

Christopher Sims, "Efficiency in the Construction Industry." Technical Studies, vol. II of the Kaiser Committee Report, pp. 145-175.

Robert J. Gordon, "A View of Real Investment in Structures." Review of Economic Statistics (November 1968), p. 423.

Robert Ball and Larry Ludwig, "Labor Requirements for Construction of Single-Family Houses." Monthly Labor Review (September 1971), pp. 12-14.

United Nations, Economic Commission for Europe, Economic Survey of Europe in 1969: Part I, Structural Trends and Prospects in the European Economy (New York: 1976), p. 92.

John W. Kendrick, Postwar Productivity Trends in the United States (NBER, 1973), pp. 77-85.

Jack Alterman and Eva E. Jacobs, "Estimates of Real Products in the United States 1947-1955." In Input, Output, and Production Measurement (New York: NBER, 1961), pp. 246-249.

Peter J. Cassimatis, The Economics of the Construction Industry (New York: National Industrial Conference Board, 1969), pp. 76-88.

Council of Economic Advisors, Economic Report of the President: 1968 (Washington, D.C.: U.S. Government Printing Office, 1968), pp. 120-124.

Ewan Clague and Leon Greenberg, "Discussion of Employment." In Automation and Technological Change (American Assembly, 1962), p. 120.

William Haber and Harold M. Levinson, Labor Relations and Productivity in the Building Trades (Ann Arbor: University of Michigan Press, 1956), p. 203.

Julian E. Lange and Daniel Quinn Mills (eds.), The Construction Industry (Lexington, Mass.: Lexington Books, 1969), pp. 88-90.

Table 7.2. Average Annual Growth of Various Input
Cost and Output Price Indices for
Construction Activity: 1947-1977.

Period	Input Cost Indices						Output Price Indices			
	EN-R	DCCI	Boeckh	Turner	WPI	CPI	CPI-R	CPI-H	NRS	BOC
1947-52	6.0%	4.6%	2.7%	5.0%	3.5%	3.5%	4.5%	na	4.5%	na
1952-57	4.1	2.0	2.7	3.3	2.1	1.2	2.8	2.2%	1.3	na
1957-62	2.7	-0.1	0.5	1.1	0.3	1.4	1.5	1.5	0.2	na [b]
1962-67	2.9	2.0	2.7	2.7	1.1	2.0	1.2	2.6	1.5	1.9% [b]
1967-72	9.2	6.7	6.6	9.0	3.4	4.6	3.6	7.0	5.6	5.7
1972-77	9.9	8.1	9.4	8.1	10.6	7.7	5.2	7.9	9.9	9.6

[a] 1953-57.
[b] 1963-67.

Sources: EN-R, DCCI, Boeckh, and Turner are computed from U.S. Department of Commerce, Industry and Trade Administration, Construction Review 25, 11 (December 1979).
WPI, CPI, CPI-R, and CPI-H are from U.S. President, Economic Report of the President,
1978, (Government Printing Office); NRS is from U.S. Department of Commerce, Bureau
of Economic Analysis. The National Income and Product Accounts of the U.S., 1929-74;
Statistical Tables, (Government Printing Office, 1977), and Survey of Current Business
57, 7 (July 1977) and 58, 7 (July 1978); BOC is from U.S. Department of Commerce, Industry and Trade Administration, Construction Review 25, 11 (December 1979).

Definitions:

EN-R: Engineering News-Record Index
DCCI: Department of Commerce Composite Construction Cost Index
Boeckh: Boeckh Index
Turner: Turner Index
WPI: Wholesale Price Index
CPI: Consumer Price Index
CPI-R: Consumer Price Index: Rent/Residential Component
CPI-H: Consumer Price Index: Homeownership Component
NRS: Department of Commerce implicit price deflator for new residential structures
BOC: Bureau of the Census price index for new single family homes, exclusive of lot value.

Input and Output Cost Measures

The available evidence does not permit a refined analysis of productivity in residential construction. The trends reported in Table 7.1 for contract construction include all residential, commercial, and industrial building as well as highway and heavy construction. In recent history, the residential component has varied between 30 and 48 percent of the total.(3)

A number of input cost and output price measures for residential construction are available from the postwar period. Tabel 7.2 presents a summary of trends in four of these cost indices. Inferences about the relationship between productivity and variations in these indices depend quite specifically on their definitions.

None of the four cost indices presented include land inputs. The Engineering News-Record index (EN-R) combines construction labor and materials input prices for fixed proportions. Since input prices are not adjusted for changes in productivity or technology, this index ignores technological change within the sector. The Boeckh index weighs materials and equipment prices for brick and frame residences by wage rates, adjusted to reflect variable labor efficiency in each of 20 locales. Consequently, some technological efficiency gains are implicit in its values. The Turner index is computed from bid estimates returned to the Turner Construction Company of the cost of standardized projects. Presumably, each firm fully accounts for inputs and labor efficiency changes in its bids, so technological advance should be fully reflected in this cost index. Unfortunately, only a few of the standard projects that underlie this index are residential in nature. Finally, the Department of Commerce Composite Index (DDCI) incorporates a number of construction cost indices (including the Engineering News-Record, Boeckh, and Turner indices). Some of its component indices account for technological change and some do not; thus it reflects, in some part, productivity advances.

A comparison of the Boeckh or the Turner index with the EN-R index implies that actual construction costs in the residential sector rose less rapidly throughout the three decades than they would have if technology were stagnant. The comparisons from 1967 on, however, suggest a reversal and a decline in residential construction productivity. A comparison of these two indices with the DCCI (which implicitly accounts for some technical change) supports the same inference.

Table 7.2 also represents a comparison of six output price indices for the same period. The wholesale price index (WPI) for industrial commodities reflects general trends in the manufacturing and mineral products sectors of the economy. The consumer price index (CPI) measures price movements in food and beverages, housing, apparel, transportation, medical services, entertainment, and other services. Two components of

the CPI's housing class also appear in the table. The rent/
residential component (CPI-R) incorporates price trends both
for apartment rent and for imputed rent of homeowners (based
on sales prices of new and existing homes). The homeownership portion of the CPI's housing class (CPI-H), introduced in
1953, combines a home purchase element with various operating
and maintenance cost elements. Also presented is the implicit
price deflator for purchases of new residential structures
(NRS) computed by the Commerce Department and the recent
Bureau of the Census price index for new single-family homes,
exclusive of lot value (BOC). Presumably the latter index is
the best indicator of output price trends.

A comparison of the NRS with the WPI or the CPI may
suggest that home-building efficiency equaled or surpassed
economywide performance until about 1967. Since 1967, however, the relative price increases of new residential structures
(NRS) or new single-family homes (BOC) have exceeded economywide price increases. Inferences based upon CPI-R or
CPI-H are more ambiguous, since they include transactions on
used homes and include the land component. Any such comparison of output prices assumes that demand fluctuations do not
change the relative prices of goods; these comparisons do,
however, measure the entire economy's efficiency in producing
housing - increases in productivity in input suppliers as well
as builders.

Recent work by Ferguson and Wheaton (1980), who analyzed the raw data underlying the BOC index, presents disaggregated trends in output prices for newly constructed
dwellings in four components: changes in the unit price of
land; changes in the quantity of land; changes in the characteristics of housing structures; and changes in the price of a
"standardized" structure. Their results indicate that improved
quality accounted for almost one-fourth of the observed increase in the prices of residential structures during the period
1972-1978.

On balance, the productivity and price evidence suggests
a pattern of modest improvement in productivity in residential
construction from 1947 through the mid-1960s, although construction seems to have lagged behind manufacturing activity.
During the more recent period, the evidence suggests little or
no improvement in productivity and a more substantial decline
relative to other sectors of the economy.

The Costs of Housing Services

A comparison of costs and productivity in the construction of
residential structures may give a misleading picture of the
importance of technical change and improved technique in the
costs of supplying housing services to consumers. Table 7.3

presents "typical" distributions of the total costs of providing newly constructed housing services as reported to the Kaiser Committee. As of 1968, only about 55 percent of the costs of new construction of single-family homes consisted of labor and materials costs. For multifamily units, about 60 percent of the cost of producing housing services was attributable to purchased inputs and labor. Development costs, including land, consisted of 25-30 percent of the costs of production.

Table 7.3. Distribution of Costs of Housing Service Provision for "Typical" Developments in 1968.

	Single-family detached house	Apartment in multifamily medium-rise building
Development Costs	31%	25%
Land	10	9
Development	15	4
Miscellaneous	6	12*
Construction Costs	69	75
Materials	37	38
On-site wages	18	22
Overhead/profit	14	15
Total	100%	100%

*Including architects' fees.

Source: McGraw Hill Co., "A Study of Comparative Time and Cost for Building Five Selective Types of Low Cost Housing." The Report of the President's Committee on Urban Housing: Technical Studies, vol. II (Washington, D.C.: U.S. Government Printing Office, 1968), p. 9.

Table 7.4 presents a "typical" distribution of the costs of consuming housing services as of 1968. The occupancy cost comparison (at 6 percent mortgage interest rates) appears quaint from the perspective of the 1980s. It reveals quite starkly, however, the importance of debt retirement in the monthly cost of consuming housing services. Even at the

Table 7.4. Distribution of Occupancy Costs of Housing
Services for "Typical" Developments in 1968.

	Single-family detached house	Apartment in multifamily medium-rise building
Debt retirement	53%[a]	42%[b]
Taxes	26	14
Utilities	16	9
Maintenance and repair	5	6
Administrative and similar costs		13
Vacancies and bad debts		9
Profit and reserves		7
Total	100%	100%

[a]Based on a 94.5 percent 30-year mortgage at 6 percent interest.

[b]Based on an 85 percent 35-year loan at 6 percent interest.

Source: McGraw Hill Co., "A Study of Comparative Time and
Cost for Building Five Selective Types of Low Cost
Housing." The Report of the President's Committee
on Urban Housing: Technical Studies, vol. II (Washington, D.C.: U.S. Government Printing Office,
1968), pp. 8-12.

typical mortgage rates of the 1960s, carrying charges represented 40-50 percent of occupancy costs. A comparison of
Tables 7.3 and 7.4 reveals that a given reduction in the cost
of materials and labor would reduce the total costs of producing housing services by only about half as much. This
would presumably be reflected in occupancy costs by reductions in the face value of mortgages. As any recent purchaser
of housing knows, however, even large reductions in the face
value of mortgages are easily offset by small changes in carrying costs.

Tables 7.3 and 7.4 are also suggestive of the importance
of exogenous factors in the production and occupancy costs for

housing services. Increased land rentals or site values observed during the past decade increase production costs, even if there are substitution possibilities between capital and land in production. A decade of increases in property taxes make occupancy costs larger, even if more services are provided in the bargain.

It appears that variations in the total costs of supplying housing services are less sensitive to technological changes in the production process per se than in many sectors of the economy. The costs of consuming these services are also less sensitive to cost reductions in labor and materials.

The Structure of the Housebuilding Industry

The residential construction industry is characterized by a relatively small scale of production as measured by gross receipts or by numbers of units completed annually. Table 7.5 reports the size distribution of multifamily and single-family builders as of 1972. For single-family builders, less than one-third of the firms reported gross receipts of one million dollars or more, or a volume of more than about 100 units; almost 40 percent of the firms reported volumes of fewer than about 20 units per year.

Table 7.5.   Distribution of Gross Receipts by Size of
Builder, 1972.

Total receipts(000)	Estimated number of units	Single-family builders	Multifamily builders
$ 0-50	0-5	8.9%	1.8%
50-99	5-10	14.6	4.1
100-249	10-20	14.8	5.6
250-499	20-40	14.7	9.0
500-999	40-100	15.6	15.3
1,000-2,499	100-200	8.4	15.6
2,500+	200+	23.1	48.7
		100%	100%

Source: U.S. Bureau of the Census, 1972 Census of Construction. (Washington, D.C.: U.S. Government Printing Office, 1974), p. 206.

In the multifamily sector, slightly less than half the firms produced an annual volume greater than 200 units and an eighth of the firms produced fewer than about 20 units in multifamily dwellings. The annual volume of the typical builder of either single-family or multifamily dwellings is quite low.

Even this description overstates the numerical concentration of builders by volume, since it only includes firms with payrolls. It is reported that, in 1967, about a third of the 110,000 home-building firms in current operation did not have a regular payroll (Baer et al., 1976, p. L-13).

Information on trends in firm size is somewhat more elusive. Table 7.6 presents trends in the size distribution of single-family builders based on membership in the National Association of Home Builders (NAHB). Inferences drawn from this table are tenuous, since NAHB has about a one-third annual turnover in its membership, both very small and very large builders are likely to be underrepresented, and the distribution of units by scale of production may vary over the business cycle. In any case, the raw data indicate a decline in the scale of the building industry during the decade of 1960s.

Table 7.6.   Size Distribution of NAHB Builders.

Units constructed	Percent of single family builders			Percent of total units constructed		
	1959	1964	1969	1959	1964	1969
1-25	57.5	64.4	69.5	10.2	15.8	21.5
26-100	29.8	27.6	24.3	25.7	32.7	36.0
101-250	8.1	5.5	4.6	21.8	22.2	23.6
250$^+$	4.6	2.5	1.6	42.3	29.4	19.0

Source: National Association of Home Builders, A Profile of the Builder and His Industry (Washington, D.C.: 1970), p. 108.

Trends since 1969 reveal an apparent increase in the size and scale of home builders. For example, it is reported that the number of firms with greater than $10 million in annual sales increased from 119 in 1968 to 369 in 1972 (figures are unadjusted for inflation).(4)   The Bluebook of Major Home-builders (1973) reports that the market share of builders with annual volumes in excess of 200 units rose by almost 11 percent between 1969 and 1972.

Table 7.7 presents the latest available information on the size distribution of housebuilders. As measured by the number of establishments, firms with less than 20 employees comprised almost 98 percent of "General Contractors-Single Family Homes," 87 percent of "General Contractors-Residential Buildings," and 94 percent of "Operative Builders." In terms of gross receipts in the industry, however, such firms comprised 78 percent, 31 percent, and 50 percent of the three industries.

The bottom part of the table indicates that firms with gross receipts in excess of a half-million dollars account for almost half of total receipts among single-family general contractors. Such firms account for almost 85 percent of receipts among operative builders.

Beyond the increasing share of the market accruing to larger firms, there is some evidence of increasing merger activity among the larger firms, at least through the mid-1970s. Merger and acquisition activity among the largest publicly held home builders has provided product-line diversification, geographic expansion and, in one-fourth of all cases some vertical integration.(5)

The rapid and sustained growth of U.S. Home, the largest American housebuilding firm since 1972, has been through merger and acquisition. Between 1969 and 1972, U.S. Home acquired 18 companies, increasing sales from $3.7M to $205M in less than three years.(6) U.S. Home merged with Homecraft in 1977, and in 1978 issued $15M in mortgage-backed securities through a wholly-owned subsidiary.(7)

Despite any trends toward increased scale, however, the economic concentration of the house-building industry is quite low. The 25-firm concentration ratio in the industry is six-tenths of one percent.

There is little recent evidence on the relationship between scale of production and the costs of production. Maisel's (1953) analysis compares production costs of builders in three size classes. He estimates that production costs, including profit and overhead, for the typical single-family dwelling are 2.6 percent lower for firms producing 25-99 units than for smaller builders, and are 7.9 percent lower for firms producing more than 100 units. More recent evidence by Cassimatis (1969) suggests that for those firms producing 200 or more units, labor and materials costs for a typical dwelling are about 12 percent lower than the costs of firms producing fewer than 50 units. C. Cook (1976) concludes on the basis of this evidence that significant economies of scale do exist. The

Table 7.7.　Size Distribution of Residential
Construction Firms in 1977
(SIC 1521, SIC 1522, SIC 1531)

Number of Employees

		Total	1-4	5-9	10-19	20-49	50-99	100-249	250-499	500+
				a.	Percentage of Establishments					
SIC	1521	100,993	72.0%	19.1%	6.6%	1.8%	0.2%	0.1%	0.0%	0.0%
SIC	1522	4,775	52.8	20.6	13.4	9.3	2.3	1.1	0.5	0.0
SIC	1531	23,477	64.1	20.7	9.5	4.0	1.1	0.4	0.2	0.0
				b.	Percentage of Total Receipts					
SIC	1521	$21.9B	33.3	25.7	19.2	12.3	4.6	2.3	0.9	1.4
SIC	1522	$ 4.5B	8.0	10.2	13.1	24.4	14.4	16.9	13.1	
SIC	1531	$22.9B	20.0	15.1	15.1	17.6	11.3	8.6	8.6	3.7

Gross Receipts (in thousands)

			0-24	25-49	50-99	100-249	250-499	500-999	1000-2499	2500+
				a.	Percentage of Establishments					
SIC	1521	100,993	14.7	15.1	21.6	27.4	12.5	5.8	2.3	0.6
SIC	1522	4,775	7.0	8.7	13.6	23.0	14.5	11.4	9.7	7.6
SIC	1531	23,477	4.7	5.6	11.7	23.4	19.9	16.4	11.9	5.7
				b.	Percentage of Receipts					
SIC	1521	$21.9B	0.9	2.6	7.2	19.9	19.8	18.1	15.5	16.0
SIC	1522	$ 4.5B	0.1	0.3	1.1	3.9	5.4	8.4	15.8	65.0
SIC	1531	$22.9B	0.0	0.2	0.9	4.1	7.2	11.9	18.7	56.9

Note:　SIC 1521:　General Contractors: Single-Family Houses
　　　　SIC 1522:　General Contractors: Residential Building
　　　　SIC 1531:　Operative Builders.

Sources: U.S. Department of Commerce, Bureau of Census,
　　　　1977 Census of Construction Industries: Industry
　　　　Studies, SIC 1521, SIC 1522, SIC 1531, CC 77-1-1, 2,
　　　　3 (Washington, D.C.: U.S. Government Printing
　　　　Office, 1980).

magnitude of the relationship between scale of production and the occupancy costs for housing services does not seem to be terribly large, however. Popular descriptions of home builders suggest that there may be significant scale economies arising from production scheduling, improved x-efficiency, and vertical integration, at least among the industry giants. For example, Fortune reports the increased stability in annual production made possible by high capitalization among the giants (e.g., Centex, Ryan Homes, and Kaufman and Broad), by mortgage-backed securities, and by increasing "professionalization" of management.(8)

One difference in production techniques by firms at the largest annual volumes is their reliance on prefabricated parts, or the output of the home manufactures industry. For example, the Department of Housing and Urban Development's analysis of 511 major home builders revealed that the 25 largest builders used "major" prefabricated parts in 52.3 percent of units completed. For other builders, the proportion of units with "major" premanufactured parts ranged between 27.2 and 35.9 percent.(9)

A comprehensive survey of the home manufacturers industry is reported by Field and Rivkin (1975). They estimate that by 1970, national production of manufactured homes (including significant use of precut, panel, or modular construction) was more than 310,000 units, and included about 21 percent of the market for new units.

For 1978, it was estimated that manufactured housing output was at about the same level, 304,000 units and a somewhat smaller market share (Mahaffey, 1979). Home manufacturers tend to operate at a larger production scale than conventional builders, but even among these firms, about 30 percent produce fewer than 100 units annually (Field and Rivkin, 1975, p. 23).

There are at least four detailed comparisons of the relative costs of housing production using conventional and home-manufacturing techniques.

Weiner (1968) considers production costs for a typical small, single detached house with 1,000 square feet of living space. He compares conventional production at a volume of 150-200 units with off-site modular construction at differing scales. According to engineering estimates, excluding land and development costs, off-site modular construction at a scale of 5,000 units a year would reduce costs by 15 percent.

Several estimates were prepared for the Douglas Commission for "typical" single-family houses (McGraw-Hill, 1968). It was estimated that the off-site production of panel walls reduces costs by less than 4 percent. Off-site construction of sectional and modular components is estimated to reduce costs, again according to engineering assumptions, by as much as 20 percent.

Rowland (1969) compared production costs for low-rise garden apartments. The cost savings attributable to fully modular construction, comparing a production scale of 12 conventional units with 1,200 manufactered units, amounts to 9.3-13.7 percent, again excluding land and development costs. Finally, a comparison of high-rise construction using precast walls and partitions with similar construction using masonry and dry-well partitions indicates a labor and materials cost saving of 16 percent (Rothenstein, 1969).

The cost savings estimated in these studies arise from two sources: the reduction in the number of worker hours required to complete a given component of the final product and the substitution of cheaper and lower-skilled labor. The nature of cost reductions is thus similar to technical progress in other sectors of the economy. However, the magnitude of cost savings depends crucially upon the benchmark for wage comparisons. The estimated cost savings noted above arise from a comparison between the unionized construction sector and the industrialized sector, not between the existing construction sector and other industry. For the construction of single-family dwellings, for example, it has been estimated that less than a third of the labor input is unionized (C. Cook, 1976, pp. 6-20).

Whether these cost savings are large or small depends upon one's perspective. First, these comparisons are based upon engineering estimates and extrapolations, not upon a comparison of actual production runs. Second, as noted previously, labor and materials inputs into structures account for roughly 40 to 60 percent of the cost of producing housing services. Third, these comparisons were made more than a decade ago. Field and Rivkin, who are firmly convinced of the potential for cost reduction through home manufacturing, admit: "We must take it on faith that economies will result from industrialization of home building because conclusive evidence of lower costs does not exist. Presumptive evidence from other industries that have undergone industrialization implies that [manufactured] home building will produce substantial savings in cost" (Field and Rivkin, 1975, p. 10).

Innovation and Research

Some inconclusiveness in the importance of home manufacturing as an alternative technique to "conventional" home building does not imply that these latter methods have been static.

Industry observers believe that the current usage of "major" industrialized housing components in conventional construction is already quite high. When such components as prehung doors and preassembled windows are included, it has been estimated that about 90 percent of all new dwelling units built by conventional builders include major industrialized housing components compared to an insignificant fraction just after World War II. In addition, it is observed that before World War II, labor comprised 70 percent of on-site construction costs compared with roughly 30 percent today.(10)

Besides the substitution of preassembled and manufactured components for on-site techniques, innovation in construction includes new materials, new techniques for assembling materials on site, new tools for implementing given techniques, and perhaps improved management x-efficiency.

Engineering changes in residential construction methods have been relatively minor, in terms of their overall incidence or their contribution to cost reduction.(11) The use of brick for both structural and veneer purposes has increased since World War II as has the proportion of post-and-beam, "California-style" construction. Better engineering knowledge about concrete products has allowed single-slab (basementless) homes to appear more frequently in cold northern climates, where they were previously unknown. Electrical wiring has been moved from baseboard raceways to the interiors of framed walls (largely because better insulation materials have made the practice safe).

These process changes do not appear to have resulted from innovation in construction methods. Wider use of brick has apparently stemmed from a shift in the relative price of brick and wood products. Post-and-beam construction is among the oldest known structural engineering methods; its increased use is attributable to changing tastes - consumer preference for "open" houses - and to the development of double-glazed insulating glass. The northerward filtration of slab-built homes has resulted from better materials, stronger and lighter concrete products, not from construction-method innovation.

Other postwar innovations in construction methods per se do entail substantial efficiency gains: the use of 2" x 3" rather than 2" x 4" studs and plates in nonload-bearing partitions, and the employment of a 24-inch framing module instead of the traditional 16-inch one. Adoption of a 24-inch module allows a somewhat less than one-third reduction in the number of studs and a corresponding decrease in the labor required for wall framing. The use of 2" x 3" lumber decreases the cost of interior partitions by about 25 percent. As with the other changes in bulding method, these innovations do not represent fundamentally new assembly concepts. Instead, they stem from the fairly recent development of lumber-quality grading (supervised by the Commerce Department's American Lumber Standards Committee) and from better engineering knowledge about lumber stress characteristics, which has established that these new practices entail little or no added safety risk (Mayer, 1978). Interestingly, the 24-inch framing module may represent better engineering than the 16-inch module (see Ventre, 1973).

Somewhat more important than innovation in construction methods, according to industry sources, have been the improvements in power tools and the greater use of heavy equipment during the past two decades. Circular handsaws, powered mechanical hoists, compressed-air jackhammers, and nailguns have all increased the productivity of laborers. Though no estimates of the cost savings attributable to tool improvements have been found, one conjecture is that nailguns alone decrease framing time by about 20 percent. Power handsaws may have generated savings of similar magnitude. Bulldozers, backhoes, and other heavy equipment have decreased the time and cost of site preparation and excavation.

It appears that the most important technical changes in residential construction have been innovations in materials and the preassembly techniques discussed earlier. When three industry experts were each asked to list the five most important postwar cost-saving innovations in construction,(12) only one response (the use of 2" x 3" studs) did not involve new materials or preassembly. The other responses were:

- prefabricated roof trusses (3 responses)
- plastic drain, waste and vent piping (3)
- other prefabricated components (2, both of the respondents mentioned roof trusses separately first, then cited other components: prehung doors and windows, prefabricated stairways and panel construction)
- speciality plywood (2)
- gypsum wall board (2)
- insulating materials (1)
- heat pumps (1)
- molded bathroom facilities (1)

The importance of materials and preassembly innovations in technical change is emphasized by other industry experts. Johnson's enumeration of "important innovations" in residential construction during the two decades after World War II includes some 120 items, more than 70 of which are materials improvement or preassembly. The most important innovations noted by Johnson (1968) include:

- gypsumboard
- improved plywood and plywood products
- particleboard
- prefinished siding and door and wall coverings
- light gauge steel I-beams and adjustable columns
- plastic piping
- molded plastic bathroom fixtures
- washerless and single-level faucets
- improved electric heat pumps
- improved gas, oil and electric furnaces

- ready mix concrete
- insulating glass
- polyethylene vapor barriers
- improved construction hardware
- acoustical ceiling tile
- indoor-outdoor carpeting

In addition to improvements in wood products - particleboard, plywood, etc. - the introduction of plastics into home building has reduced total costs.   The most well-known products are ABS (acrilonitrile-butadiene-styrene) and PVC (polyvinyl-chloride) plastic drain, waste, and vent piping. Industry sources suggest that ABS and PVC piping are employed at cost savings of about 25 percent (see Ventre, 1973, for a discussion of the cost advantages of plastics).   Polyethylene is widely used as a vapor barrier under slabs.   It is estimated that this practice has a 40 percent cost advantage over the former technique, hot-mopped felt (Johnson, 1968). Molded plastics have found increasing use in one- and multipiece bathroom components, "significantly" reducing costs.

Hard evidence on the cost savings or increased output attributable to these innovations does not exist, and any numerical estimates are merely well-informed opinion.

How well do the details of industry innovation correspond to the aggregate productivity trends of the sector?   To the extent that these innovations represent cost savings on small individual tasks and that, in the aggregate, these tasks amount to less than half of the costs of producing housing services, the effect of technological change may be rather small indeed.   Since output quality at this level of detail is quite impossible to standardize, however, some fraction of the returns to innovation may not be fully reflected in productivity measures at all.

Innovation in house building arises from formal and informal research and development that may be undertaken by house builders, suppliers, trade associations, and government.

Individual house-building firms conduct little in the way of research and development activity.   Moreover, it is reported that "there is great reluctance on the part of builders and even housing manufacturers to experiment with new products and techniques, since innovations are perceived to be risky under many market conditions."(13)

The number of research scientists and engineers employed in the construction sector suggests that resources devoted to R&D are quite small.   In 1966, the Bureau of Labor Statistics reported 800 scientists and engineers (including those with bachelor's degrees) doing research in the construction sector. In 1970, the figure reported was 1,800.   The National Science Foundation survey of scientists and engineers reported that 409 individuals with doctorates considered themselves working "principally" on housing.(15)

Some measure of the research supported by trade associations (in this case, the National Association of Home Builders, NAHB) is provided by the scale of operation. Willis (1979, p. 244) reports that in 1978 the NAHB research foundations employed a staff of "fewer than 25 people, including secretaries, that only one quarter of its work is for the general benefit of members, and that the other three quarters is proprietary work." Much of its work is testing products of suppliers to provide independent verification of their properties.

It appears, therefore, that a large fraction of the innovation in house building is the result of R&D activity by suppliers or by government. Public-sector involvement is discussed in the next section. The fraction of R&D by manufacturers and materials suppliers devoted to housing is not known (and in many cases cannot be allocated). In contrast to other potential innovations in home building, however, it appears that the economic returns to R&D are more easily appropriable by the innovator when they are in the form of identifiable materials and not improved techniques. Willis reports impressionistic evidence that suppliers' R&D efforts devoted to housing are low. For example, in interviews conducted in 1978, members of the Producers' Council (the trade association of manufacturers of building products) asserted that "very few of the large suppliers devote any of their R&D effort specifically to housebuilding" (Willis, 1979, p. 247). Research facilities of particular supplier associations such as the Brick Institute of American and the American Plywood Association, are quite small.

Important to the profitability calculus of R&D in building, even by suppliers of new materials who can capture the returns to successful innovation privately, is the profile of market penetration of a successful product. It has been estimated that a potential innovator must be prepared to wait eight to ten years after product development before reaching an appreciable fraction of the market for new dwellings.(16) Presumably, the diffusion rate of new products is sensitive to their relative reduction in production costs. But if most potential innovations are evolutionary and reduce costs for a small component of the building-production process, this suggests that the rate of adoption by builders will be low. This can be expected to affect the ex ante R&D decisions of suppliers and their level of investment in innovative activity.

III. PUBLIC POLICY, COSTS, AND EFFICIENCY

Cyclical Sensitivity and Organization

The position of the residential construction sector as a large but volatile component of total investment activity has provoked much analysis of the transmission of that volatility and of its impact upon the economy as a whole. Until recently, however, there has been little analysis of the relation between instability in final demand and the micro-behavior of firms. Two recent works have related the cyclical sensitivity of the sector to the organization of competition and the performance of the sector.

A short paper by Manski and Rosen (1978) presents a verbal analysis of the micro-economics of an industry characterized by large random variations in demand. The authors deduce five general propositions based on the general assumption that: those conditions for profit maximization--relating to production technology, output size, market area, and choice of output product itself--which are optimal when demand is stable are different from those that are optimal when demand is unstable.

First, given a choice between a production technology that is efficient within a narrow range of output and is quite inefficient outside that range and a production technology that is "reasonably" efficient over a wide band of output, but best at no output level, there will be a tendency for firms to choose the latter process if demand is unstable.

Second, given a choice between hiring labor on a long-term basis and hiring workers by the job, there will be a tendency to choose the latter when demand is unstable. (Presumably, if demand is unstable, firms will also be less likely to invest in on-the-job training for workers, even if it is specific training.)

Third, given a choice between producing, at equivalent cost, a high-quality perishable product and a lower-quality storable product, the latter choice will be made if demand is unstable.

Fourth, given a production choice between an output that performs a narrow range of functions well and others poorly, and an output that performs a broad range of functions "adequately," the latter choice will be made if demand is unstable as long as net fluctuations can be dampened.

Fifth, given a choice between developing a small market intensively and operating in a less concentrated manner in a larger area, the latter choice will be made if net fluctuations can be reduced.

The basic conclusion of the Manski-Rosen analysis is that demand instabilitiy, under these conditions, creates a tradeoff between static economic efficiency and flexibility in response to temporal variation. Flexibility and diversification makes the individual firm more able to mitigate the shocks of random changes in demand.

The model indicates that, when demand is unstable, the average price paid by consumers is higher. Importantly, however, the profits of an individual firm need not be lower in a world of demand instability than in one of perfect stability - since instability raises costs for all firms and the industry demand curve need not be perfectly elastic. Thus, while demand instabilitiy may be costly to consumers as a group, it need not be costly to any single producer.

Manski and Rosen discuss this view of cyclicality in demand in the context of six telephone interviews with suppliers to the residential construction industry. They conclude with the remarks: "The contribution of industry studies to an understanding of the behavioral implications of instability is more potential than actual. Studying the detailed structure and operations of specific industries should offer a direct and fruitful approach to the question of instability. Unfortunately, we know of no industry studies which have tried to grapple with the instability question in a major way" (1978, p. 224).

Thus, it is worth noting the international comparison of residential construction and house building recently completed by Mark Willis (1979). Willis develops a simple model of the firm facing unstable demand which is a direct extension of the Manski-Rosen analysis. Instead of postulating an industry populated by identical firms, however, Willis considers the entry and exit of marginal firms as demand increases and declines. This model predicts, for residential construction, that: (1) the industry will be highly fragmented, with a large number of in-and-out firms; (2) firms will use nonspecialized inputs in the construction of new dwellings; (3) construction firms will be unlikely to use production processes with high fixed costs; and (4) the industry will oppose public programs that would jeopardize current market shares. Willis interprets his results as implying that fewer resources will be devoted to R&D, that the selection of R&D projects will be distorted, and that firms will resist new products and processes of a labor-saving variety.

Of more interest that the theoretical refinement of this model, however, is the empirical evidence presented by the author. Willis presents a detailed comparison of aggregate house-building characteristics in the United States, England, and France, and the results of a series of interviews with builders and suppliers in the three countries.

Because housing starts have been more stable in England than in the United States and have been more stable in France than in England, a detailed international comparison provides some evidence about the link between demand conditions and industry structure. Willis' rich statistical and anecdotal evidence does indicate that firm sizes tend to follow the anticipated pattern, that French firms tend to be more capital intensive than British or American firms, and that productivity trends in construction show that increases in output per worker-hour have been significantly larger in France than in England, and somewhat larger in England than in the United States.

Willis also presents sketchy evidence on private R&D activity in the three countries. Although this evidence is far from satisfactory, the author concludes that resources devoted to R&D are relatively lower in the United States.

Willis presents a persuasive argument that these, and other comparisons of performance, are causually related to demand instability. In considering the evidence presented, however, it must be recognized that both the extent of public housing and the relative size of contracts for public housing are larger in France than in Enlgand or the United States; moreover, the historical pattern of French regional-planning activity has facilitated growth of a few large firms. Finally, for the essential inferences between demand stability and the progressivity of residential construction, the analysis has two degrees of freedom.

Historically, Savings and Loan Associations (S&Ls) have provided 40 to 60 percent of new home mortgage funds, and maximum interest rates offered by S&Ls have been limited by regulation Q. As a result, when market interest rates have exceeded ceiling rates, there have been substantial outflows of funds from S&L deposits to other forms of savings. Indeed, during the period 1965-1980, net flows into savings and loan associations have been strongly and negatively correlated with the "spread" between passbook and regulation Q ceilings.

Thus, in some part, the extreme sensitivity of mortgage lending and new construction to interest rates has been the result of public regulation. It is worth noting, therefore, that this source of cyclicality in house building will be removed by the Depository Institutions Deregulation and Monetary Control Act (PL96-221) signed into law on March 31, 1980. Under Title II of the act, regulation Q and other limitations on S&L activity will be phased out over the next six years.(17) Although the impact of interest rates on new construction activity depends more directly on the interest elasticity of demand than on specific regulation, it is forecast (indeed, it is intended by the act) that the reforms of 1980 will increase the flows of deposits into savings and loan associations and will make mortgage lending more stable.

The arguments of Manski, Rosen, and Willis indicate that the indirect effect of these reforms will be to foster productivity gains in residential construction and to stimulate innovative activity.

Geographical Fragmentation and Local Regulation

Because transport costs are an important component of materials costs, because the average size of building firms is small, and because (it is often alleged) local tastes vary, the geographic market served by most house building firms is quite small. Among the giant firms, the geographic coverage is not large. HUD's survey of the 25 largest builders revealed that they operated, on average, in 6 states, while a sample of smaller firms (26th through 100th in sales volume) operated in 3 states.(18) Today, the largest single builder, U.S. Homes, operates in 17 states, compared with 10 in 1977.

In any case, the production process is, as a result, affected by a diverse set of public policies, highly localized in nature, with differential impacts across smaller firms and with more complicated effects within the markets served by larger firms.

These local regulations, derived from the police powers of the individual states, and justified in terms of health and safety responsibilities delegated to local authorities, include: zoning controls, growth control and environmental regulations, subdivision regulations, and building code provisions.

Zoning, growth control, environmental, and subdivision regulations (19)

The classic justification for zoning regulation, which allocates particular land uses geographically, is to internalize any spillover effects arising from nuisance land uses. The spatial allocation of land uses achieved by zoning removes or reduces these externalities, increasing land values in the residential sector (and perhaps in nonresidential uses as well).

Since, however, most locally raised revenues are derived from property taxes, the fiscal motive for zoning in suburban jurisdictions may be quite strong. If public services are provided on a basis of rough equality per household, local authorities have an incentive to insure that the marginal dwelling provides more housing services (and hence local property tax revenues) than the average house. Thus, in practice, zoning regulations often specify minimum lot sizes or floor areas for single-family housing and regulate or prohibit multifamily dwellings.

The effect of such regulation on housing costs per unit of output depends upon the impact of local ordinances on the cost of land as an input into housing, as well as any additional administrative or holding costs incurred. If zoning does reduce the allocation of land to residential uses, then raw land costs may be expected to rise.

Theoretical analyses of the effect of zoning upon land allocation and input prices to housing have been undertaken by Burstein (1975), Stull (1974), Hamilton (1976), and Ohls et al. (1974). Not surprisingly, the impact of zoning upon raw land prices depends upon the amounts of devlopable land in residential and nonresidential sectors, the demand for development in alternative uses, and, most importantly, on the substitutability of demand across civil divisions with differing regulations. To the extent that the metropolitanwide system of land use regulation does reduce the supply of developable land relative to demand, we may expect prices of land inputs into housing to increase.

Empirical evidence on the effect of zoning regulation on land prices is broadly consistent with the hypothesis of land price escalation. Numerous studies have concluded that zoning ordinances increase the value of otherwise identical dwellings. In many cases, however, this effect of zoning may be attributable to the externality impact of regulation.[20]

Of more importance to the supply cost of new housing, however, is the effect of density restrictions on the price of vacant land or new housing. Sagalyn and Sternlieb's (1972) analysis concluded that large-lot (low-density) zoning increased the unit price of land for new single-family housing built in New Jersey suburbs. Gleeson's (1979) analysis of Brooklyn Park, Minnesota, estimated that two-thirds of the intracity variation in land prices (about $1,500 per acre) was attributable to zoning designation and density restriction. Peterson's (1974) analysis of northern Virginia suburbs found that density restrictions had a significant and quite large effect upon land prices. Peterson's results are consistent with a land price premium in response to zoning restrictions which varies with accessibility to downtown. At a distance of 10 miles from Washington, D.C. (Fairfax County, Va.), for example, parcels zoned one-half, 1, 2, and 10 units per acre were selling for $5,800, $7,900, $13,700, and $32,000 per acre, respectively, in 1974.

Reliance upon complex environmental and growth-management programs has increased substantially in the past decade. For example, in 1973, one jurisdiction in the San Francisco-Oakland area had growth-control regulations; three years later 31 civil divisions had such regulations (Dowall, 1980). Dowall reports an increase of 1,200 percent in the number of communities imposing environmental and/or growth management restrictions in California during the period 1972-1977 (Dowall, 1980, p. 113).

Growth-control and environmental management programs include "open space" set asides, growth timing ordinances, urban service areas, permit limitations, building moratoriums, and environmental impact review and compliance procedures. Localities typically justify these controls in terms of the benefits of environmental quality, lower municipal service and capital costs, lower property taxes, and the preservation of community "character."

Ellickson's (1977) analysis of growth management restrictions is conceptually similar to the analysis of zoning. He concludes that any effective growth management policy is likely to reduce the supply of new construction, to increase the price of vacant land, and to increase values of existing properties. Some empirical evidence is available on the magnitude of price increases. Case studies of San Jose, Santa Rosa, and Petaluma, California, all conclude that the prices of existing standardized dwelling units have increased with the adoption of such ordinances (Katz and Rosen, 1980). More important for our purposes is the effect of such tools on the supply prices of newly constructed dwellings. The San Jose analysis estimates that during the 1968-1976 period, the price of one builder's standard unit increased by 121 percent, and 43 percent of this increase was attributable to growth control (Katz and Rosen, 1980, p. 26).

Clearly, the effect of such restrictions varies with the metropolitanwide level of their imposition and with the level of demand for new units. Thus, it is worth noting that a recent survey of the San Francisco area, where housing demand has been increasing rapidly, indicates that half of the jurisdictions surveyed had imposed some type of absolute moratorium on new construction at some point since 1970 (Gabriel et al., 1980).

In addition to the effects of such ordinances on land input prices in construction, there may be substantial administrative and carrying costs imposed on developers and builders by such regulation. For example, Frieden (1979) reports that, by 1965, more than half the states imposed some form of environmental impact review for new construction. The environmental impact statement is typically the responsibility of the developer and is often prepared by consultants engaged by the developer. If the developer has purchased the land and has engaged in planning studies (as Frieden claims is typical), then a lengthy review process imposes overhead and property tax costs as well as the carrying costs for land. James and Mueller (1977) estimate that the costs of report preparation and time delays amount to 4-7 percent of total cost of new units (Dowall, 1980). In Hawaii, comparable figures are $325-$450 per unit per month of delay (Rands et al., 1980). Delay costs for Edmonton were estimated at $700-$900 per month.

Subdivision regulations can also increase the unit costs of producing new housing. Subdivision ordinances often require a complex package of off-site investments by developers, including streets, paths, lighting, landscaping, and sewers. For the San Francisco metropolitan area, Rands et al. (1980) report a range of development fees of $800 to $5,919 for a single detached unit in 1979 and a range of $3,948 to $15,301 for a seven-unit, multifamily dwelling. In the San Francisco area, Gabriel et al. (1980) report that median development fees

were \$1,907 per unit in 1979. Rands et al. (1980) report a
median development fee for single-family houses of \$2,800 (or
3.5 percent of median new home prices). The private provision of public open space, bike paths, bus shelters, parking,
and lighting are often the rule.

Finally, there is some evidence on the costs of delays
implied by development review procedures. It is estimated
that, in Houston, the process adds between \$400 and \$600 to
the cost per dwelling unit (Dowall, 1980).

The net effect of this pattern of local regulation upon
efficiency in the production of a standardized unit of residential services depends upon several factors.

First, to the extent that zoning removes or mitigates
harmful externalities, increases in land values reflect higher
levels of residential services consumed.

Second, to the extent that zoning is successful, new
housing costs per unit of service are increased and resources
are redistributed toward owners of preexisting residential
capital.

Third, to the extent that environmental and subdivision
regulations increase land and development costs by providing
additional amenities in accordance with willingness to pay,
output of residential services is increased.

Fourth, to the extent that these regulations add costs
beyond those required for health and safety, or beyond those
reflected in consumers' evaluations, they increase unit housing
costs. It has been frequently alleged that the overall effect of
these latter regulations is highly inefficient; indeed, one
estimate suggests that "unnecessary improvements" increased
development costs by almost \$900 per unit or about 2.5 percent
in northern New Jersey (Seidel, 1978).

Fifth and last, in residential construction, interest costs
and carrying charges are enormously important. Thus the real
costs imposed by delays in lengthy compliance reviews and
increases in the elapsed time of production add to the unit
cost of new housing services and are deadweight losses to
society.

Building codes

Despite the existence of a model building code (or perhaps
owing to the existence of at least five "model" building codes),
there is only a modest level of uniformity among the approximately 8,000 local ordinances that set standards for the
construction of residential housing. In addition to differences
among the codes themselves, there are differences in the
administrative application, enforcement procedures, and the
discretion given to building officials, as well as the avenues of
appeal to review boards and arbitrators. Local building codes
include three types of information: definitions, licensing re-

quirements, and standards. Definitions specify, for example, what constitutes plumbing, while licensing provisions specify who may install plumbing. Finally, standards specify the minimum quality or physical characteristics of plumbing materials or their performance characteristics.

One role of local building ordinances, therefore, in addition to the promotion of health and safety, is the promotion of job security or competition among labor groups. In addition, however, local codes ratify innovative activity by permitting new techniques, materials, or equipment to be used in construction. For the evaluation of new products and techniques, testing laboratories (such as Underwriters' Laboratories) play a key role, but no testing results are binding. Thus, approval by any testing laboratory need not imply product acceptance by any jurisdiction. The difficulty of specifying performance standards instead of input standards means that the innovator must, in principle, submit his product for testing at the local level. The criteria for acceptance may vary with the statutory provisions of the code and with the competence of local officials. As a result, it may be a long time before a cost-saving or quality-enhancing innovation achieves wide usage in the market.

A number of states, however, have adopted mandatory state codes for some types of construction. For 11 years, the state of Connecticut, for example, has had a uniform code, and has required that local officials be certified by the state. There is, however, considerable anecdotal evidence that enforcement is far from uniform. It should also be noted that some strides in uniformity of state codes have been made in the area of industrialized and prefabricated parts. For example, in California, a prefabricated unit that receives certification under state law at the factory is deemed to satisfy any local requirements in the state.

Nevertheless, to the extent that the pattern of permissable materials and techniques at the local level lags behind best-practice technology, increased unit costs of housing result.

There is conflicting evidence on the magnitude of excess costs attributable to variations in building codes. Several studies have suggested that the direct effect of building codes upon construction costs is small. For example, Maisel's (1953) early study of the San Francisco housing market concluded that an increase of less than 1 percent in the costs of newly constructed housing was attributable to "known code inefficiencies."

Burns and Mittelbach (1968), in their report to the Kaiser Committee, analyzed a survey conducted by House and Home (the leading trade journal) in 1958, and suggested that if the ten most "wasteful practices" required by building codes were eliminated, the average cost saving for single-family housing would be from 5 to 7.5 percent. "By assuming the provisions [of building codes] are randomly distributed and by taking account of their varying role in communities," the authors conclude that "the estimates represent from 1.5 to 3 percent of the price of an average house" (Burns and Mittelbach, 1968, p. 102).

Several other analysts have come to different conclusions, however. In expert testimony presented to the Kaiser Committee, Johnson (1968, p. 57) concludes that "in large urban areas, it may be possible to achieve on the order of a 10 to 15 percent reduction in direct construction costs [or 5 to 8.25 percent of selling price by Johnson's calculations] . . . if the constraints of codes and restrictive labor practices are removed and if the industry is allowed to produce as efficiently as it knows how." Survey evidence gathered by the Douglas Commission (1968, p. 262) indicated some real cost reductions achievable by mass production under more uniform building codes. The estimates indicated that if 21 "excessive requirements" - not all of which are necessarily in effect in any particular jurisdiction - were eliminated, $1,838 would be cut from a typical $12,000 FHA-insured house. This represents a 15.3 percent reduction in construction cost (or roughly 13 percent in sales price, if one-fifth of the selling price is the land component). The commission report also noted the problems of one home manufacturer who estimated that producing a standard product acceptable to the jurisdictions within his six-state market area would increase costs by $2,492 or almost 21 percent.

Information on the cost increases attributable to excessive code provisions gathered more recently is also inclusive. On the one hand, Muth and Wetzler (1976) presented regression estimates relating prices for newly constructed dwellings to a dummy variable indicating a locally modified building code. Their results suggest that the average effect of local code variation on housing prices is only about 2 percent. On the other hand, Babcock and Bosselman (1973), on the basis of interviews with builders in Ohio, concluded that codes could more than double the cost of producing residential structures.

A recent study of housing costs, undertaken by the U.S. General Accounting Office (1978), identified 64 specific building practices or methods as "unnecessarily costly," in comparison with methods or materials both less expensive and acceptable under HUD's Minimum Property Standards. The GAO then surveyed the building code restrictions of 87 jurisdictions, and estimated a median savings of $1,700 on construction costs - which could be achieved if all practices were permitted in each jurisdiction. Potential savings varied widely throughout the sample jurisdictions, from $0 to $7,300; potential savings also varied widely for jurisdictions within the same metropolitan area (from $500 to $3,100). The most widely

prohibited "best-practice" methods were: 3" (rather than 4")
concrete basement floors, a savings of $141 (prohibited by 58
of the 64 communities); 2" x 4" studs spaced 24" (rather than
16") on exterior walls, $119 (56 communities); no exterior
sheathing, $255 (50 communities); plastic piping in hot or cold
water supply (rather than copper), $130 (49 communities).

An analysis of the diffusion of innovation in homebuilding
was undertaken by Oster and Quigley (1977). For a sample of
jurisdictions, they considered the provisions of local codes that
permitted or barred a number of "best-practice" methods or
techniques (including, for example, 2" x 3" studs and 24 inch
framing in nonload-bearing partitions). Their analysis indicated that many proxies for the competence of local officials
and for the importance of local interest groups affected the
speed of diffusion greatly. In an earlier version of this paper
(Oster and Quigley, 1976), they estimated logistic diffusion
paths for several innovations. These curves suggested that
the interval between the year when 10 percent of jurisdictions
permit an innovation and the year when 90 percent grant permission may be as long as 30 years.

More important than the static excess cost inefficiencies of
building regulation, therefore, may be the dynamic effects of
these barriers upon the aggregate level of R&D effort and its
allocation. With relatively long payback periods and with
important local interests at stake, the ex ante profitability of
research in building materials is probably reduced when compared to other research activities, and the allocation of
activity between labor-saving and capital-saving innovation may
be affected.

It is difficult to estimate the aggregate effect of these
types and patterns of local regulation upon the supply cost of
housing. To some extent, the overall pattern of these regulations, no doubt, promotes health and safety or reflects
willingness to pay for improved housing services. To that
extent, associated increases in housing costs represent, not
inefficiency but increased output of housing services. To a
large extent, however, these regulatory patterns represent
attempts at redistribution from new residents and/or construction firms to owners of existing properties or to other local
interests, such as craft labor.

To the extent that this redistribution is successful, it
increases construction costs and generates additional losses
through excess carrying costs. Finally, it may affect both the
level and distribution of private research and development
activity.

Federal Support of R&D Activity

As late as 1960, the Housing and Home Finance Agency, the direct predecessor to the Department of Housing and Urban Development, had an annual research budget of $15,000 (Nelkin, 1971, p. 76). The Building Research Advisory Board (BRAB) had been in existence for 11 years. BRAB, a committee of the National Research Council (NAS) had been established in 1949 as a nongovernmental agency to stimulate and coordinate research and technology in the construction industry. One reason for BRAB's establishment, it is asserted, was to limit any federal role in housing research contemplated as a result of the 1949 Housing Act (Nelkin, 1971, p. 22). The 1949 Housing Act has authorized research on housing codes and technology, but following industrial opposition, appropriations were suspended in 1953. By 1960, some small fraction of the activities of the National Bureau of Standards was also devoted to building-related activities.

In 1962, the Civilian Industrial Technology Program (CITP) was proposed by the Kennedy administration – a Department of Commerce effort to foster technical change in lagging industries, notably housing and textiles. Congressional and industry opposition prevented the CITP program from being adopted, but from BRAB's opposition to CITP came a proposal for an expanded role for building research in the National Bureau of Standards (NBS). The present Center for Building Technology, a division of the Insitute for Applied Technology, NBS, is a descendent of the BRAB proposal.

The Center for Building Technology is the closest thing to a U.S. national research laboratory for the construction and housing industries, analogous to national laboratories in Scandinavia, France, and England. The principal difference is that the U.S. testing facility in NBS has no authority to promulgate or enforce standards itself. In 1978 the Center employed a staff of 250, including 170 professionals, had a budget of $14 million, and was engaged in a limited variety of testing and research activities (Willis, 1979, p. 89).

Currently, the standards evaluation and testing role of NBS is supplemented by the National Institute of Building Sciences, a nongovernment advisory board authorized by the Housing Act of 1974, but not established until 1976 (P.W. Cooke, 1976).

Before the establishment of HUD in 1965, federal research on building technology was virtually nonexistent. By 1969, HUD's research budget was less than $.5 million; in 1970 it increased twentyfold, and by 1980, it reached a level of $53 million.(21) Only a small fraction of these funds are allocated to building research, per se. In FY 1977, for example, the largest fraction of HUD's research budget was spent on housing assistance research (principally on housing allowances themselves and on analyses of the behavior of recipients); 17 percent was allocated to community development and neighborhood preservation research, and 11 percent was spent on state and local government research. Roughly a quarter of the budget is spent on housing energy conservation, safety, standards, management, and maintenance research.

Table 7.8 indicates the level and distribution of HUD-administered federal research funds from FY 1974 through 1980. HUD-sponsored research has declined modestly in nominal terms, more substantially in real terms, during the recent period. In contrast to other federal research activities, housing research has represented 0.22 to 0.25 percent of federal research funds. The HUD research budget is roughly 10 percent of the Department of Agriculture research budget;

Table 7.8.   Level and Distribution of HUD-Administered Federal Research Funds, 1974-1980

(millions of current dollars)

	1974	1975	1976*	1977*	1978	1979*	1980*
Housing assistance research	$16.2	$15.6	$15.6	$15.8	$12.6	$9.8	$9.8
Safety and standards	2.9	4.1	4.8	6.1	3.7	3.2	2.9
State and local government and research	7.8	8.1	5.4	8.6			
Program evaluation and support		2.7	3.8	4.3	5.3	5.9	8.6
Other HUD research	33.3	36.6	32.3	36.2	39.7	39.0	31.7
Total HUD research	$60.2	$56.6	$61.9	$71.0	$61.3	$57.9	$53.0
Energy conservation and standards (DOE transfer)					32.5	21.7	6.4

*Estimated.

Source: Department of Housing and Urban Development, HUD Statistical Yearbook (Washington, D.C.: U.S. Government Printing Office, 1974-1980).

the Department of Defense research budget is about 20 times as large.(22)

Of course, the HUD research budget does not represent the only federal resources devoted to residential construction technology. As noted in Table 7.8, substantial research on residential construction is funded by the Department of Energy and more limited research is sponsored by the Department of Defense (and the Corps of Engineers), as well as OSHA, EPA, CPSC, GSA, and the National Science Foundation (P.W. Cooke, 1976, pp. 253-259). The exact split between basic and applied research, between research on techniques and regulation is unknown, and in contrast to most Western European nations, there is no centralization of research activity.

It is instructive to consider the major attempt by the federal government to foster an improved production technology, to rationalize regulatory standards, and to create a more stable environment for residential construction.

Operation Breakthrough (23)

The housing act of 1968 expressed as a goal the completion of 26 million additional dwelling units in a ten-year period, an average annual figure that was 40 percent larger than the average annual number of housing completions during the previous 15 years. In response to the report of the Douglas Commission, which had included optimistic projections on the possibilities for industrialized, mass produced housing, the act included Section 108 to "encourage the use of new [construction] technologies." This section authorized the secretary to select plans for the development of housing using new technologies, to construct at least 5,000 dwellings a year for five years using five different technologies, to evaluate the technologies, and to report the findings to Congress.

Governor George Romney became Secretary of HUD in January 1969, without a program, but with a clear mandate from the previous Congress to increase the supply of housing quickly. "Operation Breakthrough" was announced at a press conference in May 1969 and formed the basis for much of the new secretary's testimony before the Senate that month.(24) Section 108 of the housing act had been written rather narrowly; it was intended to test whether economies of scale existed for certain promising technologies, and to report the results to Congress. According to the secretary's testimony, however, the design of Operation Breakthrough was an attempt to use off-site factory methods - "new technologies" - to increase aggregate housing production rapidly. Such a rapid increase in production required a thorough understanding of institutional factors - the cyclical nature of demand and the pattern of regulation - as well as a successful test for the presence of economies of scale along the way.

Operation Breakthrough "attempted to increase the efficiency of the market mechanism for housing output by reducing the institutional barriers among the various segments of the industry (localized building codes, zoning laws, etc.). Such action was ultimately intended to increase the market incentives for privately funded R&D, the results of which would permit the industry to respond in a timely and appropriate fashion to [secular] changes in supply or demand conditions. The breakthrough program gave heaviest emphasis to . . . the more specific R&D policy category" (C. Cook, 1976, p. L-28).

Operation Breakthrough would be implemented in three phases: Phase I, Design and Development, on cost-plus contracts with an expected duration of 2-4 months; Phase II, Prototype Completion, also on cost-plus contracts with production in another 12 months; and Phase III, Volume Production, to last indefinitely.

Initially, about 1,000 design prototypes developed during Phase I were to be constructed during Phase II on widely varying geographic sites. These prototypes would serve as sales models for Phase III production. During this period as well, NBS would conduct laboratory and field tests to verfiy the acceptability of design prototypes. Certificates of acceptance would be issued, and the producers would then manufacture their systems for sale at a private profit. Originally, each producer would install five to seven different housing systems to increase the chances of successful marketability.

Phase II construction required the selection of site planners, site developers, and site locations, as well as the selection of housing manufacturers. In addition, during Phase II, HUD would support state and local studies to identify sites for full-scale production.

Note the design of this ambitious program. It would not be until several years after volume production had been underway that the congressional mandate (to test economies of scale in the market) would have been fulfilled. Note also that the Operation Breakthrough program originally planned to subsidize only 1,000 units before beginning volume production. Section 108 authorized instead a test of 25,000 subsidized units before submitting a feasibility report to Congress.

Apparently, Operation Breakthrough, as originally conceived, would produce: houses; and factories to produce houses; and institutional regulatory reform; and research and development of new technologies; and, in addition, would provide a demonstration. Within HUD, the Office of Research and Technology was elevated to assistant-secretary level and two former NASA officials were recruited to the program. The Research and Technology office emphasized community development, analysis of the entire delivery system, and the potential for modern management techniques.

Phase I Requests for Proposals (RFPs) were issued in June 1969. The RFP stated: "Operation Breakthrough has as its primary objective the establishment of self-sustaining mechanisms for rapid, volume production of marketable housing at progressively lower costs for people of all income levels, with particular emphasis on those groups and individuals which have had difficulty in obtaining satisfactory housing in the past."(25)

Firms had 90 days to respond. More than 600 proposals were submitted (instead of the 50 to 100 that were expected), and HUD had 5 months to evaluate their technical and cost characteristics.(26) The 22 winning firms, announced in February 1970, included several firms new to the housing industry (e.g., Republic Steel) and four aerospace contractors. Ten of the systems selected were of modular design, nine were panel designs, and three used component assemblies.

Eleven sites were selected for Phase II in response to 218 proposed by communities in 36 states. With the exception of New England, they represented broad geographical coverage. Funding cutbacks subsequently eliminated two of these sites. Finally, eleven site planners and developers were selected by June 1970.

At Secretary Romney's request, the appropriations of the Office of Research and Technology were increased twentyfold, from $.5 million to $10 million. Policy decisions to emphasize integrated community development increased design and evaluation costs for a fixed Operation Breakthrough budget of $60 million.(27)

With three months to respond to the Phase I RFP, it was clear that potential entrants were forced to rely on "off-the-shelf" technologies, which would then be tested and refined during the two-to-four month development effort. The development of evaluative criteria was entirely HUD's responsibility, since HUD's certificate of acceptance would certify health, safety, habitability, and (perhaps implicitly) marketability of the dwellings.

A hard-nosed decision to design appropriate performance specifications and to conduct tests relative to performance was required if the prototypes were to be marketed at all in other localities with restrictive code provisions, and if subsequent R&D was to be stimulated. This proved to be a difficult undertaking, requiring time and money, and ultimately the redesign of more than half of the prototype plans. Phase I was scheduled for completion by August 1970, but was not in fact completed until one year later.

After issuing the RFP for Phase I, HUD contracted with the National Bureau of Standards for the development of criteria to review, evaluate, and finally to accept the developed housing systems. The NBS development of performance-based codes was reported in five volumes in December 1970. These

codes, "Guide Criteria," contained novel provisions concerning
the habitability and durability of dwellings. The performance
standards in the codes necessitated some "reasonable engineering judgments" (much as building codes themselves often do in
practice). Some ambiguity was introduced between the devel-
oper/designer interpretations and the NBS interpretation. The
National Academy of Sciences (1973, pp. 7-8) review of the
Guide Criteria indicated that the innovative set of standards
was "not written in the format and terminology customarily
used in, and felt by many to be essential to, legally enforce-
able codes." Moreover, "while many of the Guide Criteria
provisions . . . could serve as a basis for furthering the
development of existing regulatory documents, the process of
translation and selection would have to be done with professional care if existing processes were not to be unnecessarily
complicated." Finally, the review panel (National Academy of
Science, 1973a, p. 7) concluded that "the Guide Criteria contains [sic] certain provisions that are not sufficiently clear
and precise to avoid the need for excessive interpretation by
the reader."

More important than the technical ambiguity of the standards, however, were the ambiguities introduced into the
interpretations by FHA underwriters and potential leaders.

As precious time was lost during the initial phase (and as
it was feared that momentum would be lost with further delays), the strategy of parallel R&D was introduced. Parallel
R&D had been successfully employed for a decade at NASA in
producing pure hardware.

Apparently the strategy of parallel R&D proved very
costly. Four divisions - technical, site planning, "market
aggregation" (i.e., subsequent marketing under Phase III),
and financing - each conducted development activities simultaneously. The relationships among these activities were not well
defined ex ante; the implications of alternative development
activities in any one division for the other three divisions were
hardly understood. As a result, valuable "time was used up
redesigning housing systems and reallocating them across sites
to meet financial commitments arranged before the sites and
systems had been completely designed and evaluated" (C.
Cook, 1976, p. L-61).

Substantive changes had to be made in more than half of
the housing systems, which increased costs and removed innovative components. More importantly, these changes left little
time for dispassionate evaluation of the redesigned systems.
By the time the implications of these changes were understood,
it was simply too late; site development and mortgage financing
for Phase II had been locked in.

Phase II contracts were signed with 21 of the 22 building
firms and with the site developers. For legal reasons, these
contracts were ultimately negotiated on a fixed-fee basis,

which increased the risk to manufacturers. More importantly, however, HUD was in the position of being unable legally to acquire any comparative cost data from Phase II.

Given budgetary realities, Phase II could only be financed by private mortgage financing backed by the FHA. The FHA had already seen its primacy within HUD eclipsed by the elevation of the Office of Technology and Research. The assistant secretary for Housing Production and Mortgage Credit, a former president of the National Association of Home Builders, allegedly interpreted Operation Breakthrough as an attempt to "federalize" residential construction.(28) In addition, however, housing components were designed to the new "Guide Criterion" performance standards, not the input-related (and FHA-established) Minimum Property Standards (MPS). Finally, the MPS were themselves under review, and many in the industry were quite nervous that MPS would be replaced, in an instant, by the NBS performance criteria.

Industry comment on the performance standards indicated extreme concern about "the possible adoption of the Guide Criteria as a whole or of its more controversial provisions for regulatory purposes." Industry spokesmen also objected to "the lack of opportunity provided during development of the Guide Criteria for contribution of knowledge and advice from the many capable individuals outside the federal government, including those who might be most affected by the use of the Guide Criteria" (National Academy of Science, 1973b, p. 9).

In any case, applications for the financing of Phase II and Phase III were not expedited at local HUD/FHA offices. FHA financing arrangements required complicated, lengthy, and costly procedures.

The first Phase II prototype (in Kalamazoo, the secretary's home state) was completed in March 1972. Most of the other sites were about a year behind schedule. At this point, given the lengthy delays and the loss of momentum, it was decided to permit Phase II and Phase III operations in tandem, subject to the condition that Phase II prototypes be "sufficiently advanced." According to the NAS (1974, pp. 40-56) review, levels of quality assurance were quite low, especially when compared to the design tests which had been imposed in Phase I.

As Phase II and Phase III proceeded in tandem, federal rent subsidies and Section 236 subsidies were offered for Phase III units to speed production of Phase II prototypes. For the 17 producers who intended to proceed to Phase III the inducement to complete the prototypes was quite strong. Section 236 set-asides of 1,000 units per producer were offered. For the other four producers, this provided no added inducement to complete the "experiment" in a timely fashion.

As a result of the difficulties with the FHA, HUD authorized the redesign of Phase III to accomodate local building codes and the MPS regulations.

On January 16, 1973, President Nixon imposed an indefinite moratorium upon new allocations of Section 236 subsidy moneys.

The rest, as they say, is "history." The original allocation of 1,000 subsidy units per producer was honored, but no additional units were authorized. Producers were forced to substitute "standard" components and procedures for "innovative technologies" to comply with MPS, at increased site and off-site costs.

In all, about 25,000 Phase III units were completed in 150 different developments using Section 236 set-asides. Only 1,500 units were completed for unsubsidized occupancy at market interest rates (Real Estate Research Co., 1976).

By 1976, only 5 of the 22 Operation Breakthrough housing systems were still being marketed by their manufacturers (U.S. General Accounting Office, Nov. 2, 1976).

As of 1977, less than 7,000 innovative units had been marketed outside of Operation Breakthrough by these firms at market interest rates.

No factory came close to completing a single volume run.

The cost to the federal government was $72 million, or $12 million more than had initially been budgeted.

IV.   LESSONS FOR PUBLIC POLICY

The analysis presented in this chapter documents the course of technical progress in the production of residential dwellings in the postwar period. The evidence indicates that there have been a large number of individual changes in the production process - changes that have increased the quality of new construction at given costs or have reduced the cost of dwellings of given quality. The course of technical progress in residential construction has been slow and incremental in nature. No single improvement in material or method has altered basic construction techniques or construction costs in a dramatic way. In comparison with other sectors of the economy, especially the manufacturing and agricultural sectors, the rate of technical progress in house building has been slower, and aggregate resources devoted to research and development have been smaller. The overall level of public support for innovation has been quite modest and it appears that the ex ante profitability of innovation in the private sector is equally small, especially for innovations in building techniques.

In part, this pattern of technical progress may arise from factors inherent in the industrial processes. Labor and capital are small fractions of the total costs of producing new dwellings and are a smaller fraction of the occupancy costs by prospective consumers. In part, observed rates of technical progress may be attributed to the low capitalization of the industry and the predominance (at least historically) of small firms producing at low volumes. This pattern of market organization is seen to be a quite economical reaction to the sensitivity of the industry to monetary conditions and to cyclicality in the demand for newly constructed units.

In part, the structure of the industry and its lower level of capital intensity are an indirect consequence of aggregate financial policies and the regulation of depository institutions. As these federal regulatory constraints are removed over the next few years, one source of the cyclical sensitivity of new construction will be eliminated. Nevertheless, even when housing construction is able to compete equally with other investment activities in the credit market, the level of new construction will surely remain sensitive to aggregate economic conditions. Unless housing construction activity were subject to explicit legislative control, or specific "counter-counter-cyclical" subsidy policies, the level of industrial activity will remain sensitive to monetary policy. It is, of course, hard to argue that productivity considerations in the housing sector are sufficiently important to warrant special treatment for housing during periods of monetray restraint.

The supply and occupancy costs of new housing have escalated as raw land prices have increased, as property taxes have risen, and as developers have assumed more of the responsibility for provision of infrastructure. In part, these increases in the costs of output have arisen from the conscious intent of local regulation based upon fiscal and "environmental" concerns. To the extent that metropolitanwide land use controls are effective, the supply of land available for residential development is reduced and the input cost is raised. Further, to the extent that environmental regulation and subdivision restrictions affect the development process, there are three potential effects: (1) a possible wealth transfer from new residents to prior residents; (2) a possible increase in the residential quality of newly constructed dwellings; and (3) a potential for large deadweight losses from time delays, bureaucratic negotiation, and from the resulting "amenities" of new dwellings that exceed willingness to pay. Many developers do not hesitate to point out another consequence of a regulatory pattern that increases the costs and reduces the supply of newly constructed dwellings: monopoly rents to the builder who can offer newly constructed units on the market.

Given the importance of property taxes in local government finance and the land use powers delegated to local governments, there is little reason to expect that these trends will be reversed. To the extent that reliance upon local property taxes can be reduced, we may expect reductions in the cost of newly constructed dwellings. The principal arguments for decreased reliance on the property tax, however, rest quite

properly on other grounds - progressivity of the tax structure, economy of administration, horizonal equity across
individuals in different communities, and so on. Reduced
housing prices are merely a small, indirect part of this larger
issue.

It should be noted that the prohibition of land use
restrictions, density requirements, and the like, by local
governments would have direct consequences in terms of stimulating housing output, reducing costs, and promoting economically integrated local communities. It is highly unlikely, of
course, that these powers of local government will be withdrawn. It is worth noting, however, that any weakening of
these restrictions would have favorable effects upon housing -
mandatory "fair-share" zoning requirements, required compensatory payments to land owners whose properties have been
"upzoned," etc.

National experience with the federal "Operation Breakthrough" program provides several lessons for policy design.

First, despite the failure of the program to make rapid
advances in construction technology or to increase housing
output in a significant way, the program provided some tangi-
ble benefits - some improvement in coordination of building
codes, especially preemptive legislation certifying manufactured
housing components in several states; greater uniformity in
state regulations governing transport of large and bulky loads,
etc.

Second, the history of the program demonstrates the
difficulty of conducting a research, demonstration, and production program simultaneously, especially one that requires a
commitment over a reasonably long period of time. In part,
the program illustrates the difficulties of coordinating the many
interests in the building process, especially in the face of
technical and financial uncertainties. In part, it illustrates
the difficulty of providing a political commitment to housing
production or innovation over an extended period, encompass-
ing macroeconomic change in the economy.

It does appear, however, that the history of Operation
Breakthrough says practically nothing about the potential for
rapid technical change in residential construction, or about the
role of federal policy in fostering innovation. It would seem to
be quite mistaken to interpret this history as evidence that a
substantial public commitment to technical change in housing is
doomed to "failure." On the contrary, the disappointing
results of Operation Breakthrough say nothing about the innovative and experimental program authorized by Section 108
of the 1968 Housing Act. Such a program was never attempted, and it remains to be seen whether federally sponsored R&D
in construction technique can lead to cost savings and marketable innovations. A federal commitment to produce, say, 1,000
experimental dwelling units per year for 10 years would be

relatively cheap, even when viewed as a fraction of federally financed new construction under existing programs. A long-term commitment would stimulate competitors to plan in advance and to submit truly experimental designs. Public sponsorship would facilitate the dissemination of experience and the diffusion of successful techniques. If combined with performance testing by a federal laboratory, described earlier, with statutory authority to certify improvements as meeting code standards, such a federal program could increase the potential market for improvements in methods and materials.

The experience of Operation Breakthrough does indicate that two distinctions are critical to the desgin of public policies to improve productivity in residential construction. First, policies to stimulate innovation in design, materials, or technique must be distinguished from the longer-run objectives of volume production for the market. A program that combines R&D with production goals is unlikely to achieve either objective. The realities of producing a significant numerical target of innovative dwellings in a timely fashion are such that potential innovations are unlikely to be developed fully; the complex relationships among physical design components in the structure and among important interest groups in the market dictate that the implementation of R&D results proceed cautiously. Second, policies that foster cost reductions in producing housing must be interpreted more broadly than those that foster factory production. Since the federal government itself purchases an insignificant number of dwellings, any guarantee of the scale of output required for economical factory production must come from market forces.

Thus, a small-scale federal program of intensive R&D would not answer two questions important to builders, developers, and components manufacturers. It would not establish the extent of scale economies in the production of innovative housing and it would not establish the acceptance by consumers of design or materials innovations. It would seem, however, that the profitability of housing innovations in the aggregate is best established by entrepreneurs in the marketplace.

NOTES

1. See National Commission on Urban Problems (Douglas Commission), Building the American City (Washington, D.C.: U.S. Government Printing Office, 1969).

2. The goal of a "decent home" was originally espoused by the Housing Act of 1949; similar language appears in the acts of 1954 and 1959.

3. See, for example, Federal Reserve Bulletin 65, (January 1979).

4. The Professional Builder as cited in U.S. Department of Housing and Urban Development, Housing in the Seventies (October 1973), pp. 7-20.

5. U.S. Department of Housing and Urban Development, (October 1973), pp. 7-15.

6. "U.S. Homes Management Revolution," Fortune 98, 7-13 (December 4, 1978): 68-78.

7. These mortgages are currently rated AAA by Standard and Poors.

8. See "U.S. Homes' Management Revolution," Fortune (December 4, 1978), pp. 68-78; and "Defying those Interest Rates," Fortune (November 3, 1980), pp. 6-9.

9. U.S. Department of Housing and Urban Development, Housing in the Seventies, chapter 7.

10. Both these statistics are cited without attribution in at least two sources: U.S. Department of Housing and Urban Development, October 1973, pp. 7-23; and Johnson (1968).

11. Much of this discussion is based upon B.A. Gillis, Research assistant to author, "Interview Notes," with Ron J. Morony (HUD), March 11, 1980; Lee Fisher (NAHB), March 13, 1980; John Eberhard (NBS), March 13, 1980; Tom Faison (NBS), March 13-15, 1980; Joan Finch (BRAB), March 12, 1980.

12. Interviews with Morony and Faison, and Fisher.

13. U.S. Department of Housing and Urban Development (October 1973), pp. 7-24.

14. U.S. Department of Labor (1973).

15. National Science Foundation (1976).

16. U.S. Department of Housing and Urban Development, October 1973; see also Falk, 1976.

17. These implications of regulatory reform are discussed in "The Depository Institutions Deregulation and Monetary Control Act of 1980," Federal Reserve Bulletin (June 1980), pp. 444-453.

18. U.S. Department of Housing and Urban Development, (October 1973), pp. 7-17.

19. A more detailed treatment of some of these issues appears in Katz and Rosen, 1980, and Dowell, 1980.

20. For example, Stull's (1974) analysis of jurisdictions in the Boston metropolitan area indicated that house values were

higher, ceteris paribus, in jurisdictions limiting or excluding various forms of nonresidential activity. Lafferty and French (1976) and Peterson (1974) report similar results; in all cases the methodologies relate the value of single-family housing to a collection of housing, neighborhood, and (in some cases) public-service characteristics - and measure of the zoning attributes of jurisdictions.

These qualitative results of the impact of zoning are by no means unanimous, however. For example, Maser et al. (1977) found no evidence that zoning affected the allocation or price of resources in upper New York state.

21. U.S. Department of Housing and Urban Development (1980).

22. National Science Foundation (1976).

23. Much of this material has been taken from four sources: National Academy of Sciences (1974); C. Cook (1976); Real Estate Research Co, et al. (1976); and U.S. General Accounting Office (1976).

24. See U.S. House of Representatives (1969).

25. U.S. Department of Housing and Urban Development (June 23, 1969), p. C-2.

26. Many of these designs are presented and discussed in U.S. Department of Housing and Urban Development (1970); and U.S. Department of Housing and Urban Development (n.d.).

27. See C. Cook (1976) and National Academy of Sciences (1974) for a discussion of these budgetary decisions. The final cost of Operation Breakthrough was $72 million.

28. This bureacratic tension is discussed explicitly in C.C. Cook (1976, pp. L-70-72) and implicitly in Real Estate Research Co. et al. (1976).

REFERENCES

Alterman, Jack and Eva E. Jacobs, "Estimates of Real Products in the United States 1947-1955." Input, Output and Productivity Measurement (New York: National Bureau of Economic Research, 1961), pp. 246-249.

Babcock, R. F. and F. P. Bosselman, Exclusionary Zoning (New York: Praeger Publishers, 1973).

Baer, William et al., "An Analysis of Federally Funded Demonstration Projects." R-1927-DOC. (Santa Monica: The Rand Corporation, April 1976).

Ball, Robert and Harry Ludwig, "Labor Requirements for Construction of Single-Family Houses." Monthly Labor Review (September 1971), pp. 12-14.

Bluebook of Major Homebuilders (CMR Associates, 1973).

Burns, Leland S. and Frank Mittelbach, "Efficiency in the Housing Industry." The Report of the President's Committee on Urban Housing: Technical Studies, Vol. II (Washington, D.C.: U.S. Government Printing Office, 1968), pp. 75-144.

Burstein, Nancy. "The Interaction of Zoning and Local Public Finance." Ph.D. dissertation, Yale University, 1975.

Cassimatis, Peter J. The Economics of the Construction Industry (New York: National Industrial Conference Board, 1969).

Clague, Ewan and Leon Greenberg, "Discussion of Employment." Automation and Technological Change (American Assembly, 1962).

Chase Econometrics. Current Data Bank (September 1980).

Cook, Cheryl A. "Industrialized Housing Techniques." In "An Analysis of Federally Funded Demonstration Projects: Supporting Case Studies," ed. William Baer et al., Appendix L. R-1927-DOC. (Santa Monica: The Rand Corporation, April 1976).

Dacy, Douglas, C. "Productivity and Price Trends in Construction Since 1947." Review of Economics and Statistics 47 (November 1965): 406-411.

Dennison, Edward F., The Sources of U.S. Economic Growth and the Alternatives Before Us (Washington, D.C.: Brookings Institute, 1962).

Domar, Evsey, et al., "Economic Growth and Production in the United States, Canada, United Kingdom, Germany, and Japan in the Post War Period." Review of Economics and Statistics 46 (February 1964).

Dowall, David E., "The Effects of Land Use and Environmental Regulations on Housing Costs." In Housing Policy for the 1980s, ed. Roger Montgomery and Dale Rogers Marshall (Lexington, Mass.: Lexington Books, 1980), pp. 113-125.

Ellickson, R. C., "Suburban Growth Controls: An Economic and Legal Analysis." Yale Law Journal 86 (January 1977): 385-511.

Falk, David. "Building Codes in a Nutshell." Real Estate Review 5, 3 (1976): 82-91.

Federal Reserve Bulletin, "The Depository Institutions Deregulation and Monetary Control Act of 1980". (June 1980), pp. 444-453.

Ferguson, Ronald F. and William C. Wheaton, "The Determinants of Housing Inflation in the 1970s: Phase I." Mimeographed, 1980.

Field, Charles G. and Steven R. Rivkin, The Building Code Burden (Lexington, Mass.: Lexington Books, 1975).

Fortune, "Defying Those Interest Rates" (November 3, 1980).

Fortune, "U.S. Homes Management Religion" (December 4, 1978), pp. 68-78.

Frieden, Bernard J., "The New Regulation Comes to Surburbia." The Public Interest (1979), pp. 15-27.

Gabriel, Stuart, et al., "Local Land Use Regulation and Proposition 13: Some Fundings from a Recent Survey." Center for Real Estate and Urban Economics Working Paper 80-4 (Berkeley: University of California, Institute of Business and Economic Research, 1980).

Gleeson, M.E., "The Effects of an Urban Growth Management System on Land Values." Land Economics 55 (1979): 350-365.

Gordon, Robert J., "A New View of Real Investment in Structures" Review of Economics and Statistics 50 (November 1968).

Grebler, L., D. M. Blank, and L. Winnick, Capital Formation in Residential Real Estate (Princeton: Princeton University Press, 1956).

Haber, William and Harold M. Levinson, Labor Relations and Productivity in the Building Trades (Ann Arbor: University of Michigan Press, 1965).

Hamilton, Bruce W., "Capitalization of Intrajurisdictional Differences in Local Tax Prices," American Economic Review, 66, #5, December 1976, pp. 743-753.

Jaffee, Dwight M., "New Residential Construction and Energy Costs." Paper prepared for the Brookings Conference on Housing and Energy in the 1980s, November 1981.

James, F. J. and T. Mueller, "Environmental Impact Evaluation Land Use Planning, and the Housing Comsumer." AREUEA Journal 5 (1977): 279-301.

Johnson, Ralph J., "Housing Technology and Housing Costs." The Report of the President's Committee on Urban Housing Technical Studies, Vol. II (Washington, D.C.: U.S. Government Printing Office, 1968), pp. 53-64.

Katz, L. and Kenneth T. Rosen, "The Effects of Land Use Controls on Housing Policies." Center for Real Estate and Urban Economics Working Paper 80-13. (Berkeley: University of California, Institute of Business and Economic Research, 1980).

Kendrick, John W., Postwar Productivity Trends in the United States (Washington, D.C.: National Bureau of Economic Research, 1973).

Lafferty, Ronald N. and H. E. French, III, "Community Environment and the Market Value of Single Family Homes: The Effect of Dispersion on Land Values." Journal of Law and Economics 21 (October 1976): 381-394.

Lange, Julian E. and Daniel Quinn Mills, (eds.), The Construction Industry (Lexington, Mass.: Lexington Books, 1979).

Mahaffey, C. T., "The Current Status of the Manufactured Home Industry." Mimeographed, (Washington, D.C.: Center for Building Technology, National Bureau of Standards, 1979).

Manski, Charles F. and Kenneth T. Rosen, "The Implications of Demand Instability for the Behavior of Firms: The Case of Residential Construction." AREVEA Journal 6 (1978): 204-226.

Maisel, Sherman J., Housebuilding in Transition (Berkeley: University of California Press, 1953).

Maser, Steven, et al., "The Effects of Zoning and Externalities on the Price of Land: An Empirical Analysis of Monroe County, New York." Journal of Law and Economics 20 (1977): 111-132.

Mayer, M., The Builders (New York: W. W. Norton, 1978).

McGraw Hill Company, "A Study of Comparative Time and Cost for Building Five Selective Types of Low Cost Housing." The Report of the President's Committee on Urban Housing, Technical Studies, vol. II (Washington, D.C.: U.S. Goverment Printing Office, 1968), pp. 1-52.

Muth, Richard F. and E. Wetzler, "The Effect of Constraints on Housing Costs." Journal of Urban Economics 3 (1976): 57-67.

National Academy of Sciences, A Report on Operation Breakthrough, (Washington, D.C.; 1974).

_____. "Preliminary Analysis of HUD Operation Breakthrough Guide Criteria and Industry Comments." Report 1-31 (Washington, D.C.: 1973) (a)

_____., "Analysis of HUD Operation Breakthrough Guide Criteria." Report 2-31 (Washington, D.C., 1973). (b)

National Association of Homebuilders, A Profile of the Builder and His Industry. (Washington, D.C.: 1970).

National Science Foundation. "An Analysis of Federal R&D Funding by Function, 1969-1977." NSF-76-325 (Washington, D.C.: 1976).

_____., "Characteristics of the National Sample of Scientists and Enginners, 1974." NSF-76-323 (Washington, D.C.: 1976).

_____., A Report on Operation Breakthrough (Washington, D.C.: U.S. Goverment Printing Office, 1974).

Nelkin, Dorothy, The Politics of Housing Innovation (Ithaca, N.Y.: Cornell University Press, 1971).

Ohls, James C. et al., "The Effect of Zoning on Land Values." Journal of Urban Economics 1 (1974): 428-444.

Oster, Sharon A. and John M. Quigley, "Regulatory Barriers to the Diffusion of Innovation: Some Evidence from Building Codes." In Research and Innovation in the Building Regulatory Process, Publication 473, Patrick W. Cooke (Washington, D.C.: U.S. Department of Commerce, National Bureau of Standards, 1976), pp. 113-136.

_____., "Regulatory Barriers to the Diffusion of Innovation: Some Evidence from Building Codes." The Bell Journal 8 (Spring 1977): 361-377.

Peterson, George E., "Land Prices and Factor Substitution in the Metropolitan Housing Market." (Mimeographed) (Washington, D.C.: The Urban Institute 1974).

President's Committee on Urban Housing (Kaiser Committee), A Decent Home (Washington, D.C.: U.S. Government Printing Office, 1969).

Rands, M. et al., "The Impact of Local Development Fees on New Housing Construction in the San Francisco Bay Area" (Berkeley: University of California, Graduate School of Public Policy, March 1980).

Real Estate Research Corporation et al., "The Impact of Operation Breakthrough on the Nation's Housing Industry" (Chicago, Ill.: July 1976).

Rosefields, Steven and Daniel Quinn Mills, "Is Construction Technologically Stagnant?" In The Construction Industry, Julian E. Lange and Daniel Quinn Mills (Lexington, Mass.: Lexington Books, 1979), pp. 83-114.

Rothenstein, Guy, "System Building." In Building the American City (Washington, D.C.: U.S. Government Printing
Office, 1969).

Rowland, Norman, "Reston Low Income Housing Demonstration
Program" (Springfield, Va.: National Technical Information Service, 1969).

Sagalyn, L. B. and George Sternlieb, Zoning and Housing
Costs: The Impact of Land Use Controls on Housing Prices (New Brunswick, N.J.: Rutgers University Press,
1972).

Seidel, S. B., Housing Costs and Government Regulations.
(New Brunswick, N.J.: Rutgers University Press, 1978).

Sims, Christopher A., "Efficiency in the Construction Indus-
try." The Report of the President's Committee on Urban
Housing, Technical Studies, vol. II (Washington, D.C.:
U.S. Government Printing Office, 1968), pp. 145-176.

Stull, William J., "Community Environment Zoning, and the
Value of the Single Family Homes." Journal of Law and
Economics 18 (1975): 535-557.

_____., "Land Use and Zoning in an Urban Economy."
American Economic Review 64 (1974): 337-347.

United Nations Economic Commission for Europe, Economic
Survey of Europe in 1969: Part I, Structural Trends and
Prospects in the European Economy (New York/Geneva:
1976).

U.S. Council of Economic Advisors, Economic Report of the
President (Washington, D.C.: U.S. Government Printing
Office, 1968).

U.S. Department of Commerce, Bureau of the Census, 1972
Census of Construction (Washington, D.C.: U.S. Government Printing Office, 1974).

_____., 1977 Census of Construction Industries: Industry
Studies (Washington, D.C.: U.S. Government Printing
Office, 1980).

U.S. Department of Commerce, Bureau of Economic Analysis,
National Income and Product Accounts of the U.S., 1929-
1974 Statistical Tables (Washington, D.C.: U.S. Government Printing Office, 1977).

U.S. Department of Commerce, Industry and Trade Administration, Construction Review 25, 11 (Washington, D.C.:
U.S. Government Printing Office, 1979).

U.S. Department of Commerce, National Bureau of Standards,
Building Research Division Team. Guide Criteria for the
Evaluation of Operation Breakthrough Housing Systems,

NBS Report 10-200, 5 vols. (Washington, D.C.: U.S. Government Printing Office, December 1970).

U.S. Department of Commerce, National Bureau of Standards, Research and Innovation in the Building Regulatory Process, ed. Patrick W. Cooke. NBS Special Publication 473, (Washington, D.C.: U.S. Government Printing Office, 1976).

U.S. Department of Housing and Urban Development, Housing Systems Proposals for Operation Breakthrough (Washington, D.C.: U.S. Government Printing Office, 1970).

U.S. Department of Housing and Urban Development, "Operation Breakthrough - Application of Improved Housing Systems Concepts for Large Volume Production." HUD-RFP-H-55-69, (Washington, D.C.: U.S. Government Printing Office, June 23, 1969).

U.S. Department of Housing and Urban Development, Feedback: Phase I Design and Development of Housing Systems (Washington, D.C.: U.S. Government Printing Office, n.d.).

U.S. Department of Housing and Urban Development, Housing in the Seventies (Washington, D.C.: U.S. Government Printing Office, 1973).

U.S. Department of Labor, Bureau of Labor Statistics, Employment of Scientists and Engineers. 1950-1974. Bulletin 1781, (Washington, D.C.: U.S. Government Printing Office, 1973).

U.S. Federal Reserve Board, Federal Reserve Bulletin 65, PA 50 (January 1979).

U.S. General Accounting Office, "Operation Breakthrough - Lessons Learned About Demonstrating New Technology." PSAD-76-173, (Washington, D.C.: GPO, November 2, 1976).

_____., "Why Are New House Prices So High, How Are They Influenced by Government Regulations, and Can Prices be Reduced?" CED-78-101 (May 11, 1978).

U.S. House of Representatives, Committee on Banking and Currency, Hearings of the Subcommittee on Housing, 91st Congress, 1st session, 1969.

Ventre, Francis T., "Social Control of Technical Innovation: The Regulation of Building Construction." Ph.D. Dissertation, Springfield, Mass.: Massachusetts Institute of Technology, 1973.

_____., "Innovation in Residential Construction." Technology Review 82 (1979): 2-11.

_____., "On the Blackness of Kettles: Inter-Industry Comparisons in Rates of Technological Innovation." Policy Sciences 11 (1980): 309-328.

Weiner, Neil, "Supply Conditions for Low Cost Housing Production," (Springfield, Va.: National Technical Information Service, 1968).

Willis, Mark A., "The Effects of Cyclical Demand on Industry Structure and the Rate of Technical Change." Ph. D. dissertation, Yale University, 1979.

8
The Motor Vehicle Industry
Lawrence J. White

The motor vehicle industry ranks high in importance among American industries. In 1977, the motor vehicle and parts industry alone employed 1.6 million workers and had net sales of $162 billion.(1) A study of the interaction between public policy and innovation in this industry is interesting in its own right, since it tells us something about a significant fraction of industrial activity in the United States. It also yields insights into the problems of public policy when government agencies face oligopolies and into the strengths and weaknesses of regulatory policies that try to induce technological change.

When this study was initiated, the Carter administration was trying to develop, with the automobile industry, ground rules for a Cooperative Automotive Research Program (CARP). The program has been dropped by the Reagan administration; no one in the auto industry rose to defend it. This case study, then, also sheds light on some political aspects of government R&D support.

The discussion of innovation in the motor vehicle industry will deal with both product innovation and process innovation. At the beginning, it is important to distinguish between the two. By product innovation, we mean changes in the final products that consumers buy. By process innovation, we mean changes in the methods by which the products are manufactured. These two categories are not wholly separable; changes in product characteristics frequently require (or follow from) changes in manufacturing techniques. And, at the limit, discovering how to make the same quality automobile with fewer inputs and discovering how to make an improved quality automobile with the same inputs has a great deal of conceptual similarities. Still, the distinction is useful. Both kinds of innovations have been important in the motor vehicle industry, but regulatory policy has been largely aimed at product innovation. This could be contrasted with, for example, the electric utility or steel industries, in which regulatory policy (external regulation of air and water pollution) has largely affected process innovation.

The remainder of this chapter is organized as follows: section II describes the general industrial organization of the motor vehicle industry. Section III discusses the general character of technical progress in the industry, covering both product innovation and process innovation. Section IV reviews the major government policies that have influenced technical progress in this industry. Section V analyzes the questions that can be raised concerning government policies which affect technical progress in this industry, and section VI offers some brief conclusions.

II. THE INDUSTRIAL ORGANIZATION OF THE MOTOR VEHICLE INDUSTRY

The current structure of the domestic motor vehicle industry has a number of important features.(2) The major companies are large; they are few; the barriers to de novo entry are extremely high; in the automobile segment of the industry the companies are dealing largely with unsophisticated buyers in a market in which replacement demand is dominant and brand loyalty important; and lead times are long, large sums must be spent, and large swings in demand are possible, all of which combine to create large risks.

General Motors, Ford, and Chrysler are the leading companies in both the automobile and truck markets. They were, respectively, the second, fourth, and sixteenth largest industrial companies (by sales) in the United States in 1979. The fourth largest producer, American Motors, was the 109th largest industrial company in 1979; the fifth largest truck producer, International Harvester, was the 27th largest industrial company in the U.S.

These very large companies have tended to dominate the auto and truck areas. Table 8.1 provides the average North American production shares for the years 1976-1979.(3) Table 8.2 provides the average United States sales shares for these same years. Imports have gradually taken a larger share of the U.S. automobile market over the past 25 years. In 1979, the import share was 23 percent and in 1980 it rose above 25 percent. Whether the 1980 figure is a temporary surge or the beginning of a permanent plateau is a subject of current debate (and could be affected by current public policy decisions with respect to tariffs and quotas). Even with the inclusion of the imports in market share figures, it is nevertheless clear that the three large domestic producers still

Table 8.1.   Average North American Production Shares,
1976-1979.

	Automobile	Truck
Average Production (numbers)	9,940,000	3,923,000
Shares:		
American Motors	2.1%	4.9%[a]
Chrysler	13.5	13.1
Ford	26.7	33.0
International Harvester		3.4
General Motors	57.0	43.3
Other	0.7[b]	2.3

[a]Includes Jeep
[b]Includes Checker, Volkswagon of America, and Volvo of Canada.

Source: Motor Vehicle Manufacturers Association (1980).

Table 8.2.   Average U.S. Sales (Registrations),
1976-1979.

	Automobile	Trucks
Average Sales (numbers)	10,464,000	3,500,000
Shares:		
American Motors	1.8%	3.9%[a]
Chrysler	10.7	12.0
Ford	22.1	33.0
International Harvester	-	3.2
General Motors	46.9	40.7
Other	18.5[a]	7.2

[a]Includes Jeep.
[b]Includes Checker, Volkswagon of America, and imports.

Source: Automotive News (1980).

dominate the market. But it is also clear that the American market has become a more international market over the past 25 years, and even the American manufacturers are more prepared to import production components and even complete automobiles.

The barriers to entry for a new manufacturer of motor vehicles are quite high. The only successful entrants into the U.S. market in the past 30 years have been overseas manufacturers that already had a substantial manufacturing base and an established product.

All of the major domestic manufacturers are characterized by extensive vertical integration. All assemble their own vehicles and produce all or most of their major sheet-metal stampings, castings, and drive-train components: engines, transmissions, axles, and so on. All produce some of the other parts and components of their vehicles and buy the remainder from parts suppliers. Despite this extensive vertical integration in terms of processes and components, however, when the motor vehicle industry's vertical integration is measured by the ratio of value added to sales, it is only at or below the average for all manufacturing. In 1978, General Motors' ratio of value added to sales was 48.5 percent; for Ford it was 39.0 percent; and for Chrysler it was 33.2 percent. For all manufacturing in 1976, this same ratio was 43.1 percent.

The parts suppliers with whom they deal range from large companies such as Bendix, Motorola, and TRW, which are also in the Fortune 500, to small machine-tool manufacturers whose names are unfamiliar to anyone outside the motor vehicle industry. The industry is also a major customer of the steel, aluminum, rubber, and chemicals industries.

The automobile market is largely one of technically unsophisticated buyers. Replacement demand dominates the market, and brand loyalty is an important phenomenon; i.e., if a manufacturer loses sales because it has produced an unappealing product, it will have a difficult time winning them back.

Finally, the risks are high. New models require four to five years of lead time. Hundreds of millions of dollars must be spent long before a new model is introduced. Buyers are clearly fickle and, because demand is largely for replacement, can delay purchases and retain their existing cars longer. Swings of 15 percent or more in annual industry sales are not uncommon, and even larger swings in individual company sales are quite possible.

The implications of this industry structure for innovation are profound. The high barriers to entry mean that if an independent innovator has a "better idea" for a vehicle, a major component (e.g., engine or transmission), or a manufacturing process, his only hope for eventual success lies in convincing one among a literal handful of manufacturers of that innovation's worth. He has virtually no hope of establishing himself in the motor vehicle industry so as to produce the innovation himself. This situation could be contrasted with that in, say, farming, retailing, or apparel manufacture, in which efficient scale is comparatively small, entry is comparatively easy, and someone with a "better idea" could realistically expect to bring it into production.

Further, product change is necessary to attract the replacement demand. But product change is risky, and the more fundamental the change, the riskier it is.(4) Given the lack of technical sophistication of automobile customers, a strategy of relying primarily on product technology changes would be quite risky. Buyers might not respond in any event, and a serious technical failure could be quite costly and gain the company a reputation of poorly engineered cars. A strategy that instead relied primarily on styling model changes with modest technological changes would be less risky; a poorly designed model would, of course, lose sales but there would be no long-run reputation involved; in principle, better-designed models in the future could regain the sales (subject, of course, to the drags of brand loyalty). Even process changes though promising cost savings, carry the risk of causing defective parts that can be expensive to replace and can earn the company a bad reputation; again, gradualism is likely to be the favored strategy.

We now turn to the actual experience in the motor vehicle industry.

III. THE CHARACTER OF TECHNICAL PROGRESS

Product Innovation

It is difficult to provide a quantitative statement of the nature and extent of product improvements in the motor vehicle industry, and we shall not try to do so here.(5) Instead, we will offer a more qualitative description of the character of product innovation.

General Product Development

In the eight decades of their existence, cars and trucks have experienced substantial product innovation. Some aspects of the product have remained constant: the internal combustion engine is still the primary means of propulsion; four wheels are still standard; and the driver has a seat and a steering mechanism. But most other aspects of vehicles have undergone substantial changes: the size, shape, and efficiency of engines and their emissions (pollution) characteristics; the nature of transmissions; the size, weight, comfort, and safety of the vehicle; and the materials used in the vehicle.

This claim that there has been substantial change does not contradict the conclusions at the end of Section II. First, the changes have taken place over 80 years. Second, the current structure of the industry took shape only in the mid-1920s. And there has been a distinct time pattern to the industry's innovation behavior.

The first two decades of the 20th century, prior to World War I, were years of great fluidity for the industry. Overall growth was rapid, entry was easy, many firms did in fact enter and exit, and market shares fluctuated extensively. As of 1910, there were 52 companies in the automobile industry; as late as 1921 there were 88 firms in the industry. New ideas accompanied the new firms, and the electrical starter motor, the V-8 engine, the closed passenger compartment, and significant improvements in tires, lights, and electrical systems were all introduced during these decades.

The two decades between the two world wars can be seen as a transition period. Entry was more difficult. The necessary manufacturing facilities were more expensive; a reliable dealer organization was difficult to assemble. A few firms tried to enter; more exited. By the end of the 1920s the same three companies that dominate today's motor vehicle market had a 72 percent combined market share of the auto market. In the early 1920s General Motors developed its basic auto marketing strategies: "A car for every purse and purpose," which meant blanketing the market with models in every price range, and an annual model change that would encourage replacement purchases of new cars. Walter Chrysler, a General Motors "graduate," revived the ailing Maxwell-Chalmers Corporation in the early 1920s, became president in 1923, brought out the Chrysler 6 the following year, and changed the company's name to his own the year after that. He rapidly adopted marketing strategies that were similar to those of General Motors. Ford took longer to adopt them but eventually did so in the late 1920s.

With entry more difficult and with an increased marketing emphasis on styling and model changes for autos (and also with increasing maturity of the product itself), the pace of innovation in autos appears to have slackened somewhat. But refinements continued to be made. Cars became larger, heavier, and more powerful. By the 1930s bodies were all-enclosed and entirely of steel. "Aerodynamic" streamlined designs replaced the square, boxy designs of the 1920s. Automatic transmissions, power brakes, and power steering were first developed for larger trucks and buses. At the end of the 1930s automatic transmissions were beginning to be applied to automobiles, but the outbreak of World War II brought all

automobile production to a halt. It is worth noting that the
smaller companies in the industry appear to have accounted for
a disproportionate share of the innovations in the industry
prior to World War II.(6) They may have been more willing to
take the risks of product change because they were less able
to match the larger companies' styling changes.

The two decades between the late 1940s and the late 1960s
were clearly a period in which the auto market focused on
styling and model change. Technological advances consisted
primarily of the refinement and spread of the major prewar
innovations - automatic transmissions and power equipment -
and high compression engines. A good auto mechanic of the
late 1940s would have had little difficulty in understanding a
car of the late 1960s. Parts suppliers played major roles in
many of the technological developments that did occur. The
Gemmer Manufacturing Company and Bendix controlled the key
patents on power steering. Kelsey-Hayes and Bendix developed power brakes, and Bendix and Budd, disc brakes. The
Dana Corporation controlled the key patents on nonslip differentials; Motorola made the principal breakthrough on a silicon
rectifier for alternators. Ball-joint front suspensions were
developed by Thompson Products, and the first automatic
speed control, by the Perfect Circle Company. Bendix and
American Bosch did the early development work on fuel injection systems. The use of teflon for front suspension joints
was developed by du Pont and the American Metal Products
Company. Motorola and Electric Auto-Lite, before its absorp-
tion by Ford, did much of the advance work on transistorized
ignition systems. U.S. Rubber developed the toothed, glass-
fiber reinforced nylon and rubber timing belt that made Pon-
tiac's overhead camshaft engine possible. The materials
suppliers - metals, glass, plastics, paint, etc. - provided
many of the technological advances that allowed such improvements as aluminum engines, curved and tinted glass, and
lighter-weight castings. Technological advances were frequently introduced on small-volume expensive models and then,
if successful, gradually expanded to other models - a strategy
that clearly reduced risks.

The Small Car Story

The auto companies' policies with respect to one particular
"innovation" - the small car - is worth recounting in some
detail, especially since there has been much popular misper-
ception of the events of the late 1970s. The U.S. federal and
state governments have had policies that have kept the price
of gasoline relatively low. They have refrained from imposing
the substantial gasoline excise taxes that have been imposed
by many governments in Europe and Japan, and, between 1973
and 1979, the federal government explicitly tried to keep the

domestic prices of petroleum and petroleum products below those of world markets. The federal and state governments have avoided the high horsepower taxes that a number of overseas governments have imposed. And the federal and state governments have built an extensive network of high-quality highways.

These policies have had a direct influence on the pattern of automobile design in this country. Faced by cheap gasoline and good roads, most automobile buyers wanted, and got, the standard American automobile of the 1950s and 1960s - a large, high horse-power vehicle.

Even in the 1950s, though, there was a minority of buyers who wanted smaller vehicles. The three large manufacturers were reluctant to provide this minority with small cars; they saw large cars as more profitable and feared the diversion of sales from large to small cars.(7)   Only the smaller U.S. manufacturers - American Motors and Studebaker - initially provided small cars to this market. Demanders of small cars increasingly turned to imports, especially Volkswagen, in the late 1950s. Finally, reasoning that low profits on small-car sales were better than no profits or sales to imports, all three large manufacturers brought compact cars to the market in late 1959: the Chevrolet Corvair, the Ford Falcon, and the Plymouth Valiant.

These compact cars were initially successful and were followed by other compacts. The imports' share of the U.S. market declined. But the American compacts grew in size during the mid-1960s, and so did the imports' share of the market. In the late 1960s and early 1970s, Ford and General Motors brought out a new set of compacts and subcompacts.

After the Arab oil embargo of late 1973, the gasoline lines of 1974, and the large increase in gasoline prices at that time, all three companies made decisions to "down-size" all of their models and to introduce smaller models in the late 1970s. But larger cars continued as the mainstay of their production efforts; and in 1977 and 1978 this strategy appeared to be the correct one, as gasoline again became relatively cheap and the public's demand for larger cars surged.

In early 1979 the Iranian crisis led to more gasoline lines and another sharp increase in gasoline prices. The public's demand switched sharply to smaller, more fuel-efficient cars. The U.S. manufacturers, having hedged their bets somewhat, had some smaller models to offer, but many buyers turned to imports, particularly to Japanese vehicles. These cars had been designed for their home markets, where gasoline had always been much more expensive than in the United States and thus were appropriate for the changed American market.

A popular cry heard in 1979 and 1980 was that Detroit was not building the cars that the American public wanted to buy. This complaint was largely misleading. The cars had been appropriate for 1977 and 1978; the companies had hedged their bets to some extent; and the long lead time (four to five years) meant that the U.S. companies would require time to respond to the sharply changed market. Perhaps the companies should have hedged their bets to a greater extent; but this is much easier to say with 20-20 hindsight. In 1974, when the models for 1979 had to be planned, perfect foresight was not available to the auto companies' decision makers. And the consequences of guessing wrong and producing small cars for a market that wanted larger cars could have been almost as serious as guessing wrong on the other side.

Federal Involvement in Product Innovation

Prior to 1965, the federal government's role in influencing product innovation was largely restricted to the indirect pressures arising from its policies with respect to gasoline pricing, horsepower, and highways. (The only exception was the development work on torque converters which occurred on military vehicles during World War II.) In 1965, however, the federal government's policy of noninvolvement came to an end. In that year, Congress authorized the setting of emission control standards for vehicles, and the following year it authorized extensive safety standards. (A more detailed discussion of these policies will be provided below in Section IV.) These standards first took effect in the 1968 model year. (Emission controls had been required earlier in the 1960s in California.)

The standards did not have a serious impact on the industry until the early 1970s. The 1970 amendments to the Clean Air Act, however, required substantial reductions in auto emissions by 1975 and 1976, and the industry (and some of its suppliers) began focusing a substantial amount of research on emissions reduction.

The industry in the early 1970s settled on a catalyst technology to control emissions. This has been supplemented in the early 1980s by electronic exhaust sensors and microprocessors to control fuel and air mixtures. In the early 1970s, though, two overseas manufacturers, Honda and Toyo Kogyo (Mazda) chose alternative engine designs as the way to meet the emission standards.

Also in the early 1970s, the Department of Transportation's National Highway Traffic Safety Administration (NHTSA) first tried to establish standards that would require passive restraints. The safety requirements, and especially the passive restraint requirements, have meant a refocusing of research efforts toward areas that the companies would not otherwise have pursued as vigorously.

Finally, in December 1975, Congress passed the Energy Policy and Conservation Act which established fuel economy

standards for automobiles for 1978-1980 and 1985 (and authorized NHTSA to set standards for the 1981-1984 years); the
1985 standard of 27.5 miles per gallon for the sales-weighted
average of new cars sold in that year by each company is
roughly double the fuel economy achieved by the new car fleet
in 1973. For a period in 1978 and 1979, it appeared that the
fuel economy standards would be seriously binding on the
domestic manufacturers and would force them to take technological actions that they would not otherwise pursue. But the
sharp increase in gasoline prices in mid-1979 has caused car
demand to shift sharply toward smaller, more fuel-efficient
cars, and the current standards through 1985 are unlikely to
be binding. The shift in demand induced by the high fuel
price by itself is shifting the sales-weighted average of miles
per gallon substantially upward, and this shift in demand is
apparently providing more than enough inducement for the
companies to develop models and technologies that will yield
yet greater fuel economy.

In summary, then, the emissions and safety regulations of
the early 1970s and the sharp increase in the price of gasoline
in 1979 has led to a substantial change in the pattern of
innovation in the motor vehicle industry. Much more effort is
being devoted to meeting the regulatory requirements and in
increasing fuel efficiency. The cycle of model changes has
been considerably stretched out, as compared with the pattern
of the 1960s. The industry's product innovation attention has
clearly been focused in a new direction.

Process Innovation

Unlike product innovation, process innovation in the motor
vehicle industry is susceptible to quantitative investigation, at
least in an indirect manner. We can examine indexes of motor
vehicle industry output per worker and retail automobile prices
and compare their trends over time with those in other sectors
in the economy. As we shall see, the performance in these
areas by the motor vehicle industry has been relatively good.
The claim that most engineers in Detroit would be willing to
sell their grandmothers for the opportunity to save 25¢ per car
may be an exaggeration, but it is clear that the industry has
been quite cost and cost-reduction conscious.

Also, in examining the labor productivity and price
indexes, we can try to shed some light on the question of
whether government regulation has caused a slackening of
productivity improvements in this industry.

Before the quantitative work is discussed further, one
qualification should be added. Since 1959, the Bureau of
Labor Statistics has been adjusting the new car price index for
product quality improvements.(10) Thus, the relative pattern

of car prices via-a-vis other prices in the economy is a product of both process and product improvements. Similarly, because the output indexes are derived by deflating value indexes by price indexes, the labor productivity measures similarly reflect a mix of both process and product innovations. This author's strong impression is that most of the net advantage of the motor vehicle industry relative to the rest of the economy in these indexes is due to process improvements, but there is no satisfactory way to verify this assertion.

Labor productivity indexes

The time pattern of an index of output per worker will reflect not only innovations but also simple substitution of capital (and, if the degree of vertical integration is not held constant, other materials) for labor. Thus, changes in the index are likely to overstate the extent of pure innovation. Still, labor productivity improvements are a major concern of public policy, and they probably are indicative of broad cost tendencies.

In an earlier study,(11) this author found that for the years 1949-1967 labor productivity in the motor vehicle industry improved at an average rate of 3.96 to 4.33 percent per year. That study used relatively crude measures of output (the Federal Reserve Board's indexes of industrial production) and a set of labor indexes that were not matched exactly to the output series.

It is now possible to update and refine those figures. The Bureau of Labor Statistics specifically compiles an index of output per employee-hour for the motor vehicles and equipment industry (SIC 371). The output and labor indexes are matched to each other, and the data are collected on an establishment basis, so the gross problems of changes in the degree of vertical integration have been eliminated.(12) The data extend back to 1957 and also include a split between production workers and non-production workers.

Table 8.3 provides the average annual increases in output per employee-hour between 1957 and 1978 and for a number of subperiod splits. Over the entire period, labor productivity improved at a rate of 3.5 percent per year in the motor vehicles and equipment industry; the rate of increase was about the same for production and nonproduction workers. By contrast, in all manufacturing, labor productivity rose by an average of only 2.7 percent per year, and in the entire private sector, labor productivity rose by only 2.4 percent per year.

Data for the time-period splits can be used to test the proposition that there has been a significant slowing of the rate of productivity increase and that government regulation might be a cause of this slackening. The "switching of re-

Table 8.3.   Average Annual Percentage Increases in
Output per Labor Hour*

Motor Vehicles and Equipment Industry

	All Employees	Production Workers	Nonproduction Workers	All Manufacturing	All Private Business
1957-1978	3.5%	3.5%	3.6%	2.7%	2.4%
1957-1965	5.1	4.8	6.2	3.6	3.3
1966-1978	3.2	3.2	2.8	2.2	1.6
1957-1966	4.8	4.5	5.8	3.6	3.3
1967-1978	3.3	3.3	3.2	2.2	1.6
1957-1967	4.5	4.3	5.1	3.3	3.3
1968-1978	3.3	3.3	3.3	2.2	1.5
1957-1968	4.4	4.2	5.0	3.3	3.3
1969-1978	3.6	3.5	4.0	2.3	1.6
1957-1969	4.1	4.0	4.7	3.2	3.2
1970-1978	3.7	3.5	4.4	2.3	1.5
1957-1970	3.7	3.7	3.9	3.0	3.0
1971-1978	3.0	2.9	3.3	2.0	1.3
1957-1971	3.8	3.7	4.0	3.0	2.9
1972-1978	3.3	3.1	3.8	1.9	1.2

---

*All rates of increase are the slope coefficient of an ordinary least squares regression of the logarithm of output per labor hour   on time.

gimes" methodology (13)   provides a method for testing those propositions.   The methodology calls for an examination of alternative splits of the data to find the split that yields the largest difference in "regimes."

As can be seen in Table 8.3, the data for the splits by period indicate that labor productivity rose less rapidly in the latter part of these 22 years than in the former part; the difference is significant.   But the pattern of the splits indicates that the slower growth had begun by the mid-1960s and did not get any worse in the 1970s.   If we use the "switching of regimes" framework, we find that the maximum degree of difference in regimes is not found in the splits that focus on the 1970s as one "regime."   But government regulations began to have a serious impact on the industry only in the early 1970s. Thus, this set of data would not support a claim that regulation was responsible for the slackening of productivity growth in this industry. (Note, though, that for all manufacturing and for the entire private sector, as the splits focus more on the 1970s, the growth in productivity does slacken. The causes of this general deceleration in productivity growth are, of course, widely debated. Regulation may be one of them.)

This last conclusion is reinforced by the data for production and nonproduction employees. One of the claimed consequences of regulation is that more employees must spend more time filling out reporting forms and engineers and technicians must spend more of their time trying to devise ways to meet government regulation. If this were occurring in a serious fashion in the 1970s in the motor vehicle industry, we would expect to see a greater slackening in the rate of increase of output per nonproduction worker than in the rate of increase of output per production worker. The opposite appears to have been the case.

Relative rates of price increases

A second way of trying to measure relative rates of innovation is to measure relative rates of price increases. Of course, price increases also reflect increases in the costs of inputs and changes in profit margins, as well as innovations. In this respect, though, a comparison of price increases is probably biased against a favorable showing by the motor vehicle industry, since the cost of one of its major inputs, labor, has been rising more rapidly than the cost of labor in most other sectors. (Offsetting this to some extent, however, is the fact that the automobile price index is regularly adjusted for quality improvements, whereas prices in other sectors sometimes are and sometimes are not adjusted for quality improvements.)

Table 8.4 provides the average annual rate of increase of the new car component of the consumer price index between 1955 and 1979.(14) For comparison purposes, the average increases in the durable goods component of the CPI and in the overall CPI are also provided, as are the relative rates of increase of the new car price index as compared to the other two indexes. As can be seen, the rate of increase of new car prices has been appreciably below that of the overall CPI and even of the prices of durable goods generally. The former result is not surprising, since the overall CPI includes the prices of services (whose rate of productivity increase would be expected to lag behind that of manufactured goods and hence their prices would be expected to rise more rapidly) and the prices of petroleum products and other energy items

Table 8.4.   Average Annual Percentage Increase in Prices*

	New Car Component of CPI	Durable Goods Component of CPI	Overall CPI	New Car ÷ Durable Goods	New Car ÷ CPI
1955-1979	1.7%	2.7%	3.8%	-0.9%	-2.1%
1955-1965	1.0	0.7	1.6	0.3	-0.6
1966-1979	3.8	5.1	6.2	-1.3	-2.4
1955-1966	0.7	0.7	1.6	0.0	-0.9
1967-1979	4.0	5.3	6.4	-1.3	-2.4
1955-1967	0.5	0.6	1.6	-0.1	-1.1
1968-1979	4.3	5.6	6.6	-1.3	-2.4
1955-1968	0.5	0.7	1.8	-0.2	-1.3
1969-1974	4.6	5.8	6.8	-1.3	-2.3
1955-1969	0.5	0.8	1.9	-0.3	-1.4
1970-1979	4.9	6.1	7.1	-1.2	-2.2
1955-1970	0.5	1.0	2.1	-0.4	-1.6
1971-1979	5.3	6.5	7.4	-1.1	-2.1
1955-1971	0.6	1.1	2.3	-0.5	-1.6
1972-1979	6.1	7.0	7.7	-0.9	-1.6

*All rates of increase are the slope coefficient of an ordinary least squares regression of the logarithm of the price index on time.

(whose prices have risen sharply since 1973). But the differential vis-a-vis the durable-goods component (of which new automobiles themselves are about 12 percent) is less expected and hence more impressive.

Again, we can use the "switching of regimes" methodology to examine differences in behavior within the longer 1955-1979 period. As is indicated in Table 8.4, the new car price increases were greater in the latter period than in the earlier period.(15) But the latter period was one of greater inflation generally. The negative differences between the new car price index and the durable price index and between the new car index and the CPI were greater in the latter period than in the former. But, as the splits focus more on the 1970s, the negative differences diminish.

How can we reconcile this finding of a narrowing in the 1970s of the relative performance of prices with the previous evidence of a widening of the relative performance of labor productivity?   First, it might be caused by a more rapid increase in the costs of inputs into motor vehicles as compared with inputs into other sectors of the economy.   Unfortunately, input cost indexes are not available.   But we can rule out this possibility for labor inputs.   Table 8.5 presents the average annual rates of increase of Ford's U.S. hourly labor costs (including fringe benefits) and, for comparison, the rates of increase in all manufacturing and in all private business. Over the period 1959-1979, the United Automobile Workers did relatively well for its members.   But as the time period splits focus on the 1970s, the relative increase in Ford's labor costs appear to have narrowed vis-a-vis the rest of the economy; i.e., relatively accelerating labor costs cannot explain the narrowing price performance.   Unfortunately, it is not possible to make a similar determination for other inputs.

Table 8.5.   Average Annual Increase in
Hourly Compensation
(including fringe benefits)*

	Ford Motor Co. (U.S.)	All Manufacturing	All Private Business	Ford ÷ Manufacturing	Ford ÷ Private Business
1959-1979	7.8%	6.1%	6.6%	1.7%	1.2%
1959-1965	4.6	3.4	4.3	1.2	0.4
1966-1979	9.4	7.6	7.6	1.8	1.8
1959-1966	4.8	3.5	4.5	1.3	0.3
1962-1979	9.6	7.7	7.7	1.9	1.9
1959-1967	5.1	3.6	4.6	1.5	0.5
1968-1979	9.9	7.9	7.8	2.0	2.1
1959-1968	5.2	3.8	4.8	1.4	0.4
1969-1979	10.1	8.1	8.0	2.0	2.1
1959-1969	5.3	4.1	5.0	1.2	0.3
1970-1979	10.0	8.3	8.1	1.8	1.9
1959-1970	5.5	4.3	5.2	1.2	0.3
1971-1979	10.0	8.6	8.3	1.5	1.7
1959-1971	5.9	4.5	5.4	1.3	0.5
1972-1979	10.2	8.8	8.5	1.4	1.7

*All rates of increase are the slope coefficient of an ordinary least squares regression of the logarithm of the price index on time.

Alternatively, the labor productivity indexes may be capturing mostly the effects of input substitutions, and the relative rate of increase in total factor productivity in motor vehicles vis-a-vis other sectors may have slackened in the 1970s. For example, General Motors' inflation-adjusted accounting indicates that the real amount of capital per employee (worldwide) increased by 65 percent between 1967 and 1979. Unfortunately, little other data are available.

Thus, we are left with a puzzle. The motor vehicle industry's relative performance in labor productivity improved in the 1970s; its relative performance in price increases deteriorated. Both measures represent imperfect ways of capturing the effects of process innovation, and we cannot tell which is closer to the true concept we are seeking. We are left concluding that regulation certainly did not affect rates of labor productivity increases, but we cannot conclude anything generally about the effects of regulation on process innovation.

Further comments on process innovation

The motor vehicle industry has also been subject to the standard array of regulations that have confronted most other industries: e.g., Environmental Protection Agency regulation of air and water pollutant emissions from factories, Occupational Safety and Health Administration regulation of workplace practices, Department of Labor regulation of pension funds, and so on. The impact of these other regulations has been minor, as compared with the exhaust emissions and safety regulations. The relative magnitude of the most important of these other regulatory areas, industrial air and water pollution control, can be determined from General Motors' expenditures in this area as compared with that company's expenditures on motor vehicle emissions and safety. These figures are found in Table 8.6. The relative importance of industrial air and water pollution control has been growing, but it is still below 20 percent of the company's expenditures on motor vehicle emissions and safety.

A few final comments on process innovation are warranted. First, since the 1920s an important trend in the industry's manufacturing processes has been the substitution of capital for labor. This is exemplified by the highly automated transfer-machine technology of engine plants and the current "robotization" of assembly line processes. In part this capital deepening has simply been "natural" substitution of capital for labor as relative wage rates have increased. But innovations in capital equipment have permitted and encouraged this substitution. This pattern of innovation is to be expected, though the literature on "induced innovation" is surprisingly muddled on this point.(16) The proper interpretation of this literature shows that, ceteris paribus, the inputs

Table 8.6. General Motors' Expenditures on Industrial
Air and Water Pollution Control and on Motor Vehicle
Emissions and Safety.

	Industrial Air and Water Pollution Control (1)	Motor Vehicle Emissions and Safety (2)	(1) ÷ (2)
1968	$17 million	$445 million	3.8%
1969	65 ·	503	6.8
1970	35	520	6.7
1971	55	578	9.5
1972	58	745	7.8
1973	69	963	7.1
1974	74	868	8.5
1975	58	532	10.9
1976	65	542	12.0
1977	97	670	14.5
1978	154	912	16.9
1979	220	1,113	19.8

Source: General Motors' 10-K reports.

with the largest cost shares will receive the most attention
with respect to process innovations.(17)   Labor has been and
continues to be, by far, the largest single cost component in
motor vehicle production.   For General Motors, for example,
labor costs were 30 percent of the costs of all inputs in 1978.

Second, it appears that the motor vehicle industry has
not been the major discoverer or developer of most major new
manufacturing processes.   Rather, supplier firms have generally taken the lead.   This has been true in areas like cold
extrusion of metals, cold rolling of splines on shafts, electrical
discharge machining of dies, and numerical control of die
cutting.(18)   But, as the data in this section have indicated,
the motor vehicle industry has been quite good at adapting
and adopting these innovation for use in its home territory.

Third, William Abernathy has argued that the full cycle of
product and process development is likely to contain the seeds

of its own stagnation.(19)   Early in the development of a
product (e.g., the internal combustion engine), production
processes are flexible; most technological improvements occur
through product changes, which can be accommodated by the
flexible production technology. But over time, the production
processes become more refined and more rigid; the processes
are less tolerant of changes in product.   This rigidity encourages more of the research effort to focus on process
refinements rather than on product changes, which increases
rigidity even further.(20)   But there are diminishing returns
to these process improvements.   Hence an industry may find
itself confronting a "productivity dilemma." Abernathy argues
that engine production processes have displayed just this
pattern, and he contrasts the rigidity of the (capital-
intensive) engine machining plant with the flexibility of the
(more labor-intensive) assembly plant.

Both Abernathy's argument and his evidence are open to
question.(21) First, if there has been an increase in rigidity,
it may will be due to the general trend toward greater capital-
intensity, caused by rising labor costs, rather than to the
specific innovation model described by Abernathy.   Second, as
we argued in Section II, the basic structure of the industry
probably biases it away from fundamental product change; any
production rigidity that occurs, either from Abernathy's model
or from the general process of capital deepening, is probably
only pushing the industry slightly farther in a direction in
which it was already headed.

Third, his comparison of engine plants and assembly
plants is not convincing.   Are engine lines in fact less flexible
than assembly lines?   (Computerized numerical control processes have surely increased the flexibility of machining
processes.)   Are there fewer basic varieties of engine types
than there are basic varieties of car models?   Has it been more
difficult to develop new engines than it has been to develop
new car models?   Also, Abernathy's evidence indicates that the
patterns of labor productivity improvements in engine plants
and assembly plants have been remarkably similar.(22)

Fourth, both the automobile industry and the general
economy have experienced a deceleration of productivity
improvement in the 1970s. The data in Table 8.3 indicate that
the automobile industry's productivity growth decline in the
1970s has, if anything, been smaller than that of the general
economy. The "productivity dilemma" is a general one for the
entire U.S. economy and remains a puzzle to many observers.(23)   It is highly unlikely that the consequences of
Abernathy's specific innovation model could have simultaneously
struck most of the major sectors of the U.S. economy.   Finally, electronics and on-board microprocessors may well provide
major product improvements for autos in the 1980s, which
would seem to contradict the Abernathy thesis.

Research and Development Expenditure

Thus far, we have examined innovations - the outcome of the process of technological change. One other measure that is frequently examined is expenditures on research and development - the inputs into the process. This measure cannot tell us anything about innovation unless there is a strict one-to-one relationship between inputs and outputs in this process, but it may be able to tell us something about efforts at innovation.

The data on R&D expenditures as a percentage of sales for the leading three motor vehicle manufacturers are provided in Table 8.7. Unfortunately, the data extend back only to 1967. Before discussing the implication of the data, we should offer some caveats. First, the data that are reported by the companies pertain to worldwide operation; the North American motor vehicle data are only estimates. Second, there are no strict accounting standards for what is counted as "R&D expenditures." Thus, different companies may include different items; and, since R&D is a high-prestige activity, there is probably a general tendency toward overstatement.

The data in Table 8.7 indicate somewhat similar trends. The Chrysler and Ford worldwide R&D efforts were relatively unchanged in the late 1960s and early 1970s, rose in 1973 and 1974, declined subsequently, and then peaked in the late 1970s; their more limited North American data show a similar pattern. General Motors' worldwide R&D effort rose earlier in the 1970s, peaked in 1974, then declined, and only rose moderately in 1979; the more limited North American data do not show a rise in the early 1970s but show the same peak in 1974.

Unexpected sales shortfalls (lowering the base of the percentage) may have been at least partly responsible for the rise in 1979; all three companies had disappointing sales in that year, and Chrysler had disappointing sales in 1978 as well.

The last column of Table 8.7 puts these numbers in perspective. The average R&D sales percentages for all manufacturing in the 1970s was 2 percent. Thus, Ford and General Motors have clearly been above the average, while Chrysler recently has also exceeded the average.

Has regulation influenced this pattern of R&D? Though regulation clearly had a greater impact on the motor vehicle industry in the 1970s than in the 1960s, it is difficult to tell if its effects were greater at the beginning or the end of the 1970s. The Clean Air Act's stringent emissions requirements for automobiles were originally scheduled for 1975 and 1976, but they were subsequently delayed to the early 1980s, and stringent standards for trucks were added to the schedule for the early 1980s. Passive restraints were originally scheduled

Table 8.7. Research and Development Expenditures,
as a Percentage of Sales.

	Chrysler		Ford		General Motors		Motor Vehicle and Equipment Industry	All Manufacturing
	Worldwide	North American Motor Vehicle[ab]	Worldwide	North American Motor Vehicle[a]	Worldwide	North American Motor Vehicle[a]		
1967	1.8	–	3.1[a]	3.5	3.0[a]	–	–	2.1
1968	1.8	–	3.0[a]	3.0	3.0[a]	2.6	–	2.1
1969	2.3	–	3.0[a]	3.2	3.0[a]	2.5	–	2.2
1970	1.9	–	3.1	3.2	5.3[c]	3.7[c]	–	2.1
1971	1.8	1.2	3.1	3.3	3.6	2.4	–	2.0
1972	1.9	2.0	3.1	2.9	3.5	2.4	2.8	2.0
1973	2.1	2.2	3.5	3.4	3.5	2.4	2.9	2.0
1974	2.2	2.4	3.5	3.4	4.3	3.0	3.2	2.0
1975	1.7	2.0	3.1	3.3	3.1	2.7	3.0	2.0
1976	1.8	–	3.2	3.2	2.7	2.2	2.7	2.0
1977	2.0	–	3.1	–	2.6	–	2.7	2.0
1978	2.5	–	3.4	–	2.6	–	–	–
1979	3.0	–	4.0	–	2.9	–	–	–

[a] R&D expenditures estimated by Carroll and Schneider (1979).
[b] Sales estimated by Kaiser (1979).
[c] Abnormally high, because of strike.

Source: 10-K reports filed by the companies with the Securities and Exchange Commission; U.S. National Science Foundation (1979); Carroll and Schneider (1979); Kaiser (1979).

for 1976 but then delayed to the early 1980s. Prior to the sharp increase in gasoline prices in 1979, it appeared that the fuel economy standards of the early 1980s would require extensive innovation by the companies. Thus, a good case could be made for either end of the 1970s as having required heavier R&D expenditures so as to meet impending regulatory requirements.

Some additional light on this question is yielded by one extra data series. General Motors, in its 10-K reports, has listed its expenditures on "research, engineering, reliability, inspection, and testing" for emissions control and safety regulation from 1968 onward. (It is interesting that the company has not made a separate listing for fuel economy improvements; apparently, it has considered the improvements to be largely market-motivated rather than required by regulation, even during 1978 and 1979 when the fuel economy standards of the 1980s seemed most likely to be binding.) These expenditures, as a percentage of North American automotive sales, are listed in Table 8.8. It appears that regulation was imposing heavier requirements at the beginning of the decade than at its end. This probably explains the pattern in the overall General Motors R&D series in Table 8.7. The Ford and Chrysler rises in 1973 and 1974 were probably due also to regulatory requirements. The General Motors evidence, though, makes it likely that the Chrysler and Ford peaks at the end of the decade were due to a combination of decreased sales and the pressures of developing more fuel-efficient vehicles and not to emissions and safety regulation.

Table 8.8. General Motors' Research and Development
Expenditures on Emissions Control and Safety,
as a Percentage of North American Automotive Sales.

1968	1.7
1969	1.8
1970	2.4*
1971	1.7
1972	1.8
1973	1.7
1974	1.8
1975	1.4
1976	1.1
1977	1.1
1978	1.2
1979	1.3

*Abnormally high, because of strike.

Source: General Motors' 10-K reports.

A Summary on the Character of Innovation

It seems clear that government regulation has had a major effect on innovation in the major vehicle industry. This regulation has had its effects largely in the product modification area. It does not appear that regulation can be held responsible for the modest slackening in productivity improvements that has occurred. But regulation appears to have had a quantitative effect on the R&D budgets of the major companies in the early 1970s.

We now turn to a more complete description of this regulation and the other government actions and programs that have possibly affected innovation in the motor vehicle industry.

IV.   GOVERNMENT POLICIES

There are three sets of policies by the federal government that have been relevant to innovation in the motor vehicle industry: regulation of air-pollutant emissions, safety, and fuel economy; the 1969 antitrust suit attacking joint behavior with respect to emissions control; and direct government funding of research. The policies operate in quite distinctive ways and will be described separately;   the   controversies   concerning   their effects on innovation will be left to Section V.

Regulation

Responsibility for regulation of air pollutant emissions is lodged with the Environmental Protection Agency (EPA); the Department of Transportation's National Highway Traffic Safety Administration (NHTSA) has responsibility for safety and fuel economy regulation. The history and method of each regulatory program is distinct.

Air pollutant emissions regulation

The federal government first became involved in emissions regulation in 1965.(24)   The Motor Vehicle Air Pollution Act of 1965 directed the Department of Health, Education, and Welfare (the predecessor in this area to EPA, which was established in 1970) to set emissions standards for automobiles.   The first standards applied to the 1968 model year and covered hydrocarbon (HC) and carbon monoxide (CO) emissions.   By 1970 the standards implied approximately a 50 percent reduction in exhaust emissions from uncontrolled levels.   In December of that year, the 1970 amendments to the Clean Air Act called for

a further 90 percent reduction in HC and CO emissions by
1975 (i.e., a 95 percent reduction from uncontrolled levels was
implied) and a 90 percent reduction in nitrogen oxides (NO_x)
by 1976.(25)

In 1973 the deadlines were delayed a year through administrative decisions by EPA. In 1974, Congress delayed them
for another year, and in 1975 EPA delayed them yet another
year. Finally, in the late summer of 1977 Congress passed the
1977 amendments to the Clean Air Act which delayed the HC
and CO requirements to 1980 and 1981, respectively,(26) and
eased the NO_x reduction to 75 percent with 1981 as the new
deadline.(27) The 1977 amendments also specified that HC and
CO emissions from trucks should be reduced by 90 percent
from uncontrolled (1969) levels by 1983 and the NO_x emissions
should be reduced by 75 percent by 1985; the 1970 amendments had simply required that EPA regulate truck emissions,
without specifying the levels, and EPA had set standards that
were considerably less stringent than those that were required
of automobiles. Finally, the 1977 amendments called for EPA to
set standards for particulate emissions for vehicles; this was
aimed primarily at diesels. EPA has subsequently set standards for automobiles and light-duty trucks that call for a 40
to 50 percent reduction in particulate emissions by 1982 and an
80 to 85 percent reduction by 1985.(28) EPA is currently
developing particulate standards for heavy-duty trucks.

There are a number of important characteristics of emissions regulation. First, for the categories of automobiles and
of light-duty trucks, the emissions standards are set in terms
of the maximum allowable emissions of each grams per mile for
each vehicle. The standards apply uniformly within each
category to all new vehicles sold; no averaging is allowed, and
small Hondas and large Cadillacs are expected to meet the same
standards (but the grams-per-mile standards are more lenient
for the class of light-duty trucks than for automobiles). In
effect, the requirements assume that all automobiles serve the
same purpose and hence should meet the same absolute regulatory requirements; the same has been assumed to be true for
light-duty trucks. For heavy-duty trucks, however, it has
been more obvious that this assumption could not be made, and
the emissions requirements instead are stated in terms of
maximum allowable emissions per brake-horsepower-hour.

Second, selling a vehicle that does not meet the standards
is a violation of the law and carries a fine of up to $10,000
per vehicle. The fine is understood by all parties to be
prohibitive and not to operate as an emissions or effluent fee.
The only exception to this is that the 1977 amendments allow a
"nonconformance penalty" for heavy-duty trucks, which could
operate like an effluent fee. EPA has not yet made this
nonconformance penalty operational but has indicated that it
intends to do so.(29)

Finally, the requirements apply only to the first five years or 50,000 of a vehicle's life, whichever comes first.(30) Sample vehicles are tested over 50,000 miles prior to production, and since 1976 EPA has tested samples from assembly line production. But there are no current federal requirements on actual emissions from in-use vehicles, though a number of states and localities currently have in-use emissions limits and EPA has plans for more comprehensive inspection and maintenance programs in the 1980s.

Safety regulation

The federal government first became involved in vehicle safety in 1962.(31) In that year, Congress directed the Department of Commerce (the predecessor in this area to NHTSA, which was established in 1970) to set standards for hydraulic brake fluid. Standards for seat belts (which were offered voluntarily by the auto companies) were set the following year. In 1964 Congress directed the General Service Administration (GSA) to prescribe safety standards for vehicles bought by the federal government. De jure, this changed very little, since the GSA had always had the power to set specifications for the vehicles it bought; de facto, it indicated that Congress expected more safety. In 1965 GSA set 16 safety standards (and one air pollution control standard) for the 1967 model cars it would buy.

In 1966, in the wake of Ralph Nader's Unsafe at any Speed (32) and the revelation that detectives hired by General Motors had harassed Nader, Congress passed the National Traffic and Motor Vehicle Safety Act of 1966. It directed the Department of Commerce to set safety standards for all vehicles. The first standards were set for the 1968 model year, and further standards were set for subsequent years; in 1970 NHTSA assumed responsibility for safety regulation.

The regulations that have proved most controversial have been the passive restraint standard for automobiles, a bumper impact protection standard for automobiles, and a brake standard for heavy-duty trucks. The passive restraint standard has attracted the most attention. In response to surveys that indicated that only 15 to 20 percent of car occupants were actually using the seatbelts that were mandatorily provided in all cars, NHTSA first proposed in 1972 that passive restraints (which would automatically protect car occupants in the event of a crash, without their having to take positive actions) be required on all automobiles by 1976. At the time it was thought that airbags were the only way to meet the requirement. The regulations were challenged in the courts and overturned in late 1972.(33) (The electronic interlock system, which would not allow a car to start unless the front seat occupants had buckled their belts, was an interim measure for

1974 and 1975 which survived the court challenge; it did not, however, survive the wave of consumer unhappiness in 1974 when many cars' systems failed to work properly, and Congress specifically repealed the interlock requirement in 1974.)

NHTSA went back to the drawing boards, and in 1977 the agency again proposed that passive restraints be required, this time to be phased in during the 1982-1984 model years. This time the regulations survived court challenges.(34)   By 1980, however, it appeared that automatic belts (which automatically enclose a front seat occupant when the front door is closed), rather than airbags, would be the devices installed in most or all cars in order to meet the requirements.   This angered some members of Congress and officials of NHTSA during the Carter administration, who feared that many motorists would disconnect the automatic belts (whereas it would be more difficult to disconnect airbags).   In 1981, the newly elected Reagan administration decided to rescind the passive restraint requirement.

Like the emissions standards, there are separate safety standards for automobiles, light-duty trucks, and heavy-duty trucks, and the standards apply uniformly to all new vehicles sold within each category.   Selling a new vehicle that does not meet the standards is a violation of the law and carries a penalty of up to $1,000 per vehicle; again, the penalty is meant to be punitive and not to operate like an effluent fee. Finally, there is no federal program for in-use inspection, but NHTSA has actively encouraged the states to establish safety inspection systems.

Fuel economy regulation

In December 1975, Congress passed the Energy Policy and Conservation Act, which established fuel economy standards for automobiles: 18 miles per gallon for 1978, 19 mpg for 1979, 20 mpg for 1980, and 27.5 mpg for 1985.   The act also instructed NHTSA to set standards for the 1971-1974 interim years for autos, which the agency did in July 1977, and to set standards for light-duty trucks, which the agency has also done and first applied to the 1979 models.   Standards for heavy-duty trucks were not required (again, one suspects, because the claim that all vehicles within the class could be regulated uniformly could not be sustained).

Unlike the emissions and safety standards, the fuel economy standards do not apply to each individual vehicle but rather to the sales-weighted average of each manufacturer's vehicle sales in the appropriate category.   Further, the original law contained a one-year carry-forward, carry-back provision, and in 1980 the carry-forward, carry-back allowance was extended to three years.   If a manufacturer fails to meet the standard (and cannot take advantage of the carry-back,

carry-forward), the company is subject to a penalty of $5 per
0.1 mpg that his fleet average falls short of the standard, to
be applied to all vehicles sold by that manufacturer in that
category.(35) Thus, if General Motors sold four million cars
in a model year and the sales-weighted average fuel economy
of that fleet fell short of the standard 0.5 mpg, General
Motors would pay a penalty of $100 million. This clearly is a
nontrivial penalty, but it is not draconian; it provides
incentives that are comparable to those of an effluent fee.

The standards apply to the first 50,000 miles of a vehicle's life; the fuel economy of each model vehicle is determined
as a by-product of EPA's preproduction certification tests for
air pollutant emissions. There are no requirements that apply
to in-use vehicles.

For about two years - roughly the period between
NHTSA's establishment of the interim year standards on July
1977 and July 1979, when the full effects of the sharp rise in
gasoline prices had been felt - it appeared that the fuel
economy standards for autos would be binding and would force
the companies to take actions that the market would not otherwise have motivated.(36) But consumer response to the increase in gasoline prices has shifted the sales-weighted
averages of the manufacturers sharply toward smaller, more
fuel-efficient vehicles; similarly, the likely market response to
future fuel-saving innovations appears much more favorable
and would motivate the companies to pursue fuel efficiency
vigorously, even in the absence of the current standards.
The current debates over fuel economy standards, then, focus
on whether and to what extent the automobile fuel economy
standards should be tightened after 1985 and on the light-duty
truck standards for the mid-1980s.

One other, less well known fuel economy provision should
be mentioned. In the fall of 1978 Congress passed the Energy
Tax Act of 1978, which contained a set of "gas guzzler" taxes.
These are excise taxes that imply no violation of the law.
They are wholly independent of the fuel economy standards
just described and apply to each car sole that falls below a
certain level. For example, for the 1980 model year, each car
that failed to achieve 15 mpg was subject to a tax of 200 to
550 dollars, depending on how short of that figure it fell; to
this author's knowledge, no cars were required to pay the tax
in 1980.(37) By 1986, the minimum acceptable level will be
22.5 mpg and the tax will range from $500 to $3,850.

Antitrust Policy

In January 1969, the Justice Department brought an antitrust
suit against the individual motor vehicle manufacturers and
their industry association.(38) The suit charged that a 1955

cross-licensing agreement among the manufacturers and other joint behavior had constituted a "contract, combination . . . . or conspiracy in restraint of trade" that had delayed the development of pollutant emissions control technology and was therefore a violation of Section 1 of the Sherman Act.(39) It is worth noting that in the 1920s and 1930s the industry had had a cross-licensing agreement that had covered patents in all areas of automotive technology. That agreement had been allowed to lapse in 1957; the last patents covered were those acquired in 1939.

The suit was settled by a consent decree in September 1969 with no admission of guilt by any of the defendants. They agreed, however, to end the cross-licensing agreement, to avoid exchanging proprietary information with each other, and, in essence, to refrain from any joint behavior with respect to the development of emissions control technology. Ten years later, in May 1979, the Justice Department agreed to allow General Motors to sell technical assistance on emissions control and passive restraint technology to Chrysler, which was ailing financially; the department feared that Chrysler simply could not afford the necessary research and might fail to meet the requirements, with uncertain consequences. The department had previously allowed General Motors to do the same for American Motors. In July 1979, Ford complained that it was now unfairly the "odd man out," the only company still required to observe the prohibition on exchanging information. In April 1979, the District Court supervising the consent decree had decided to extend the decree's prohibition on information exchanges and joint reports to government agencies for an additional ten years; in July, sympathetic to Ford's plea, the court reversed itself and ended the prohibitions (40); an appeal to the Court of Appeals for the Ninth Circuit followed, with an inconclusive outcome. But in February 1982 the Department of Justice announced a major revision in the consent decree, loosening some of the provisions and calling for an expiration of the entire decree in 1987.(41) From that date on, only the normal provisions of the antitrust laws (without the additional strictures of the consent decree) will be applicable.

Federal Funding of Research

A number of federal agencies have conducted their own research and funded outside research on motor vehicles.(42) The motivating forces behind this research have varied from agency to agency. The Army and the Postal Service have conducted research as part of their vehicle procurement programs. EPA and NHTSA have conducted and funded some research to gain a basis for regulatory requirements; since there is always an explicit or implicit feasibility test that regulations must meet, the agencies need their own sources of information, to serve at least as a partial check on what the motor vehicle companies are claiming is feasible. NHTSA research on safety has been aimed at demonstrating safety possibilities and goading the industry into further research efforts. The Urban Mass Transit Administration, also within the Department of Transportation, has funded research on bus design as part of its efforts to help develop and fund urban transit systems. The Energy Development and Research Administration (ERDA), within the Department of Energy, has funded research on the development of electric vehicles and on turbine engines as part of its mission to encourage more efficient use of energy and to encourage alternatives to petroleum use.

The research described thus far would be characterized as applied research or even as development efforts. ERDA also conducts and funds basic research on automotive engines and combustion processes.

The current amount of research and development spending by the federal government on motor vehicles is not large – probably no more than $250 million per year, with only about $40 million of this constituting basic research.(43) By comparison, the three largest motor vehicle manufacturers alone spent $4 billion on research and development in 1979, roughly $3 billion of which was spent on North American motor vehicle research. If the spending of the other vehicle manufacturers and of the parts and materials suppliers were included, the North American total would be well above $4 billion. And the federal government currently spends about $29 billion on all research, of which $14.2 billion is spent for nonmilitary purposes. Thus, the federal sums spent on motor vehicle research are not large in comparison either to motor vehicle industry R&D or total federal R&D expenditures.

Of particular interest to this study is the Cooperative Automotive Research Program (CARP). CARP had its origins in a December 1978 speech by Secretary of Transportation Brock Adams, in which he called for an effort to "reinvent the automobile" so that a goal of 50 mpg could be met by the 1990s. It should be noted that the automobile companies themselves had not been brought into any governmental discussions of such a program, prior to Adams speech. Their responses varied from lukewarm to hostile. Adams' proposal was eventually transformed into a design for a jointly funded basic research program on fundamental aspects of motor vehicle construction, design, and operation, with the federal government and the automobile industry to split the costs equally (and the manufacturers splitting their share of the costs among themselves in proportion to sales). The program, in essence, specified a quota of research funding that each of the parties would separately undertake (above a specified baseline level) from an agreed-upon list of research topics. The research results were expected to be disseminated to all participants. A CARP oversight committee was to monitor the program, to make sure that all of the parties lived up to their part of the bargain. For the 1981 fiscal year, the federal government appropriated $12 million; the eventual goal was a $50 million per year federal contribution, with an equal contribution by industry.

The firms in the industry continued to be reluctant partners. When the Reagan administration eliminated CARP, the industry offered few protests.

V. QUESTIONS CONCERNING GOVERNMENT POLICIES

Regulation

Most of the debate concerning motor vehicle regulation focuses on the stringency of the regulations; i.e., it is a debate over the costs and benefits of the regulations and whether the levels of stringency should be increased or decreased. We will not review this debate here (44) but will instead concentrate on the consequences of the form of the regulations for innovation in the motor vehicle industry.

First, the current form of regulation may be trying to induce too much innovation from the motor vehicle industry and not enough changes in behavior from motorists. The primary approach of Congress and the regulators has been to confront the motor vehicle industry and, in effect, say, "This is your problem; you should fix it." Partly, this stance reflects Congress' belief in the boundless technological ingenuity of American industry; partly, this is an easy and popular political position. But, for any desired level of overall achievement in regulatory areas, a better balance of company action and motorist action would surely reduce social costs. Higher prices for gasoline are surely the low-social-cost way of reducing fuel consumption. As the experience of the increase in gasoline prices in 1973-1974 and again in 1979 has shown, higher prices do lead to less driving, a shift in demand toward more fuel-efficient vehicles, and greater manufacturer interest in fuel efficiency. A system of in-use inspection of motor vehicles emissions and effluent fees levied on motorists would induce them to maintain their cars properly and to seek out low-emissions models.(45) Incentives to encourage motorists to wear seat belts would surely be less socially expensive than the mandatory installation of passive restraints.(46)

Second, the inflexible nature of standards that take the form of "every vehicle must meet the standard or else . . ." (e.g., the emissions and safety standards) has a serious effect on innovation. It discourages research on innovations that may be low cost but cannot quite achieve the standards; "a miss is as good as a mile." It discourages research on innovations that are very good at meeting some standards but have difficulty in meeting others. For example, diesel engines are naturally low emitters of CO and HC; diesel vehicles usually have emissions below the "cleanest" comparable gasoline vehicles. But it is difficult to reduce the NO_x and particulate emissions of diesels to the levels achieved by gasoline vehicles. The inability to trade off good achievement in the one area against not-so-good achievement in the other area has, at various times in the past decade, discouraged research in diesels.

Further, it is clear that the original 1975 and 1976 deadlines embodied in the 1970 amendments to the Clean Air Act created a mixture of incentives for the companies: on the one hand, if they thought the act was really going to be enforced, they needed to find a quick, low-risk way of meeting the requirements, even if this meant a high cost technology. On the other hand, it was highly unlikely that the few large motor vehicle companies would be shut down or severely penalized for failure to meet the standards, as long as the appearance of a good faith effort was maintained; the credibility of the enforcement of the punitive penalties was quite low. Hence, some surreptitious foot dragging would have been worthwhile. It appears that both kinds of incentives came into play at various times for various companies.(47) Neither set of incentives provides the proper motivation for research and development. An effluent fee system would do so.

Third, because the regulatory structure can only impose standards that are perceived to be feasible and feasibility (to a great extent) depends on research and development information generated by the regulated companies themselves, the regulation itself may retard research and innovation in the regulated area. This is most likely to happen when the regulator faces a monopoly or a tight oligopoly. A monopolist's interests in restricting information are quite clear. In such instances, a regulatory agency would have to rely on its own research or on that of third parties, but both are clearly inferior to the kind of information that the regulated industry itself is capable of generating. (By contrast, in a competitive industry, each firm would try separately to develop the feasible technology in the hope that the regulator would then adopt the appropriately stringent standards and the successful firm could watch its rivals wither or could make large profits from licensing its technology to them.) An effluent fee (or similar incentive) approach would reduce this problem, since even the monopoly firm always experiences a direct gain in discovering low-cost ways of reducing its emissions yet further.

Fourth, the fleet averaging approach embodied in fuel economy standards must allow greater flexibility and encourage more socially desirable innovation than does the uniform standard for all vehicles approach embodied in the emissions and safety standards.(48)

Antitrust

The arguments for and against joint research efforts are fairly straightforward. Joint research can avoid costly duplication. It can encourage an interchange and interplay of ideas that may lead to new ideas that might not otherwise occur. But the joint interests of an industry may lie in suppressing some innovations that individual competition might pursue. This point is especially clear in the context of externality regulation, but it can hold equally validly for innovations in an unregulated market. This need not be another version of the apochryphal tale of the oil companies suppressing the invention of the pill that turns water into gasoline; instead it is a logical extension of the proposition that a monopoly could find it worthwhile not to offer some varieties or qualities of a good that a competitive industry would offer.(49)

Accordingly, one's assessment of the wisdom of joint research efforts depends on one's views of likely duplication, idea exchanges, and joint oligopolistic innovation suppression. There is little question that the motor vehicle industry has experienced more competition from overseas sources in the past 20 years than was true in previous decades. Nevertheless, regulatory actions are still based heavily on perceptions of what the domestic industry is able to achieve. And the degree of competition from abroad is always subject to the whims of the political process.

It is this author's judgment that the gains from joint research are not likely to be great and the risks are probably greater, especially in regulatory areas.

Federal Funding of Research

The arguments for and against federal funding of research are also fairly straightforward. Research, especially basic research, has the familiar property of externalities. The firm doing the research is unlikely to be able to capture all of the gains of the output of the research. Thus, the social benefits from research exceed the private benefits. Also, private firms may be less inclined to take risks than would be socially worthwhile, or they may have too short a time horizon and use too high a discount rate. For all of these reasons, a profit-maximizing firm is likely to conduct too little research from a social perspective, and this problem becomes progressively worse as the research becomes progressively more basic. Thus, there is a case to be made for some kind of social funding or assistance that will supplement private research efforts. Also, if a monopoly firm in an industry is not conducting research on products that it fears may simply divert demand from its more profitable items,(50) a case for government research can be made. Further, as noted in Section IV, in a regulatory context, government agencies need to conduct research so as to have a check on what the regulated industries are claiming is feasible. And, in a purchasing context, government agencies may need to conduct their own research so as to assess better the products they purchase and perhaps to suggest alternatives to vendors.

There are two major problems with government funding of research that is meant to supplement private efforts. First, the government-funded research may supplant rather than supplement the private research; i.e., government-funded research becomes a substitute for, rather than a complement to, private research. This is progressively more likely as the research becomes increasingly applied and development oriented. Thus, the net addition to total research is smaller than expected or, at the limit, nonexistent, and there are clear distributional (equity) consequences that flow from government funding rather than private funding.

Second, because government agencies do not face a market test, there is less assurance that the research that is funded by government will ultimately prove to be socially worthwhile, at least as judged by markets. (Of course, if government is funding research on externality problems, which the private sector would otherwise ignore, a market test is inappropriate - unless a market test is created, through devices such as effluent fees.) Government agencies have neither the profit-maximizing motives of private firms nor the competitive push of fear of survival in markets. Thus, inappropriate research and waste become more likely, and this problem becomes more severe as the research comes closer to the market--that is, as it becomes more applied and development oriented.

The arguments both for and against government funding point in the same direction: government funding that is directed at supplementing private research should focus as much as possible on the basic research end of the spectrum. Only the monopoly-limitation, regulation, and purchasing arguments point toward more applied research.

By these standards, the federal programs get a mixed rating. Some basic research is being funded ($40 million), but much applied research (e.g., on electric vehicles) is also

occurring.   NHTSA conducts a modest amount of research
($60-65 million), much of it on accident causation; EPA con-
ducts very little research (at best, a few million dollars).

The CARP program seemed reasonably well designed to
avoid the pitfalls of federal funding.   It focused on basic
research.   It tried to induce added research from all parties,
above some baseline.   And its design avoided most of the joint
collusion problem by specifying that each party undertake the
research separately but disseminate the results.   A program of
$100 million per year would not have been large when viewed
against total motor vehicle R&D budgets in the $4 billion area,
but it surely would have constituted a much larger fraction of
the total basic research that is being conducted.

In the end, the CARP program might have had more value
as a symbol - as an indication of a nonhostile attitude by the
federal government toward the industry - than as a program
that achieved great breakthroughs; but it was well designed,
it was inexpensive, and it was unlikely to do any harm.   Most
economists would surely wish they could say the same about
many other government programs.

The Special Problems of a Public Policy
in the Presence of Tight Oligopoly

At a number of points in this review of public policy, the
automobile industry's market structure has been an important
consideration.   It may be worth reviewing that structure and
reflecting on its implications.

The domestic industry is a tight oligopoly, with the three
largest firms accounting for over 95 percent of domestic
automobile production and almost 90 percent of domestic truck
production.   The domestic companies do face substantial competition from imports, but this competition is always subject to
limitation by legislation; and regulators are much more sensitive to the technological capabilities of the domestic companies
than to those of foreign companies shipping cars to the United
States.   The domestic firms are large in absolute size and
employ large numbers of workers, with their supplier firms
employing yet larger numbers.   This employment is concentrated in a number of states, mostly in the Midwest.   Technology
information flows fairly freely among the companies; there are
few important technological secrets in the industry.

In this context, public policy faces a number of serious
limitations.   The political process appears unwilling - or at
least greatly reluctant - to allow any of the companies to cease
operations; the Chrysler loan-guarantee program bears adequate witness to this phenomenon.   The companies may thus be
in position to bluff and "game" the regulators; each company
may feel that the federal government will not allow its rivals to

achieve a large technological jump that would threaten that
company's existence.   Chrysler may well have deliberately
slowed its emissions-control efforts in 1972.(51)   Thus, in
their efforts to promulgate and enforce regulations (and
achieve technological change through regulation), regulators
(and Congress) may be able to harass, embarass, and mildly
penalize the companies; but regulators are unlikely effectively
to threaten the companies' closure for failure to meet regulations, as long as the companies have provided the appearance
of a "good faith" effort to meet the regulations.   EPA Administrator William Ruckelshaus learned this in April 1973, when he
established interim HC and CO emissions standards for 1975,
rather than enforce the original, more stringent 1975 standards
that none of Chrysler's models could meet.(52)   Subsequent
legislation by Congress and delays by EPA have reinforced
this lesson.   Substantial progress has nevertheless occurred in
reducing automotive pollutant emissions - but considerably
slower than had been originally envisioned by the Clean Air
Act of 1970.

The powerful forces of the profit motive - the desire to
"make a buck" and to cut costs - should be harnessed to
serve the ultimate goals of the regulatory policies.   Congress
and the regulatory agencies should pursue regulatory strategies that would provide undiluted incentives for the companies
to improve their technologies in socially beneficial ways.   "Meet
the standard or else we will close you down" is not such a
strategy.   Effluent fees, properly structured incentive penalties, and fleet averaging are such strategies.   They should
prove beneficial, even if the industry is a tight oligopoly or
monopoly.   Further, vigorous antitrust scrutiny of the industry and its practices should be maintained, to discourage any
oligopolistic tendencies that may lead to true collusion.   And
any policies that sanction joint research efforts should be
structured (as was CARP) to reduce the likelihood of antisocial
collusion emerging.

CONCLUSIONS

As this chapter is written, the U.S. motor vehicle industry is
going through its most troubled period in history.   All of the
major companies ran losses in 1980; Chrysler's loss established
a U.S. corporate record.   The domestic industry has suffered
a substantial reduction in sales - the product of a soft national
economy, high nominal interest rates, and a sharp switch in
consumer demands toward smaller, more fuel-efficient vehicles
that overseas manufacturers have been in a better position to
satisfy quickly.   The industry's competitive problem vis-a-vis
Japanese imports has also been exacerbated by a domestic wage

structure that currently yields average U.S. auto worker wages that are substantially above the average wage in the United States and, at present exchange rates, are about twice those of auto workers in Japan.

The political process has not ignored the domestic industry's plight. The federal government has provided loan guarantees to bail out Chrysler. Import protection measures continue to receive serious consideration. The Reagan administration in April 1981 relaxed a number of automotive safety and pollution regulations and promised to pursue further revisions. (Though the regulatory changes affect all vehicles sold, regardless of source of manufacture, the changes are likely to benefit the domestic manufacturers differentially. For example, as a rough approximation, pollutant emissions are proportional to the size of the engine, and efforts to meet an emissions standard expressed in grams per mile will be more difficult for larger engines; since the domestic manufacturers have tended to build larger vehicles with larger engines, they have had greater problems in meeting the standards.)

The causes listed at the beginning of this section (plus Chrysler's peculiar history of poor management) seem adequate to explain the domestic industry's problems. Radical technological programs do not seem warranted. Sensible restructuring and scaling back of regulatory programs, modest research encouragement, and vigilant (but not overly zealous) antitrust policies are necessary. And, in the end, one should remember that nowhere is it written in stone that the United States should always maintain a comparative advantage in motor vehicle production such that only 15 percent of autos should be bought from abroad. Motor vehicle production is unlikely to disappear from the United States. And even if exchange rates, wage rates, and technologies were to point in that direction, should the trend be resisted? A growing, flexible economy should accommodate (and ease) change, not expend large amounts of public monies and private welfare (through tariffs and other trade restrictions) resisting it.

NOTES

1. See Motor Vehicle Manufacturers Association (1980), p. 70.

2. For further details, see White (1971, 1977a).

3. Since the U.S.-Canadian Automotive Agreement of 1965, there has been free trade in motor vehicles and components for the companies, so they have been able to rationalize their production across both borders.

4. This point is emphasized by Heywood et al. (1974) and Linden et al. (1976).

5. Hedonic price indexes are a means of providing a limited quantitative measure of product change. See Griliches (1971).

6. See U.S. Senate (1958), pt. 7, pp. 3812-3813.

7. For a more complete discussion, see White (1971), chap. 11.

8. See Chrysler v. Department of Transportation, 472 F. 2d 654 (1972).

9. See Pacific Legal Foundation v. Department of Transportation, 593 F. 2d 1338 (1979).

10. The adjustment are not based on any hedonic price equations but are, rather, based largely on the estimated costs of the product changes.

11. See White (1971), pp. 256-258.

12. Problems still remain if the vertical integration of the processes within an establishment increase or decrease.

13. See Goldfeld and Quandt (1976).

14. The Bureau of Labor Statistics began to measure the actual transaction retail prices, rather than list prices, in mid-1954, so 1955 seems to be the safest place to start.

15. It is worth remembering that the costs of safety and emissions-control equipment are adjusted out of the new car price index by the Bureau of Labor Statistics.

16. See White (1977c).

17. See White (1977c).

18. For details, see White (1971), pp. 221-222.

19. See Abernathy (1978).

20. Note that this does not mean that product innovations are being suppressed, but rather that the technology creates incentives discouraging product innovation.

21. See White (1979).

22. See Abernathy (1978), pp. 178-180.

23. See Dennsion (1979).

24. The State of California was involved earlier, in the 1950s. For further details see White (1971, chap. 14) and Mills and White (1978).

25. Controls over evaporative emissions of hydrocarbons from carburetors and gas tanks were also regulated.

26. EPA has the power to grant waivers of the CO requirement for the 1981 and 1982 model years, which it has done for some models.

27. EPA has the power to grant waivers for American Motors for two years and for diesel automobiles for four model years. EPA has granted some waivers for some models for some years.

28. For further details, see White (1981), chap. 7.

29. For further details, see White (1981), chap. 6.

30. For heavy-duty gasoline trucks, the requirement is for the first five years, 50,000 miles, or 1,500 hours of engine operation; for heavy-duty trucks, the requirement is for the first five years, 100,000 miles, or 3,000 hours of engine operation.

31. For further details, see White (1971), chap. 14.

32. See Nader (1965).

33. See Chrysler Corp. v. Department of Transportation, 472 F2d 659 (1972).

34. See Pacific Legal Foundation v. Department of Transportation, 593 F 2d 1338 (1979).

35. Penalties up to $10 per 0.1 mpg per vehicle are theoretically possible but are unlikely.

36. For further details, see White (1981), chap. 8.

37. It is worth noting that the official mpg figure for regulation purposes is a weighted average of a highway driving cycle (55 percent weight) and a city driving cycle (45 percent); the former is usually an appreciably higher number than the latter. Because both (and the average) apparently overstate the actual experience of most drivers, EPA now reports only the city figure as the likely mpg of each model.

38. See U.S. v. Automobile Mfrers. Assn., 1969 Trade Cases, no. 72907.

39. For some of the Justice Department's evidence, see U.S. Senate (1973, pp. 445-456).

40. See U.S. v. Motor Vehicle Mfrers. Assn., 1979-2 Trade Cases, no. 62,759.

41. See Federal Register, 47 (February 19, 1982): 7529-7541.

42. See Heywood et al. (1974) and Linden et al. (1976).

43. The numbers are difficult to pull together comprehensively and partly depend on one's definition of what constitutes research relevant to motor vehicles and also what constitutes "research and development." The $250 million encompasses fairly broad definitions and would include such things as test-

ing. One compilation of federal R&D spending (U.S. National Science Foundation, 1979) indicates that in fiscal year 1977, the federal government financed $411 million of research and development conducted by the motor vehicle and equipment industry. But a large part of that R&D was defense related, with little direct connection with motor vehicles.

44. For arguments that the emissions standards have been too stringent, see Mills and White (1978) and White (1980); for arguments that the safety standards have been ineffective, see Peltzman (1975).

45. For a further discussion of the feasibility and desirability of an in-use emissions fee systems, see Mills and White (1978).

46. It appears, though, that state inspection systems may not be effective in reducing accidents. See Crain (1980).

48. See White (1981), chap. 7.

49. For a demonstration of this proposition, see White (1977b).

50. See White (1977b).

51. See U.S. Senate (1973), pp. 211-43, 46-51.

52. See U.S. Senate (1973), p. 25.

REFERENCES

Abernathy, William J., The Productivity Dilemma: Roadblock to Innovation in the Automobile Industry (Baltimore: Johns Hopkins University Press, 1978).

Automotive News, 1980 Market Data Book Issue (April 1980).

Carroll, John M. and Richard P. Schneider, "Historical Financial Data, Domestic Automobile Manufacturers." Report by Arthur D. Little, prepared for U.S. Department of Transportation (January 1979).

Crain, W. Mark, Vehicle Safety Inspection Systems: How Effective? (Washington, D.C.: American Enterprise Institute, 1980).

Dennison, Edward F., Accounting for Slower Economic Growth (Washington, D.C.: Brookings, 1979).

Goldfeld, Stephen M. and Richard E. Quandt (eds.), Studies in Nonlinear Estimation (Cambridge, Mass.: Ballinger, 1976).

Griliches, Zvi, Price Indexes and Quality Change (Cambridge, Mass.: Harvard University Press, 1971).

Heywood, John B., Henry D. Jacoby, and Lawrence H. Linden, "The Role of Federal R&D on Alternative Automotive Power Systems." Report submitted to the Office of Energy R&D Policy, National Science Foundation, by the Energy Laboratory, MIT (Cambridge, Mass.: November 1974).

Kaiser, R., "Historic (1971-1975) Cost-Revenue Analysis of the Automotive Operations of the Major U.S. Automotive Products Manufacturers." Report by H.H. Aerospace Design Co., prepared for U.S. Department of Transportation (January 1979).

Linden, Lawrence H., John B. Heywood, Henry D. Jacoby, Howard Margolis, "Federal Support for the Development of Alternative Automotive Power Systems." Working Paper submitted to the Office of Energy R&D Policy, National Science Foundation, by the Energy Laboratory, MIT (Cambridge, Mass.: March 1976).

Mills, Edwin S. and Lawrence J. White, "Government Policies Toward Automotive Emissions Control." In Approaches to Controlling Air Pollution, ed. Ann F. Friedlaender (Cambridge, Mass.: MIT Press, 1978).

Motor Vehicle Manufacturers Association, Facts and Figures '80 (Detroit: MVMA, 1980).

Nader, Ralph, Unsafe at Any Speed (New York: Grossman, 1965).

Peltzman, Sam, "The Effects of Automobile Safety Regulation." Journal of Political Economy 83 (August 1975): 677-725.

U.S. Department of Transportation, Office of the Assistant Secretary for Policy and International Affairs, The U.S. Automobile Industry, 1980 (Washington, D.C.: January 1981).

U.S. National Science Foundation, Research and Development in Industry, 1977 (Washington, D.C.: 1979).

U.S. Senate, Committee on the Judiciary, Subcommittee on Antitrust and Monopoly, Hearings on Administered Prices, Automobiles, 85th Congress, 2nd session (1958).

U.S. Senate, Committee on Public Works, Subcommittee on Air and Water Pollution, Decision of the Administrator of the Environmental Protection Agency Regarding Suspension of the 1975 Auto Emissions Standards, Hearings, 93rd Congress, 1st session (1973).

White, Lawrence J., The Automobile Industry Since 1945 (Cambridge, Mass.: Harvard University Press, 1971).

_____., "The Automobile Industry." In The Structure of American Industry, 5th ed., Walter Adams, (New York: Macmillan, 1977). (a)

_____., "Market Structure and Product Varieties." American Economic Review 67 (March 1977): 179-182). (b)

_____., "Does Technological Change Respond to Economic Incentives?" New York University Faculty of Business Administration Working Paper Series, #77-98 (December 1977). (c)

_____., "Review of Abernathy." Business History Review 53 (Winter 1979): 546-548.

_____., "Automobile Emissions Control Policy: Success Story or Wrongheaded Regulations?" In Government, Technology, and the Future of the Automobile, ed. Douglas H. Ginsburg and William J. Abernathy (New York: McGraw-Hill, 1980).

_____., Reforming Regulation: Processes and Problems (Englewood Cliffs, N.J.: Prentice-Hall, 1981).

9
Government Stimulus of Technological Progress: Lessons from American History
Richard R. Nelson

I. ANALYZING A COMPLEX HISTORICAL RECORD

The preceding case studies reveal a record that is rich and complex. The United States indeed has had considerable experience with policies aimed to spur or guide or constrain industrial innovation. Let me briefly review that experience as recounted in the case studies.

A Brief Review

From the beginnings of the industry, the federal government has been a major stimulator and supporter of technological advance in aircraft. Military procurement has accounted for a significant fraction of total sales of the industry. Direct government support of R&D has taken several forms. During the heyday of NACA, government funds supported R&D and testing relating to aircraft in general; during this time the generic aspects of military and commercial technologies were relatively undifferentiated and advances in understanding or design principles relevant to one usually were relevant to the other as well. Of course, the government also funded R&D on airframes and components for specific military needs, although in many cases the companies invested their own funds in hopes of winning a procurement contract. Since World War II government R&D monies have gone largely into work with specific military application in mind. It has turned out that a good portion of military technology continues also to be applicable to civil aviation, although recently these technologies have been drawing apart. The postwar era also was marked by an attempt on the part of the government to pull forth and support the development of a commercial supersonic transport, an experience that ended as an expensive abort. CAB regulation of the airlines, and the constraints on vertical integration imposed by the Airmail Act of 1934, also have been important influences on how civil aircraft technology has evolved.

There has also been a strong military, and space, interest in computer and semiconductor technology. In semiconductors, most of the early work that laid the foundations for the industry was privately financed. Government R&D funding came later. By contrast, much of the early exploratory research on computers was done under government contracts. Government procurement accounted for a large percentage of the sale of both industries in the early days. While, as the industries began to tap commercial markets, government procurement and R&D funding came to play smaller roles, the government market continues to be significant. Public monies have continued to support advanced education and university-based research relevant to these industries. Antitrust considerations have played an important role in the evolution of both industries. Had Bell Laboratories and Western Electric gone into commercial production of semiconductors, the industry likely would have taken on a very different shape from the shape it has taken. The same consent decree that blocked commercial production of transistors by AT&T also kept Bell Labs and Western Electric out of the commercial computer business, and possibly enabled IBM to come to dominate that industry. Antitrust controversy seems to swirl continuously around IBM because of the dominant position it has achieved in the commercial computer market.

For many years public funds have supported applied and basic research, higher education, and extension, relevant to agriculture. Unlike the situation in the three industries mentioned above, in the case of agriculture there has been no major public procurement interest. The farmers of the United States, however, have formed a strong political constituency demanding, and to some extent guiding, government R&D support. The public R&D system has been operated largely through the agricultural colleges and experimentation stations of the state universities. Decision making in R&D allocation has been largely decentralized to the individual station, which depend on their state legislatures for a hefty portion of their funding.

In pharmaceuticals, as in agriculture, significant federal monies have gone into basic research and into the establishment and maintenance of programs to train scientists. However, federal funds for pharmaceutical applied research and development have been fenced into the field of cancer, and to "orphan drugs" for which the commercial market is likely to be small. It is apparent that there exists a strong political constituency for basic-research funding; at the same time there are strong political constraints against significant federal encroachment into the proprietary domains staked out by the pharmaceutical companies. Pharmaceuticals is also an industry marked by a complicated regulatory regime that significantly affects the cost of R&D.

The automobile and residential construction industries have experienced neither significant federal procurement nor much federal R&D support for either basic or applied work. Regulatory regimes, however, have strongly influenced technological advance in both sectors. Both sectors have seen federal attempts to launch an R&D support program. Political support for these, however, has been weak and where programs were initiated, they were not sustained.

The Analytic Problem

How can lessons be drawn from the rich experience described in these case studies, and from other studies? In principle, we want to draw up a matrix. The rows would delineate various policy instruments, the columns certain industry characteristics, the entries would measure the feasibility and effectiveness of a policy under a particular set of industry characteristics. This is, however, easier said than done. Simply classifying the policies and the relevant industry characteristics is a challenging task; tracing cause and effect relationships is extraordinarily difficult.

In general, a wide variety of policies have impinged on each economic sector and each policy has been complex and changing over time. In both aviation and agriculture, government funds have gone into support of applied R&D, but the programs and the objectives in these two cases are very different. Regulation has meant different things in automobiles and in pharmaceuticals. The research reported in this volume permits a start, but only a start, toward a more sophisticated and sensitive way of classifying the diverse range of policies used by government.

What are the industry characteristics that determine the feasibility and likely effectiveness of various policy instruments (assuming these can be well described)? Why has major government R&D support proved feasible and effective in aviation, but not in residential construction? The question suggests that one important industry characteristic is the presence or absence of a well defined procurement interest. Perhaps so, but government R&D support has been feasible and effective in agriculture. What differentiates agriculture from housing? Simply identifying the key industry characteristics that seem to explain these differences is a challenging analytic task.

Even if we could lay out the rows and columns in an objective manner, cause and effect relationships are not easy

to discern; technological progress in an industry might be fast
or slow and take the particular directions that it did for any
of a wide variety of reasons. Given the current state of
knowledge it is not possible to estimate a policy's effect with
any precision. To what extent did public R&D money simply
replace private R&D monies in the early days of the computer
industry? In aviation? Has public R&D support really made a
difference lately in semiconductors? To what extent has
regulation deterred pharmaceutical innovation? These are
difficult questions.

In short, it is very hard to tease out from the historical
record clear-cut lessons that are applicable to future policy
decisions. It is possible, however, to make some judgments.
essentially hypotheses about the kinds of policies that are
feasible and effective in various contexts. These hypotheses
appear consistent with the historical record as revealed by the
case studies presented here, and with other evidence I know
about. Nonetheless, like any theory that fits a fragment of
evidence, this one may prove quite wrong in a number of
places, or even in broad scope.

We are interested ultimately in understanding the sources
of variation. Different policies have been applied in different
industries. Some have been smashing successes, others have
been ineffective or worse. However, in order to sort out the
characteristics, reasons for, and effects of variation, it is
important to get hold of the common elements. There are
several general characteristics of technological advance that
are apparent in all the case studies. One is the apparent
inherent uncertainty involved in technological advance. A
second is the central, but often myopic and strongly context-
dependent, role of producers and consumers in the generation
and screening of technological advance. The third is the
important role played by nonmarket elements (as well as market
ones) in the institutional structure influencing technological
advance.

All of the case studies reveal that technological advance
involves considerable uncertainty. When a person or organization begins a search for a new product or a process it is
never clear exactly what the precise outcome will be. Design
configurations and solutions take shape in the course of trying
to achieve these. The ultimate success, or failure, of the
quest is revealed only after the fact. The uncertainties take
on a somewhat different form in each technology. Thus Grabowski and Vernon describe the hunt for a new pharmaceutical
as, literally, a search. Katz and Phillips discuss the considerable uncertainty during the 1950s regarding which new
technology was going to replace the old vacuum tube in computer design. Mowery and Rosenberg point out that in the
design of civil aircraft, theoretical calculations resolve only a
small portion of the uncertainties. Some of the semiconductor

companies placed their bets heavily on integrated circuits; others hung back.

Technological uncertainties are compounded by market uncertainties. It is very difficult to predict with any accuracy which future technologies will be useful, and which will be bought at a profitable volume and price. Just as different individuals and R&D organizations lay their bets differently about which technological paths are the most promising, so they tend to differ in their assessment of the market. A number of companies that developed strong technological capabilities for the design of computers failed to anticipate a large business market. IBM made a bet that such a market existed, at the same time that it acquired the technological capabilities to cover its bet. The American automobile companies had little reason to believe that consumer demand would swing sharply toward smaller, more fuel-efficient vehicles, but it did.

Thus, while the details differ from industry to industry, in none of the cases do R&D and follow-on technological work appear to be activities that can be planned in any neat and tidy sense. The uncertainties seem to be innate. From a social point of view, effective pursuit of technological advance seems to call for the exploration of a wide variety of alternatives and the selective screening of these after their characteristics have been better revealed - a process that seems wasteful with the wonderful vision of hindsight. As the supersonic-transport case indicates, however, hindsight may be much clearer than foresight.

All of the case studies also reveal the central role of the producer-provider (usually private enterprise) and the demand-user (who may be private or public) in the generating and screening of technological advances. The producer, and the user, each have important informational and motivational advantages over other parties. Producers live with the prevailing process and product technology, and know things about it - its strengths, its weaknesses, certain potentialities for change - that people and organizations without that experience cannot know. Users have similar special knowledge about the products and services they employ. It is natural, and essential, that this special knowledge and immediate motivation for improvement play a central role in inducing and guiding the innovation process. Moreover, in a market setting it is users who ultimately will determine whether a product will be demanded, and producers whether and how it will be produced.

This said, it should be recognized that that vision may be narrow and that motivation is very dependent on context. Both the computer and semiconductor case studies reveal companies reluctant to move away from technologies with which they were familiar to try radically different ones. In the semiconductor case, it is interesting that new companies, not the old tube producers, were the successful innovators. Similarly, user-consumers, like producers, fall into comfortable habits. Had IBM waited for potential users of business computers to articulate a clear-cut demand for them before deciding that a market likely existed, the advent of the computer age would have been significantly delayed.

The motivation of producer and user is strongly influenced by the details of the technologies involved, and by the particular insitutional and legal setting. There is little gain for a for-profit seed vendor to develop better self-propagating seeds. It does pay the seed vendor to develop better hybrid seeds since the farmer, each year, has to go back to the source; he cannot create next year's seeds from this year's plants. It was a delicate, and not inevitable, legal decision that ruled that antibiotics, although natural substances, were patentable. While patents don't carry much force in the semiconductor industry, and innovations are quickly imitated, the advantages of a head start are still significant enough that firms have motive to innovate. Government regulation, much more than expressed consumer demand, has pulled innovation toward safer and less environmentally harmful automobile designs. CAB regulation, in the form of constraints on air fares, tilted airline competition toward providing more attractive service and stimulated the market for faster and more comfortable planes. It was a governmental market, not a private market, that made it profitable for Texas instruments and IBM to invest in semiconductor and computer R&D. Both building codes and fluctuations in the demand for housing significantly dampen incentives for innovation in building construction.

In sum, while producers and consumers play central roles in the innovation process - and they should - their informational advantages may be associated with myopia. Their motivations are strongly influenced by special technological circumstances and the particular legal and institutional setting and by public as well as private demands.

More generally, it is important to recognize that technological change involves nonmarket as well as market elements. In all of the industry studies presented in this volume, there was a public interest, expressed through public policies, in certain aspects of the performance of the industries. There were elements of cooperation as well as competition in research and development.

In aviation, computers, and semiconductors, there was, for obvious reasons, a public interest in how the technologies and the industries evolved that transcended the interest of particular private purchasers or producers of the products. In these cases the public interest was manifested in a governmental demand for goods and services of a quite specialized variety and in policies associated with procurement.

In the other four industries studied there was no such important procurement interest. However, a public interest in certain aspects of industry performance shows up in other policies. In the cases of pharmaceuticals, automobiles, housing, and agriculture (as well as aircraft), a public interest in safety, environmental protection, and in ensuring certain general standards was made manifest in regulations. Several of these industries are also marked by various forms of subsidy to producers or consumers. Citizens and scholars may divide on the merits and demerits of these regulations and subsidies. But this makes it no less a fact that public policies to constrain or supplement market mechanisms pervade the American economy. And the workings of these policies significantly influence the environment for industrial innovation.

Also, while in all of the industries surveyed, for-profit firms creating and taking a proprietary interest in certain technologies are a large part of the story, in all of the industries one can observe as well a system of R&D cooperation, and exchange of technological information. In some cases government policy has played a large role in building and supporting this cooperative system; in other cases its role has been smaller.

With these common elements laid out, we can now explore the differences in policies, in industry characteristics, and in the apparent viability and effectiveness of policies, revealed by our case studies. (In what follows I also will draw, where appropriate, on other studies.) There are several alternative paths to follow. One could try to assess what industries are success stories in some sense and discuss the policies and structures associated with these, and then go on to consider the failures. One could divide the industries according to some kind of structural characteristics. It has proved more straightforward to try to classify policies (instruments) and proceed to consider where they were and were not employed, and why, and how effective they have been in different contexts.

A Road Map

One rough division among instruments places those involving direct government funding of R&D in one category and those that indirectly influence R&D or other activities associated with industrial innovation in another. While this division is plausible on its face, notice that the lines between the categories are blurred, not sharp. How does one treat, for example, procurement contracts that cover the cost of R&D incurred earlier by a company, who anticipated the subsequent contract? How does one treat special tax credits for R&D? These problems notwithstanding, I shall hazard such a break.

Section II will deal with government support of R&D. The objective here is to categorize meaningfully the different kinds of government R&D support programs revealed in our case studies, to analyze the reasons for the significant differences in such programs across industries, and to make judgments as to what kinds of programs worked and which didn't. I distinguish among four kinds of government R&D support programs: (1) those associated with public procurement or other well-defined public objectives, (2) those that involve an extension of support of scientific basic research to support of research to advance generic technological knowledge, (3) programs that are aimed at meeting reasonably well defined clientele demands, and (4) attempting to support "winners" in commercial competition.

Section III considers a wide range of government policies that do not involve direct R&D support - procurement, regulation both old style and new, antitrust, policy regarding patents - to name the central ones. But simply listing these as instruments covers up some fundamental problems. Regulation, for example, has meant fundamentally different things in different industries; the thrust of antitrust policies has also been different, etc. Relatedly and equally important, the central purpose of these policies often has little to do with spurring or guiding industrial innovation. There are serious questions as to whether they should be regarded as promising instruments for that purpose. In section IV, I draw some general conclusions.

II. GOVERNMENT SUPPORT OF RESEARCH AND DEVELOPMENT

The case studies reveal dramatic differences among the industries in the extent and kind of federal R&D support. The government has been an important source of both applied and basic-research funding in the evolution of aviation, computer, and semiconductor technologies. The government also has productively supported both applied and basic research in agriculture. While the government has been an important supporter of basic research relevant to pharmaceuticals, public funding of work aimed directly at identifying and testing new drugs has been limited to the field of cancer, and certain rare diseases. The government never has been able to mount a sustained R&D program relevant to the housing and automobile industries.

It is not easy to measure the efficacy of the various government R&D support programs. In the three defense-relevant industries they certainly have bought us technological primacy. Critics have argued both that much of the bought technology has not been necessary for national security but rather has inflamed the arms race, and that many of the R&D programs have been inordinately expensive and wasteful.   It should be noted that contributions to the advance of civilian technology made by defense and space programs, while the focus of our case studies, has been a "spillover" and certainly not the principal intent of these programs.   The advance of civilian technology was the central purpose of government R&D support programs in agriculture, and of basic biomedical research.   The rate of return on the public investment in R&D for agriculture undoubtedly has been very high.   Quantitative estimates are more difficult with respect to the returns from support of biomedical research; however this is too generally regarded as a very successful research program.   The case studies also reveal two expensive fiascos: the supersonic transport project, and "Operation Breakthrough" in the housing industry.

How can one make intellectual order out of this varied experience?   I propose it is important to distinguish among four different kinds of government R&D support programs. Most government R&D support is aimed to achieve a well-defined government purpose - like the procurement of a new weapon system.   Other government programs support basic or generic research relevant to a particular technology or technologies rather than being pointed toward achieving any particular product of process; examples are research on the nutritional needs of wheat, or the properties of certain exotic materials.   Some government programs support applied research and development on products and processes that serve civilian, not governmental purposes, and whose acceptance depends in large part on market calculations made by nongovernmental actors.   Programs of this sort ought to be divided into programs in which the potential users have considerable influence on allocation, as has been the case in agriculture, and programs in which a government agency has relatively free-handed control over the setting of goals and priorities, as in the supersonic transport project.   Programs obviously differ in the range of industries for which they are politically feasible, and in the kinds of circumstances where they are likely to be effective.

R&D Support Associated with Procurement Needs or Other Well-Defined Public Purposes

In three of our case studies - aviation, computers, and semiconductors - there was a strong and recognized governmental demand for the products produced by the industry that led to a particular and focused public interest in certain kinds of technological advances.   A recognized public sector demand for certain types of technological improvement lends two important features to the policy context. First, it means that the government (or the relevant government agency) is in a position to define technological targets according to its own criteria and that it has (or at least has the motivation to have) some expertise about the technologies in question. Second, the recognized governmental need lends legitimacy to government attempts to stimulate and guide the evolution of the relevant technologies.

One should note that public procurement does not inevitably lead to active public-sector effort to mold or stimulate technological advance. The federal government procures typewriters, office calculators, automobiles, and a wide variety of products that are identical (or virtually so) to those purchased by nongovernmental users. In these cases the federal government usually has chosen simply to act as an informed shopper. In the three industries in question, however, the relevant government agencies deliberately tried to induce the development of products that were suited for their purposes. The vehicles employed included procurement contracts written so as to cover the R&D costs of the particular design (a disguised form of R&D support), direct R&D support associated with a procurement contract, and support of basic and generic research.

If public-sector needs and private-sector needs differed sharply, the procurement and applied research and development funding aspects of such policies would not facilitate the evolution of technology for the private sector. At least these three cases suggest, however, that governmental efforts to advance technology for public-sector purposes can also enhance technological capabilities to meet private needs. In the early days of these technologies, R&D aimed for a governmental purpose almost always had some commercial spillover. As these technologies matured, the governmental (military) market and the civilian market began to separate. Government-financed applied research and development associated with public procurement, and R&D financed by the companies themselves and aimed for products in the civilian market, became dissimilar. At the present time, the principal impact of the government on the evolution of civilian technology in these industries would appear to come via public support of basic and generic research. This fall-off in "spillover" has led to proposals that the government consciously fund projects that have likely civilian benefits. The supersonic transport project ought to warn against this strategy.

The key difference between government procurement-oriented R&D and government R&D on the supersonic transport, of course, is that in the former case, but not the latter, a government agency spends to further its own reasonably well defined purposes, and its evaluation of the technology emanating from R&D determines whether the new technology will be used or not. While there may be "spillover" to the private sector, the principal objective is better provision of a public good. The evaluation of the program is by a government agency, not a private market. I recognize that in some cases the well-being of an industry is viewed as a public interest. I shall discuss the special issues involved in public support of R&D to enhance the commercial capabilities of a private industry subsequently. Here I am concerned with public R&D aimed at public purposes that are more narrowly defined.

Public support of R&D to find drugs to deal with cancer and certain rare diseases has something in common with the procurement cases. Here, as with the examples of defense procurement, a government agency stands ready to see that the fruits of R&D are employed. There is a recognized public commitment to try to cure or relieve the suffering of people with serious diseases. If necessary, public monies will go into the procurement of whatever it takes to do this. Drugs to cure cancer or other life-threatening diseases are not, as it were, in the position of having to make in on a conventional commercial market. As with the case of the decision by the Department of Defense to procure a new fighter (or as with the space program) one can argue about how much tax money ought to go into the pursuit of the objective, and about whether the program is being conducted efficiently. But there is little question about the political legitimacy of the program, or about the potential ability of government decision makers to marshal the information needed to make sensible R&D decisions.

The case of pollution abatement, I propose, is similar in context if not in policy. Since the middle 1960s there has been a well-recognized public interest attached to the development of technologies that are less polluting than those currently being employed. Some public monies have gone into R&D on pollution abatement, but the Clean Air Act of 1970 marked a commitment to a strategy for achieving the objective that minimized the government's direct role in funding R&D. The strategy was to induce private funding of R&D through the imposition of regulatory requirements that could only be met by the development of new technologies. White and other scholars have argued that this has proved an inefficient and costly way of drawing forth the new technologies. Given a recognized public commitment to their achievement, the government certainly was in a position to fund R&D on its own. Of course, the effectiveness of such a program would depend on how it would be guided and administered.

Clearly, there are a wide range of technologies associated with public procurement, or public subsidy of certain kinds of private purchases, or regulation, where there is a recognized public interest in certain kinds of advances. The government has adopted a wide variety of strategies on the extent and kind of R&D it will support in these areas. I suspect that in many cases the government has been too passive and has relied too heavily on market mechanisms alone.

It would be a mistake, however, to think of R&D on all public-sector purposes as inevitably similar to defense procurement. Indeed for many public-sector needs, the DOD model is clearly the wrong one. The development of better technologies for the provision of public services, like mass transport, garbage collection, repairing city streets, and others, potentially can yield a very high rate of return on the public R&D dollar. Unlike the Department of Defense, however, when the Department of Transportation or the Department of Housing and Urban Development make R&D allocation decisions, they are not usually making them regarding items that they themselves will procure. The principal users will be state and local governments. Similarly, public financing of the R&D required by environmental and safety goals may yield high social returns and avoid the high private costs and tangled relations that come from the current regulatory strategies. However, the new technologies will ultimately be employed by private firms, not federal agencies. The institutional machinery needed to spend such public R&D monies efficiently will have to be different from that of the Department of Defense or NASA. An effective program will require strong participation from the users.

The National Cooperative Highway Research Program is a good example of a relatively successful program. In its guidance and support it has managed to marshall the attention of state government's, as well as the relevant engineering and technical societies.

It is understandable why discussion of the appropriate role of the Federal government in stimulating industrial innovation has focused largely on products sold on private markets. However, public-sector demands are large and broad and many of these receive little R&D attention. The question of the range of public demands where the federal government ought to play a significant role in R&D support merits more attention than it has received.

My emphasis has been on R&D to permit public demands to be met more efficiently, not spillover from that R&D to enhance capability to meet private demands. The case studies suggest that there will be such spillover when the technology employed to meet public and private needs are similar, or where publicly supported R&D enhances generic capabilities as contrasted with creating only a special-purpose design. If an important objective is to enhance spillover, I propose that the appropriate strategy is to give extra weight to basic and generic research.

Support of Basic and Generic Research

Absent a recognized public interest in the evolution of a particular technology, the government's ability to fund R&D efficiently faces certain constraints. In the first place a government agency has no particular claim to be able to determine R&D priorities and may be blocked from access to the information necessary to do so. Second, the legitimacy of publicly financed R&D programs, which may upset the status quo within an industry, may be questioned and such programs politically stymied. The case studies reveal these constraints to be quite stringent regarding government support of applied R&D aimed to achieve particular new products and processes. They appear to be less confining for public support of basic and generic research a step or two away from specific application.

Our case studies show the government actively involved in support of such research not only in the three industries in which there was a strong procurement interest - aviation, computers, and semiconductors - but also in agriculture and the scientific fields relating to pharmaceutical developments. The aborted Cooperative Automotive Research Program represented an attempt to extend this type of public program to the automobile industry.

To understand the nature and importance of these public programs, it is important to recognize that technological knowledge inevitably involves a public as well as a proprietary component. The public part of technological knowledge generally does not relate to the design or operational details of a particular product or process but to broad design concepts, general working characteristics of processes, properties of materials that are used, testing techniques, etc. Most of such knowledge is not patentable. Much of it is openly shared among scientists and engineers working in the field, whether they are located at universities, government laboratories, or corporate laboratories.

The kind of research that leads to such knowledge is not generally the sort that an academic scholar, pursuing fashionable questions in a standard scientific field, would explore. Rather, the research questions are posed by technological problems and opportunities and the objective is to enhance their understanding and the capability to solve practical problems. In some industries progressive private companies themselves support some of this type of research. While some secrecy is involved, it is recognized that the findings from this type of research ought to flow into the public domain. As our case studies illustrate, often the funding is at least partly public. Such a research system fits in between more fundamental research defined by the traditional sciences, and the applied research and development of the firms in the industry.

To be effective, the system has to make good contact with both sides, but avoid too much overlap and duplication.

In the judgment of Evenson and other scholars, the agricultural sciences have in general managed to define their niche appropriately. The research they do lies in between the basic academic sciences such as chemistry and biology on the one side, and on the other the research that goes on in public experimentation stations and private companies to develop, for example, better seeds or fertilizers. Both sides influence the kind of research that is done and monitor quality and efficacy. The biomedical research community is a similar system. It too is pulled from one side by the interests of practitioners (physicians) and private companies in having practical problems illuminated, and is disciplined from the other side by scientists in the more basic sciences. It is interesting that both the agriculture science and the biomedical science tend to find their home in universities, but in professional schools rather than in colleges of arts and sciences.

The government provides the bulk of support for these two research communities. The allocation of research resources, however, is guided only loosely by government agencies. The Department of Agriculture, the state legislatures, and the National Institute of Health - the principal support agencies - leave the details of allocation to machinery operated by the research communities themselves. However, in political deliberations about the level of funding and broad research strategies the focus is very much on the practical benefits that have flowed from the programs and the practical problems that future research promises to resolve.

Mowery and Rosenberg remark that the old NACA did not sponsor much in the way of basic research. In the pulling and tugging on the one hand to be applied and relevant and on the other to be rigorous and scientific, the first kind of pull clearly was significantly stronger than the second. This may well reflect the fact that NACA, unlike the agricultural experimentation stations and the medical schools, was a free-standing organizational entity not affiliated with a university or universities. Nonetheless, NACA undertook many experiments and studies that were relevant to aviation technology in general, rather than concentrating on particular aircraft designs that were being contemplated or were on the drawing board. In that sense, NACA certainly did support generic research and, as history testifies, to strong positive affect. The role of NACA diminished after World War II. In the postwar era the armed services increasingly funded their principal contractors to do the kind of research that NACA used to do.

No large and separate government-funded generic research programs mark the computer and semiconductor industries. While sometimes special government agencies were involved (for example the Advanced Research Project Agency of DOD and the National Bureau of Standards), the major governmental agencies supporting such research have been those interested in procurement, and the NSF. The funds have flowed to both university and corporate laboratories.

It is hard to draw clear lessons from the aborted experience with CARP. That experience might be read as suggesting that government programs in support of basic and generic research are politically acceptable in virtually any industry. Companies do not perceive such programs as posing sharp threats to their commercial positions, or the threats, if perceived, are seen as diffuse and not readily indentifiable as dangerous to any particular portion of the industry. Since proprietary knowledge is not needed to guide allocation, mechanisms can be established to allocate resources sensibly. Yet, General Motors was worried about the antitrust implications. More important, the industry never was particularly supportive of the program, perhaps because they played so small a role in its initiation. It is clear enough that such programs require industry support and participation if they are to be politically viable.

The key question is the efficacy of such programs. In the industry studies in this volume the verdict is positive. When private companies support little generic research, the case for public support seems specially strong. When private companies support such research, the case for public funding may be diminished, but certainly not eliminated. Thus, in the computer industry and in semiconductors, where the companies themselves do engage in significant funding of generic research, there is advocacy not opposition for government funding of research at universities. While there is a risk that public funds in such cases largely replace private funds rather than adding to them, I don't think the risk is particularly great.

I offer the following conjecture. Government funding of basic and generic research not only enhances the amount of such work undertaken, even when the companies themselves are spending significantly amounts; when the government funds such work, particularly at universities, the research is more "public" than when business finances it. Moreover, when a considerable amount is publicly financed, the treatment of research results as public is contagious. Business-hired or supported scientists are part of the communications network, and behave according to its canons.

As remarked above, in all of the industries studied there were networks of communication among scientists and engineers as well as industrial secrecy. However, in the industries where public support of basic and generic research is significant the information exchange seems to be wider and deeper. It is interesting that in all five such industries, technological advance has been especially rapid.

In some of these industries the private companies finance university research. In recent years the Semi-Conductor Industry Association has been exploring ways in which firms in the industry could cooperate in the funding of academic research. This is an interesting possible development and raises the possibilities of joint industry government funding of basic and generic research at universities. There are some serious question, however, about how the research findings from such programs will be treated. Tensions between the corporate funders and the academic undertakers of research already are apparent. To what extent should the funders have special access? What, if any, restraints should be put on publication? Universities and potential corporate sponsors will have to wrestle with these questions.

Support of Clientele-Oriented Applied Research

Public support of basic and generic research does not require program officers to form judgments about which particular commercial developments would be most valuable. Rather, the objective is to enhance understanding of relatively basic principles, to explore certain potentially widely applicable technological routes, etc. Furthermore, this kind of research seldom poses an immediate, perceived threat to the proprietary interests of particular groups of firms. The research community itself can be tapped to guide fund allocation. In contrast, government programs of support of applied research and development for an industry whose products are evaluated largely on commercial markets requires a mechanism to make commercial judgments. Such a program also may appear threatening to certain firms. The case of public support of applied research and development for agriculture indicates that, even with these constraints, a government program may be feasible and effective. It is interesting to consider which aspects of the industry, and the program, have led to success.

In the first place, farming is an atomistic industry and farmers are not in rivalrous competition with each other. Differential access to certain kinds of technological knowledge, or property rights in certain technologies, are not important to individual farmers. This fact at once means that farmers have little incentive to engage in R&D on their own behalf and opens the possibility that the farming community itself would provide a political constituency for public support of R&D.

The federal/state agricultural experimentation system established under the Hatch and subsequent acts marshalled that support and put the farmers in a position of evaluating and influencing the publicly funded applied R&D. The system is highly decentralized. The regional nature of agricultural technology means that farmers in individual states see it to their advantage that their particular technologies be advanced as rapidly as possible.

Evenson and other historians of technical change in agriculture have argued that the applied research and development efforts of the experimentation stations did not yield particularly high rates of return until a body of more scientific and technological understanding was developed. It was this combination of an evolving set of agricultural sciences based in the universities and supported publicly, and applied research and development also publicly funded but monitored politically by the farming community, that has made public support of agricultural technology as successful as it has been. Where private companies are funding significant amounts of innovative work and the industry is reasonably competitive, it is in the interest of the farmers as well as the companies that public R&D money be allocated to other things. As Evenson describes it, a reasonably well defined division of labor has emerged between publicly and privately funded applied research.

Can the experience in agriculture be duplicated elsewhere? It is apparent that many people have seen housing and agriculture as quite similar. Henry Wallace, who earlier served as Roosevelt's Secretary of Agriculture, clearly drew the analogy when, after the war, he tried (and failed) as Secretary of Commerce to initiate a major program of federal funding of building research. The efforts to revive that idea under the Kennedy administration were also explicitly based on the agricultural analogy. The analogy also was drawn in "Operation Breakthrough." It is obvious that there are important differences.

In the first place, while the building industry is atomistic, construction markets are local and therefore individual builders are, to some extent, in rivalrous competition with one another. However, since individual builders possess little in the way of proprietary knowledge, this was not a particularly important obstacle. What was more important was that suppliers of inputs and equipment to builders produce different, and rivalrous, products. Direct government support of applied research and development was viewed by many of them as potentially threatening. Had the builders of houses formed a strong constituency for government support of R&D, these resistances of input suppliers might have been overcome. No such constituency developed, however. Unlike the case in agriculture where farmers in a particular region saw it to their competitive advantage (as a group) to have their technologies advanced relative to the technologies employed by farmers in other regions, builders apparently saw so such advantages for them. It must be remarked that no special efforts have been made on the part of the proponents of a program in support of R&D in residential construction to develop an industry constituency.

Nor did there exist in housing, as there came to exist in agriculture, a scientific community that could point persuasively to promising areas for applied research and development. Residential construction lacks a broad scientific base from which to mount an effective applied research and development endeavor.

Thus, agriculture had both a constituency interested in getting applied research and development relevant to their needs undertaken, and ultimately at least a sound scientific basis beneath its technologies. Residential construction has neither. My conjecture is that programs in support of residential construction technology will not be politically feasible until the clientele is established to support and guard them, and will not be effective in the absence of some sort of underlying scientific base.

It is not clear how far the agricultural model of public support of applied R&D can be extended. There may be a number of industries, however, to which such a program is applicable. Again, the key ingredients would appear to be a group of users of the technology who are not in rivalrous competition with each other but who, together, have a significant interest in getting their technologies advanced, and a scientific base strong enough that applied research and development can be fruitful. Some of these industries are fragmented private ones, for examples, furniture. The regionally divided regulated industries are another candidate group. Indeed electric and gas utilities already do support, collectively, applied research and development on common needs under the auspices of the Electric Power Research Institute, and the Gas Research Institute. In both cases a government approved user fee, which is passed on to consumers in the form of a surcharge in utility rates, is the principal source of financing.

The above programs are cooperative applied R&D associations. The federal-state agricultural experimentation stations might also be viewed as industry R&D cooperative associations, but with an important difference from most such. Much of the policy discussions about cooperative research and development has presumed that public funds should account for only a small portion of total R&D monies and that the industry should contribute the bulk of the funds, save for, perhaps, the first few years of the program. Under such terms it has proved hard to initiate and sustain much cooperative R&D. The agricultural case suggests that the requirement for industry financing may be a mistake. In industries, like agriculture, where such programs are plausible, prices tend to follow costs. The returns to successful R&D go largely to consumers, not to producers. The limit to extending the agricultural model is not that the public at large would not benefit, but that the conditions under which this model is applicable would appear to be rather special.

Government-Guided Applied R&D with Commercial Ends

In Operation Breakthrough and in the Supersonic Transport Project, the government got itself into the business of trying to identity or develop products that would sell well on complex commercial markets. In Operation Breakthrough the Department of Housing and Urban Development was neither a major builder of houses nor a buyer of nonsubsidized housing. It thus did not have any particular technological or managerial expertise that would enable it to assess the economics of particular designs; nor did it have any basis for judging the appeal of such designs to consumers. Thus, it was easy for the department and Congress to lose track of the objectives as the program was debated politically. Similarly, the FAA was not in the business of building, or procuring, commercial airplanes. The commercial airlines were singularly discouraging when asked about their interest in a supersonic transport. The aircraft producers showed no particular interest in designing and building such a vehicle until the subsidies grew very large.

Very few of the housing designs created through Operation Breakthrough proved commercially viable, nor did they serve as a significant basis for follow-up design work. The British/French experience with their supersonic transport indicates how fortunate the United States was that the program was stopped before it resulted in a technologically (though not commercially) viable aircraft.

The lesson here is a general one, not particular to these two cases. There are many other studied cases, most of these European, in which government has tried to identify and support particular products that it was hoped would ultimately prove to be commercial successes. While there are few successes, the batting average has been very low, except when the government in question has been willing to subsidize or require the procurement of the completed product as well as the R&D on it.

This should not be surprising. In many of the industries in which this has been attempted (in Europe), the private companies also were investing in R&D, and the government was in a position either of duplicating private effort, subsidizing that effort and probably therefore replacing private R&D monies, or investing in a design that the private companies had decided to leave alone. In the last case it might be argued that there is a legitimate public role in supporting work on designs that are a generation ahead of those that the companies themselves are exploring. As the supersonic transport and a number of other like examples indicates, however, the sensible way to explore the next generation of technologies is through doing generic research, building and studying prototypes, and so on. The appropriate research program would be one that is modeled after NACA, not one modeled after the supersonic transport project.

If the United States were to drop its antitrust laws, and the objective of preserving intra-U.S. competition that those laws are supposed to embody, then it might be possible to mount a policy to help industry search for "winners." In various of the European countries and Japan, competition is viewed not so much in terms of rivalry among domestic companies, as in terms of competition from abroad. In these circumstances it is possible for the government to work with industry as a whole, and to participate in laying the bets and dividing the market. As the law exists in the United States, much of the information needed to guide a government program to help industry find and support "winner" is proprietary, not shared among firms, and not accessible to a governmental body. The experience of the European governments in trying to pick winners indicates that the costs of these American constraints are not severe; constraints are looser in Europe and the record of public policies to identify and support winners is not encouraging. The experience in Japan may or may not be different. At present, not enough is known about what the Japanese actually do to make a judgment on this. In any case, modes of government-industry cooperation in Japan are so radically different from those in the United States that it is doubtful we can learn much that will be of use to us from the Japanese experience.

It is a shame that so much of the discussion about government support of industrial R&D in the United States has swirled around the question: should the government try to pick winners? The evidence from our case studies answers that question negatively. The experience also shows, however, that there are many other potentially fruitful ways that the government can support industrial research and development.

III. POLICY AFFECTING THE CLIMATE FOR PRIVATE R&D

Much of the preceding section was spent disentangling various kinds of government R&D support, attempting to identify the reasons why such support has taken different form in different industries, and hazarding guesses as to the effects. The same kinds of analytical challenges face us in this section, in which our concern is with a variety of government policies that have influenced the climate for private R&D and innovation but do not involve direct governmental support of R&D. Regulation, for example, has had very different meanings in the various industries studied.

The fact that the policies considered here do not involve direct R&D support may not be the most important difference

between them and the policies considered in the preceding
section, however. The policies discussed above obviously were
intended to influence technological advance. Most of the
policies considered here were put in place for other purposes.
It is not clear whether or to what extent they can be realistically regarded as instruments that might be consciously
employed to influence innovation. Put another way, the problem is this: virtually every policy of government influences the
climate for innovation in some way, in greater or lesser
degree. For only a few is the influence on innovation a major
factor considered in their design and implementation. Which
policies should be considered explicitly here? Presumably,
those whose influence on innovation is significant, and whose
design might be modified by policy makers in response to
presentation of evidence about its effects on innovation. Unfortunately, evidence of magnitude of impact is hard to come
by. Therefore the focus must be, and should be, on policies
widely regarded - whether correctly or not - as having a
significant effect, and as subject to modification to make that
effect more positive or less negative. Since the case studies
contain relatively rich material on them, I shall focus on four
such classes of policy: procurement, regulation, antitrust, and
patent and other policies affecting property rights on inventions. I conclude this section with the suggestion that it is
probably mistaken to see any of these instruments as likely to
be manipulated so as to play a significantly more powerful role
in stimulating industrial innovation.

Procurement

In the undertaking of its varied activities, the federal
government procures a wide range of products. If one considers, as well, the state and local governmental activities that
the federal government helps to finance, and the regulatory
and other objectives of government, the government is a direct
or indirect purchaser of virtually everything. However, the
range of products where the government actively and consciously has attempted to spur technological advances is quite
limited. Earlier I argued that returns to extending that range
considerably might be high. Such an extension would enhance
the capability of government to meet accepted public-sector
needs, and at the same time contribute to the advance of
technologies used to produce private goods and services.
     As the case studies show, there are a number of ways in
which the government can attempt to draw forth technological
advances. At one extreme it can itself undertake or contract
for virtually all of the R&D in the area. At the other, it can
simply act as an informed purchaser. In-between, it can
advertise its interest in products with certain characteristics,

and entice and support private R&D efforts through its procurement policies, but not engage in much direct public funding of R&D.

Much has been said about the role of government in
"making a market" for certain kinds of technological advances,
with the implicit assumption that this is a very different kind
of policy from that of government R&D support of the work
leading to those advances. I find the distinction blurred, not
sharp. I suspect that, as in the three defense-related industries studied in this volume, wherever the government tries
to "make a market" for a new technology, it inevitably and
appropriately will be drawn into some R&D support. Conversely, government R&D support of public-sector technologies
does not make sense in the absence of aggressive procurement
policies which, in turn, almost inevitably, will induce certain
privately financed efforts. The mix of R&D inducement and
R&D support is a matter of tactics, not strategy. Aggressive
procurement is one aspect of a policy designed to draw forth
better technologies that are used in public-sector activities,
and that also may have nongovernmental applications. Such
procurement policies are a complement for government R&D
support policies in such arenas, not a substitute.

Regulation

If the reader of this volume commenced with any strong simple
ideas of the effect of regulation on technological change in
industry, a reading of the case studies should have disabused
him of these. The studies reveal how diverse regulation is
and how complex and subtle are its influences.

The automobile industry and, to a lesser extent, residential construction reveal what has been called "new style"
regulation at work. (As the housing example testifies, new
style regulation is not so new.) Regulation here amounted to
the imposition of certain requirements on the products or the
technology employed with the objective of assuring certain
standards of quality, or safety, or protecting the environment,
etc. Regulation, however, has had quite different purposes in
the two cases, and has had different consequences for technical advance.

In the housing case, regulation has been conservative.
Building codes and standards have stuck pretty close to prevailing techniques and materials, or simple modifications
thereof. Far from being aimed at drawing forth new materials
and methods, housing regulation has tried to monitor and
screen these and in fact has made significant innovation
expensive if not downright impossible. In contrast, in the
case of automobiles, regulation has been used aggressively to
pull forth new technologies. When the regulation were imposed

it was well understood that prevailing technologies could not meet the standards.   One can argue about whether regulation was the most appropriate or efficient method to pull forth the desired innovation.   White, and other scholars, believe that the route has been inefficient and expensive.   Although this regulatory strategy may have led to government neglect of direct R&D funding, it is certainly not the case that regulation has deterred all innovation.

Pharmaceutical regulation is something else again.   Originally concerned with maintaining purity standards and safety, in the 1960s regulation began to try to assure efficacy as well and to constrain and monitor the safety of the R&D process itself.   There are very real questions about whether the post-1960s regulatory environment has actually increased the efficacy of the new drugs that reach the market, or guarded the safety of patients and experimental subjects to any significantly enhanced degree.   As Grabowski and Vernon argue, it is not easy to pin down and separate the effect of U.S. pharmaceutical regulation on the flow of new pharmaceuticals into the cornucopia.   It is clear, however, that regulation has significantly increased R&D costs, and delayed the introduction of new drugs compared to the date of introduction in countries with different regulatory regimes.

The effects of new-style regulation show up less strikingly in the other industry studies.   However, environmental and safety regulation has in recent years come to play a significant role in influencing the fertilizers and pesticides that farmers are allowed to use and, relatedly, the tests and hurdles a new substance must overcome before it can be introduced to the market.   To my knowledge, however, no study of the effect of such regulations on the flow of fertilizers and pesticides comparable to the studies of the effects of regulations on the introduction of new pharmaceuticals has ever been made.

Of our case studies, civil aviation has been the industry that has been most strongly influenced by what has been called "old-style" public-utility regulation - regulation aimed at constraining prices and requiring certain standards of service delivery.   In this particular case the airlines, while regulated, were in rivalrous competition with each other.   Further, the industry doing most of the relevant R&D - the airframe industry - was not regulated.   The consequence of regulation undoubtedly was to spur innovation.

As has been the case in other regulated but rivalrous industries, for example railroads, airline regulation must be understood as setting floors under prices as well as establishing ceilings.   In the airline case the result was that, since rate competition was blocked on lucrative competitive runs, the airlines' competitiveness spilled over into the providing of better services, and seats on more attractive aircraft.   The consequence was that the airlines provided a strong, indeed eager, market for new aircraft. It has been often argued that old-style public-utility regulation stifles innovation; this most emphatically was not the case here. This is not to argue that the regulation of air transport was a desirable policy from a social point of view or even that the stimulus provided by regulation for the development of transport aircraft was socially desirable. It is simply to warn against the simple-minded notion that regulation generally deters all kinds of innovation.

In view of the diversity of regulation and its impact, deregulation or regulatory reform means different things in different industries. For the airlines it has meant the abandonment of rate regulation and the relaxation of CAB control on routes. While the new regime of aircraft competition may provide strong demand for certain kinds of new aircraft, it is hard to argue that the demand will be any stronger than it was under the old regulated regime, although the pattern of demand may be different. Airline deregulation is part and parcel of the deregulation movement for industries that, in the past, have been treated as public utilities even though their structure permitted considerable competition.

Reform of environmental and safety regulation involves a different set of issues and strategies. What is needed here is regulation-setting machinery that will consider costs as well as benefits, greater use of performance standards, and in some cases use of fees or marketable licenses rather than quantitative restrictions. Such a reformed regulatory regime would quite likely provide a better, if not necessarily a stronger, environment for the generation of technological advances that respect environmental and safety values. However, what is needed here is more sophisticated regulation, not "deregulation." Unfortunately, much of the apparent thrust toward modification of "new-style-regulation" is toward abandonment rather than reform.

For the pharmaceutical industry, regulatory reform largely means simplifying and speeding up the evaluation procedures for new drugs. Grabowski and Vernon argue that the current regulatory regime has significantly retarded and increased the cost of pharmaceutical innovation in the United States, and that the most effective available vehicle for spurring innovation is regulatory reform. Of the industries studied in this volume, however, pharmaceuticals is probably unique in this respect.

Antitrust

Just as with regulation, many people carry around in their heads an oversimplified and distorted view of what antitrust has meant for technological advance. The case studies reveal quite complicated and varied stories.

The pharmaceutical and automobile industries have been traditional targets of antitrust prosecution. Usually, however, the antitrust cases have not involved innovation or R&D directly but rather have been concerned with such old-fashioned matters as price fixing or other "conspiracies in the restraint of trade." In the pharmaceutical industry a few of these have involved patent licensing, and other related issues. However, neither the Grabowski and Vernon study, nor other studies of the pharmaceutical industry, has argued that antitrust has had much of an influence on innovation in the industry one way or another.

In the automobile industry, it is quite possible that concern about antitrust action has deterred General Motors from being as technologically aggressive as it might have been. On a few occasions antitrust has touched directly on issues relating to R&D and technological advance. The restrictions on patent pooling and on certain forms of cooperative R&D were noted in White's case study. The lawyers for the automobile companies certainly had misgivings about what the antitrust division would do if they joined the proposed Cooperative Automotive Research Program. Present antitrust guidelines, however, which permit cooperative R&D if the results are not treated as proprietary, would appear to leave room for programs of this sort and for most fruitful kinds of government-industry cooperative programs.

The computer industry is an interesting one for thinking through certain conundrums about antitrust and industrial innovation. The history presented in the case study stops at just about the time that IBM achieved the dominance that it now has maintained for close to 20 years. As Katz and Phillips show, IBM was successful in part because it guessed right technologically, and in part because it judged the market correctly. Other scholars have remarked that its prior dominance in the punch-card calculator business gave IBM a special advantage in the sale of computers to business users. Scholars and lawyers may dispute whether it was technological leadership, shrewd judging of the market, effective marketing, taking advantage of old ties, or behavior subject to prosecution under the antitrust laws, that enabled IBM to preserve its dominance (in large-scale civilian computers). Nonetheless, the antitrust cases have involved in an essential way complaints about the way IBM goes about designing and introducing new computers, and the remedies proposed include some that would significantly limit the freedom of action of IBM regarding R&D and innovation. It is naive to believe that the current pull-off of the government from the IBM case is the end of that story. AT&T will be part of the new chapter.

The case studies reveal at least two striking instances where antitrust and other structural policies preserved or created a competitive market structure with apparent salutory effects on industrial innovation. Although there is good evidence that AT&T had no interest in going into production for sale of transistors anyhow, the 1956 consent decree legally foreclosed that option. The evolution of the semiconductor industry might have been different had AT&T decided to enter the commercial market. One might also note that the consent decree, while most visible in our semiconductor study, stopped AT&T from going into any commercial market not directly connected with the telephone service. The evolution of the commercial computer industry might have been significantly different, absent the restraints on Bell labs and Western Electric. The future of this industry certainly will involve these organizations.

A second example of government policies that influenced an industry's structure in a way that had a profound effect on technological advance is the revised Airmail Act of 1934. This act broke up vertical integration among airlines, airline manufacturers, and engine manufacturers, and left a more open and competitive structure. Again, it is difficult to judge what would have happened if the industry had remained vertically integrated, but it is hard to imagine that technological advance would have been any faster than it was.

Patent and Related Policies

How about public policies that affect patenting and, more generally, the ability of the company to appropriate the returns of an invention it makes? Again, the picture is mixed and complex.

In the pharmaceutical industry it is apparent that the ability to patent a new drug is virtually essential if that drug is to be profitable for the company that creates it. Indeed, the whole history of the pharmaceutical industry would have been different had the courts ruled that antibiotics, as natural substances, could not be patented. However, in pharmaceuticals the question of the effective duration of a proprietary market hinges not only on patent life but on the decisions of physicians and pharmacists, and laws impinging on these decisions, regarding whether to provide a generic or brand-name drug when the former is available. Arguments against generic prescription are, in effect, arguments that protection provided by a patent ought to extend beyond its legal limit. Of course, the effective life of a patent in the pharmaceutical industry depends on the relationship between the date of patenting, and the date of commercial introduction of the product. The testing and licensing requirements mean that there is often a considerable lag between patent application and commercialization. Returns to invention in the pharmaceutical industry clearly depend on a wider set of variables than the strength of patents.

For many of the other industries studied, legal protection of proprietary rights seems to be less important than it is in pharmaceuticals. Key patents have played a role in the evolution of mechanical machinery in agriculture and in inducing new chemical compounds such as fertilizers and pesticides. However, while hybrids were judged patentable, it is not apparent that a patent adds much to the protection a seed company has for its particular hybrid. A potential competitor cannot readily discern the exact nature of the crossing that led to the particular hybrid seed. In this case the patent may be a minor rather than a major element in assuring appropriability.

In semiconductors, while firms patent their new devices, these patents do not have much force. Sometimes producers of new devices are able to hide their design from potential competitors by "potting." But in this industry imitation generally is quick. Indeed, the insistence of government and other purchasers of semiconductors on "second sourcing" in effect requires that a firm's new design be produced by another firm as well as the innovator. The profits to a successful innovator in this industry would appear to reside largely in the head start that provides a short period when the innovating firm is the sole supplier, and a ability to move down the learning curve before other firms get into production.

With a few interesting exceptions, patents appear not to have played a particularly important role in inducing, or making profitable, innovation in automobiles or civil aircraft. Indeed, in both industries there has been a tradition of relatively easy patent licensing, or even patent pooling. The reason for the lack of interest in a particular patent would appear to be that automobiles and aircraft are complex systems, and that particular patentable components do not play much of a role in determining the attractiveness of the overall system. It is the general overall engineering of the product that counts, and that is not readily patentable. The same situation seems to apply in computers. While patent suits marked the early history of the industry, IBM's prominent position does not rest on its patent holdings.

General Purpose Instruments, More Generally

It would be easy to draw on the case studies and other material to extend the list of government policies that influence the climate for industrial innovation. Some of these policies are broad in scope, although their influence differs from industry to industry. The tax codes are one of these. While the influence of the tax code is pervasive, particular features, such as the treatment of capital gains, appear to be particularly important in certain industries. Thus, it has been

argued that the higher taxation of capital gains that came with
the tax bills of the early 1970s had an especially strong
negative effect on funds to finance innovation in the semiconductor industry.   It is unlikely that these status changes had
a comparable effect on aviation.   While monetary policy is
cross-cutting, our particular monetary institutions segregate
the housing industry, and make that industry bear the brunt
of the economic fluctuations to a great extent.   Some policies
are aimed at particular industries.   Special price-support
programs certainly have influenced technological advance in
agriculture.   The trade agreement with Japan regarding the
importance of television sets especially affected the U.S.
semiconductor industry.   One could go on.   However, if our
search is for instruments that can be considered powerful tools
for a policy to stimulate industrial innovation, such extended
listing and analysis is not likely to be fruitful.   There are
several reasons.

First, the broad policies in question have been put in
place for a variety of reasons.   Arguments about their effect
on industrial innovation will carry only limited weight in
influencing the debate about their reform.   This is not to say
that such arguments have no influence.   A tax credit for R&D
was proposed by several groups as an important instrument to
spur innovation, and such a tax credit was part of the recent
Reagan tax-modification package.   But an R&D tax credit was
but a small part of that bill, and it is unlikely that the
particular proposal would have been heeded had there not been
a general thrust toward tax reductions of various kinds.

Second, the broad policies in question often differ in the
particulars of their applications from sector to sector.   Therefore, it is virtually impossible to identify any general rules for
reform of any of these instruments for the purpose of spurring
industrial innovation.   Rather, the most salient proposals
would appear to be industry specific, for example, particular
reforms of pharmaceutical regulation.

Third, while in some cases there is undoubtedly a tradeoff between stimulus of industrial innovation and other policy
objectives, our perusal of the case studies suggests that in
most instances the reforms that make sense as a stimulus to
the right kind of innovation make sense in terms of more
general criteria as well.   Thus, while regulatory reform is not
a broad panacea for stimulating faster or better-directed
technological advance, the kinds of reforms that scholars long
have proposed on grounds of general economic efficiency for
pharmaceutical regulation, and auto emissions control, probably
would affect innovation in the right direction.   Our case
studies reveal a few instances where antitrust may be acting
as a restraint on certain types of industrial innovation, but
they certainly provide no general indictment of antitrust policy
on these grounds.   The antitrust issues involved in the suits

against IBM or AT&T were complicated. As a general rule, however, it does not appear that antitrust is hobbling innovation by business. Similarly, there appears to be no general magic in reform of the patent law, or in the patent policies of government agencies that fund R&D.

Let me not be misunderstood. It may well be that establishment of a generally supportive climate for industrial R&D is the most important thing the government can do to facilitate industrial innovation. I would put particular stress, first, on support of basic science and scientific and engineering education. The strength of American universities and the education they provided has been cited by many scholars as the key to American technological primacy in the 1950s and 1960s.

I would stress, as well, the importance of strong aggregate demand, relatively stable demand growth, and predictable prices. When business conditions are good, and incomes and demand are growing rapidly and predictably, business firms can anticipate an expanded market, and make their investment and R&D plans accordingly. When demand is stagnant, or uncertain, investment in new plant and equipment is deterred, and R&D aimed to tap new markets may look like a very risky proposition. Of the industries studied in this volume, housing is the one that is most noticeably influenced by changing macroeconomic conditions. Quigley, and others, have argued that the cyclical sensitivity of residential construction is an important factor explaining the structure of the industry, and the limited incentives for innovation associated with investment in durable equipment. Virtually all industry is subject to some cyclical influences, however. The demand of farmers for new agricultural implements is cyclically sensitive. A nontrivial proportion of the demand for semiconductor is cyclically sensitive. Economic slumps hurt the airlines, diminish their ability and incentive to invest in new equipment, and reduce returns to the design and development of new aircraft.

Even were there no effects on innovation, however, it should be the objective of macroeconomic policy to achieve sustained growth, high employment, and steady prices. As with regulatory and antitrust policy, the objective of stimulating innovation carries no particular implications for fiscal and monetary policies.

It seems to be like this in general. If the specific interest is in stimulating innovation, it is a mistake to look largely to general-purpose policies. Their design can be influenced only marginally by concern about their effect on innovation, and usually policies that make sense on general grounds provide an appropriate environment for innovation in any case. If "innovation" policy is to have any meaning, search for it must be focused on more specialized instruments.

IV.   A BRIEF SUMMING UP

In the preceding section we identified a wide range of government policies that defined the climate, influenced incentives
for, and imposed constraints on industrial research and development.   In virtually all of our cases studied, one or more of
these government policies were an important part of the story.
However, the most important such policies differed from industry to industry.   While it is apparent that a number of specific
reforms might have significant benefits, the case studies do
not reveal any general and powerful guidelines for regulatory
or antitrust or patent policy reform.   If a serious mandate
reemerges to find and implement government policies that will
significantly spur industrial innovation, there is an understandable temptation to look for modification in these instruments to do the trick.   But there isn't much leverage there.
Moreover, the kinds of improvements in macroeconomic and
other policies that make most sense for stimulating the right
kind of innovation make good sense in terms of other criteria
as well.

     If the government is to look specifically for policies that
may have a significant stimulating effect on industrial innovation, the place to look is in the bag of R&D support policies.
This chapter has not attempted to give a general rationale or
justification for active government support of R&D, or to draw
up fine theoretical arguments to guide such policies.   As
stated earlier, a decade or so ago economists had much clearer
and more pointed theoretical views about these matters.   The
externalities from R&D and the uncertainties involved led,
according to the theoretical perspective prominent at that time,
to a divergence between the quantity of R&D expenditure that
firms would find most profitable and the quantity that was
optimal from a social point of view.   The firms would spend too
little.   Public support or subsidy was therefore warranted,
and ought to be focused on those kinds of R&D and on those
industries where the externalities and the uncertainties were
the greatest.   Subsequent theoretical work has led economists
to draw a more complicated picture.   A competitive regime in
which firms gain property rights on certain of their technologies draws forth some R&D that is socially wasteful.   Major
technological uncertainties call for a variety of approaches with
open knowledge of routes being explored and what is being
found along the way, and not for a big push along one particular road.   The problem with market-induced industrial R&D
allocation lies in the portfolio, the allocation of resources,
rather than in a total magnitude of effort.

     But if the problem cannot simply be characterized as "too
little" research and development, the design of appropriate
government policies requires mechanisms to identify the partic-

ular kinds of research, and sometimes the particular projects, that are being underfunded. Therein lies the problem. Government agencies are seriously constrained in the information they are able to marshal directly or indirectly to guide the allocation of public R&D monies.

The historical experience canvased in this volume suggests that there are three potentially fruitful routes that can be followed. One is to associate government R&D support with procurement or other well-defined public objective. A second is to define and fund arenas of nonproprietary research and allow the appropriate scientific community to guide R&D allocation. The third is to develop mechanisms whereby potential users guide the allocation of applied research and development funds.

A fourth kind of policy, in which government officials try themselves to identify the kinds of projects that are likely to be winners in a commercial market competition, is seductive. The evidence collected in this volume and other studies suggests, however, that this is a strategy to be avoided.

These are qualitative judgments drawn from qualitative and impressionistic case studies. While I can provide some reasoning to make them plausible, I can provide no tidy and powerful general theoretical justification for them. Perhaps the lesson that economists should draw from their earlier attempts to base prescription for government R&D policy on theoretical arguments is that this is a dangerous game. Economic reality is too complicated for any simply theory to fit well. More complicated theories generally point in different policy directions depending on the quantitative magnitude of certain key parameters. The design of good policy depends on hard empirical research, and not simply on theoretical reasoning.

There are two major weaknesses with the evidence provided in this volume supporting the above propositions about policies. First, the evidence comes largely from studies of seven U.S. industries. Second, the evidence is qualitative and judgmental, not quantitative and readily verifiable.

The first weakness is not as serious as it might seem, although this study would have been enriched if coverage had been wider. There are available a number of other industry studies, some of the United States, some of Europe. There are also several across-the-board evaluations of government policies in support of industrial innovation, particularly policies of European countries. The conclusions drawn in this chapter were influenced not only by the case studies presented here, but also by this other evidence, and are consistent by and large with both bodies of data.

The second weakness is the serious one. One can try to avoid having to base conclusions largely on qualitative and impressionistic evidence by constructing formal models and hypotheses and estimating and testing these with statistics. To some extent this kind of work has been done for agriculture. But such quantitative conclusions are no better than the models and the data on which they are based, and these contain large elements of the subjective and judgmental. Personally, I fear more the faith that lay persons, policy makers, and even scholars, often show in quantitative conclusions drawn from shaky models and data than I do conclusions that are explicitly qualitative and judgmental. When our knowledge is stronger, when we understand things well enough to have confidence in the basic form of the models we write down, when we have data that are more conformable with our operating models than is presently the case, the quantitative studies can play a greater role. I would argue that at the present time, however, the most promising route towards such stronger knowledge is through case studies of the sort presented here, and the kind of qualitative judgmental analysis developed in this chapter.

Index

About the Authors

RICHARD NELSON is Professor of Economics, and Director of the Institution for Social and Policy Studies, at Yale University. He has written extensively on long run economic change, and on problems of economic organization and public control. Much of his research has been focused on the processes of technical change, and on related issues of public policy.

RICHARD LEVIN is Professor of Economics and of Organization and Management at Yale University. He is the author of several articles on the economics of technical change, and he is currently directing a research program at Yale on the relationships between technical advance and economic structure.

DAVID MOWERY is Assistant Professor of Economics and Social Science at Carnegie-Mellon University. His current research concerns the development of industrial research in American manufacturing during 1900-1950 and government policy toward innovation in the private sector.

NATHAN ROSENBERG, Professor of Economics at Stanford University, is an economic historian. The particular focus of his research has been on technical change in U.S. industry and in Europe. He has written a number of articles and books that relate to the question of the role of government in technological innovation.

BARBARA KATZ is Associate Professor of Economics at the Graduate School of Business Administration at New York University. Her speciality is applied microeconomic analysis.

ALMARIN PHILLIPS is Professor of Law, Economics and Public Policy at the University of Pennsylvania. His primary interests are in industrial organization, with emphasis on technological change, and in regulation. He has written extensively on these subject.

ROBERT EVENSON is Professor of Economics at Yale University, and a member of Yale's Economic Growth Center. His specialty is agricultural economics, and before getting his Ph.D. he was a self-employed farmer in Minnesota. He has written widely on agricultural economics, and also on technological change and economic development more broadly.

HENRY GRABOWSKI is Professor of Economics at Duke University. He is also director of the Program on Pharmaceutical and Health Economics theory. His principal field of research has been industrial organization, and in recent years he has specialized in studying the pharmaceutical industry. He has authored many articles and several books on these topics, and has served as a consultant to several government agencies regarding them.

JOHN VERNON also is Professor of Economics at Duke University, and also is an industrial organization economist. His writing has ranged over a number of topics, but in recent years he, as Grabowski, has focused on the pharmaceutical industry.

JOHN QUIGLEY is Professor of Economics and of Public Policy at the University of California, Berkeley, and is an associate at Berkeley's Center for Real Estate and Urban Economics. He has written extensively on issues relating to housing and labor markets, urban development, and local public finance.

LAWRENCE J. WHITE is Professor of Economics at the Graduate School of Business Administration, New York University. He has written on the American Automobile Industry, and on various aspects of the regulation. During 1982-1983 he is on leave from NYU to serve as Director of the Economic Policy Office of the Anti-Trust Division, U.S. Department of Justice.